MEDICAL
INTELLIGENCE
UNIT 27

Adipose Tissues

Susanne Klaus, Ph.D.
German Institute of Human Nutrition
Potsdam-Rehbrücke, Germany

CRC Press
Taylor & Francis Group
Boca Raton London New York

CRC Press is an imprint of the
Taylor & Francis Group, an **informa** business

ADIPOSE TISSUES

Medical Intelligence Unit

Designed by Jesse Kelly-Landes

First published 2001 by Landes Bioscience

Published 2018 by CRC Press
Taylor & Francis Group
6000 Broken Sound Parkway NW, Suite 300
Boca Raton, FL 33487-2742

Copyright © 2001 by Taylor & Francis Group, LLC
CRC Press is an imprint of Taylor & Francis Group, an Informa business

First issued in paperback 2019

No claim to original U.S. Government works

ISBN 13: 978-0-367-44719-9 (pbk)
ISBN 13: 978-1-58706-040-3 (hbk)

Visit the Taylor & Francis Web site at
http://www.taylorandfrancis.com

and the CRC Press Web site at
http://www.crcpress.com

Library of Congress Cataloging-in-Publication Data

Adipose tissues / [edited by] Susanne Klaus.
    p. ; cm. -- (Medical intelligence unit)
   Includes bibliographical references and index.
   ISBN 1-58706-040-X (hardcover)
    1. Adipose tissues--Physiology. 2. Adipose tissues--Pathophysiology. 3.
Obesity--Pathophysiology. I. Klaus, Susanne, 1959- II. Series.
   [DNLM: 1. Adipose Tissue. QS 532.5.A3 A23475 2000]
   QP88.15 .A326 2000
   611'.0182--dc21

CONTENTS

EDITOR

Susanne Klaus, Ph.D.
German Institute of Human Nutrtion
Potsdam-Rehbrücke, Germany
Chapters 1, 4, 5

CONTRIBUTORS

Gérard Ailhaud
Institut de Recherches Signalisation,
   Biologie du Développement et Cancer
Laboratoire "Biologie du Développement
   du Tissu Adipeux"
Centre de Biochimie
Nice, France
E-mail: ailhaud@unice.fr
Chapter 3

Timothy J. Bartness
Department of Biology and Department
   of Psychology
Georgia State University
Atlanta, Georgia, U.S.A.
E-mail: bartness@gsu.edu
Chapters 6, 7

Michael Boschmann
German Institute of Human Nutrition
Bergholz-Rehbrücke, Germany
E-mail: boschmann@www.dife.de
Chapter 8

Saverio Cinti
Institute of Normal Human
   Morphology-Anatomy
Faculty of Medicine
University of Ancona
Ancona, Italy
E-mail: cinti@popcsi.unian.it
Chapter 2

Gregory E. Demas
Department of Biology
Georgia State University
Atlanta, Georgia, U.S.A.
Chapters 6, 7

Hans Hauner
German Diabetes Research Institute
Heinrich-Heine-University Düsseldorf
Düsseldorf, Germany
E-mail: hauner@dif.uni-duesseldorf.de
Chapter 10

Nigel Hoggard
Molecular Physiology Group
Rowett Research Institute
Scotland, U.K.
Chapter 9

Susanne Klaus
German Institute of Human Nutrition
Potsdam-Rehbrücke, Germany
E-mail: klaus@www.dife.de
Chapters 1, 4, 5

D. Vernon Rayner
Molecular Physiology Group
Rowett Research Institute
Scotland, U.K.
Chapter 9

Thomas Skurk
German Diabetes Research Institute
Heinrich-Heine-University Düsseldorf
Düsseldorf, Germany
Chapter 10

C. Kay Song
Department of Biology
Georgia State University
Atlanta, Georgia, U.S.A.
Chapters 6, 7

Paul Trayhurn
Institute of Nutrition Research
University of Oslo
Norway
E-mail: p.trayhurn@basalmed.uio.no
Chapter 9

PREFACE

The ability to store fat as a metabolic fuel is crucial for survival of many vertebrate species. They possess specialized adipose tissues adapted to the storage and release of triglycerides. In recent years it became more and more clear that adipose tissue represents not only a passive tissue merely responding to nutritional challenges, but that it is rather an organ actively involved in energy homeostasis and regulation of important metabolic functions. Furthermore, it was recognized that adipose tissue is not a homogeneous tissue: different anatomical depots display different metabolic properties and are subject to different endocrine and neural regulation.

Historically, the two major types of adipose tissue, namely white adipose tissue (WAT) and brown adipose tissue (BAT) have drawn interest from researchers of quite different fields. Scientists in medical and more specifically obesity research were interested in WAT, whereas BAT was mainly studied by zoologists focused on thermoregulation of small mammals. The discovery that brown fat plays also a role in overall energy homeostasis and possibly in the defense of body weight against excess energy intake, however, has triggered enormous basic and pharmaceutical research efforts in the last 20 years. Today we know that white and brown adipocytes share many metabolic and molecular pathways; although their physiological function, i.e., energy storage and energy dissipation, respectively, are quite opposite for WAT and BAT.

Different types of adipose tissue are actively involved in the regulation of energy and fuel homeostasis which have important implications for the development of nutrition-related diseases like obesity and diabetes. Most available reviews cover very specific aspects of WAT or BAT and white or brown adipocytes, respectively. Therefore, we think it timely to provide a comprehensive volume covering the whole range of topics of adipose biology from morphology over function, development to physiological and molecular regulation and heterogeneity. Our aim is specifically to tie together the most recent findings on the molecular mechanisms of adipocyte development and gene expression with the most important histological, physiological, and metabolic characteristics of the different adipose tissues. This includes also central nervous system innervation of the adipose, heterogeneity of white adipose tissue metabolism, and—last but not least—adipose tissue pathology in human obesity.

I am aware that all aspects related to adipose tissue and adipocytes can not be covered exhaustively in just one volume. Therefore here we try to cover the basic aspects of adipose tissue biology, discussing in more detail only obesity as a special metabolic state. Adipose functions during aging, exercise, lactation, in bone morrow, etc. are only mentioned briefly, although we tried to supply references for additional information. Due to space limitations many references are given to overview articles, and I apologize to all authors whose original contributions could therefore not be listed here.

I am very honored and proud that renowned experts from different fields of adipose biology have participated in the effort to put this book together, and I would like to thank all of the authors for their excellent contributions. I sincerely hope that this volume will provide insights into the fascinating biology of adipose tissues not only to the newly interested reader, but also for researchers and clinicians already working in this field.

Susanne Klaus
May 24, 2000

CHAPTER 1

Overview:
Biological Significance of Fat and Adipose Tissues

Susanne Klaus

From a human point of view fat or adipose tissue is nowadays often considered as superfluous and undesired, especially considering the current "emaciated" beauty ideal of the Western world. This could ultimately lead to the provocative question: "Do we really need adipose tissue for normal life?" (see also the introduction to Chapter 3 of this volume). The answer to this question can be approached from different angles. From an evolutionary perspective it was very important especially for homeothermic animals (mammals and birds) to develop the ability to store relatively large amounts of energy. Homeothermic animals are able to keep their body temperature at a constant high level which enables them to be active in cold environments, i.e., very high and low latitudes, during the winter and also at night, thus giving them an advantage over endothermic animals like insects or reptiles. However, this implies a high, continuous energy expenditure for thermogenesis. As food, i.e., an exogenous energy supply is not continuously available, a significant energy reservoir is essential for survival in times of food scarcity. From a clinical point of view, loss of adipose tissue as encountered in syndromes of lipodystrophy or lipoatrophy is associated with severe metabolic complications like diabetes, hypermetabolism, and organomegaly of several organs including the liver (for review see ref. 1). Interestingly, recently developed transgenic mice with virtually no white fat displayed very similar symptoms.[2] Together this implies that adipose tissue is indeed important for normal physiological functions and metabolism and is not only an energy reservoir for emergency situations.

Triglycerides, i.e., fat, represent an ideal energy store (see below) and are organized in adipose tissues, collectively also called "the adipose organ". It was pointed out by S. Cinti that adipose tissue truly fulfills the definition of an organ, because it consists of two main tissue types—white and brown fat—which collaborate in energy partitioning either towards storage (white fat) or thermogenesis (brown fat).[3] In Table 1.1 are listed the different types and subtypes of adipose tissue which can be categorized according to their location and function.[4] This chapter will present a general overview of the diverse biological roles of fat and adipose tissue. The most important functions and features of adipose tissue will then be presented more explicitly in the following chapters.

Fat, the Ideal Energy Store

Fat, i.e., triglycerides (TG) represents an excellent energy reservoir, firstly because of its high energy density, and secondly because it is very little hydrated in contrast to glycogen or

Adipose Tissues, edited by Susanne Klaus. ©2001 Eurekah.com.

Table 1.1. Adipose tissue types and their biological function

Type of adipose tissue	Biological function
white adipose tissue (WAT)	energy reservoir
(subcutaneous, visceral, orbital)	water source (metabolic water)
	insulation (thermal and mechanical)
	endocrine function
mammary gland fat pad	epithelial growth
	lactation
brown adipose tissue (BAT)	thermogenesis
bone marrow fat	energy reservoir
	thermogenesis ?
	hematopoiesis ?
	osteogenesis ?

protein which require up to five times their weight in associated water. As shown in Table 1.2, storage of the same amount of energy in the form of glycogen would imply an 8- to 12-fold weight which would have to be carried around. Considering that an average, non-obese man of 70 kg has about 15 kg of body fat, it is obvious that accumulation of about 150 kg of glycogen (with associated water) instead would not be very practical. Throughout the animal kingdom fat accumulation is therefore the most important mean of energy storage.

According to the energy status of an animal or human, body fat content can vary considerably. Adipose tissue is thus the organ with the greatest plasticity: it can increase its mass by hypertrophic as well as hyperplastic growth (see also Chapter 10). Figure 1.1. gives an example of the large range of body fat stores in the red backed vole (*Clethrionomys glareolus*), a small rodent common in Middle and Northern Europe. Animals which were live-trapped in the winter only had about 3% of total body fat which increased to 13% when they were kept outdoors with food ad libitum, and further increased to 22% under laboratory conditions (food ad libitum, 23 °C ambient temperature, and 16h day light).[5] Interestingly, the maximum body fat content of 22% is within the range of the body fat content of normal weight humans.

Many animals build up adipose depots as an energy reservoir for times of little or no food intake. A classical example are hibernators like the alpine marmot which feeds exclusively in summer and spends six to eight months a year underground without any food intake. Its body fat upon entrance into hibernation is usually between 20 and 30%, sometimes up to 40% of body mass. From typically 1 kg of stored fat only 100-200 g are left in the spring when the

Table 1.2. Isocaloric weight of carbohydrates and protein in comparison to fat (triglycerides)

Macronutrient / energy store	Physiological energy value	Associated water	Isocaloric weight (including water)
Triglycerides	38.9 kJ / g	ca 0.15 ml / g	1
Carbohydrate	17.2 kJ / g	3-5 ml / g	8-12
Protein	17.2 kJ / g	3-5 ml / g	8-12

marmot emerges.[6] Another example are migratory birds which cover great distances during migration without feeding. They rely almost exclusively on fat as oxidative fuel which they accumulate in a subcutaneous depot adjacent to the breast muscle. In migratory birds body fat can attribute as much as 40-50% to body mass, much more than in nonmigratory birds.[7]

Apart from the described cases of intentional fat accumulation, most animals do not continually increase their body fat even when food is abundant; they are very well able to keep a balanced energy budget. One exception is the Israeli sand rat (*Psammomys obesus*), a desert gerbil which is used as an animal model of obesity and diabetes. When these animals are kept under laboratory conditions with unlimited access to laboratory chow, they develop massive obesity and subsequent diabetes. Apparently they are adapted to extreme environmental conditions feeding naturally only on a plant diet low in energy and high in salt content which results in prompt overeating and fat accumulation when they are confronted with "normal" food.[8,9]

Fat and Adipose Tissue in Humans

Like most mammals humans also accumulate fat as the major long term energy reservoir. As illustrated in Figure 1.2., a fat reserve of 15 kg contains 590 MJ, theoretically enough to provide energy for over 50 days, considering an average daily energy expenditure of around 10 MJ.[10] Normal ranges of relative body fat mass are 12-20% and 20-30% in men and women, respectively, with the larger portion present in subcutaneous depots. A body fat content of over 25% and 33% in men and women, respectively, is considered as critical.[11] Depending on the anatomical location, adipose depots display considerable differences in metabolic activity. Subcutaneous fat is much less metabolically active than visceral fat which means that not fat mass alone but also body fat distribution can have profound effects on metabolic complications associated with obesity (see Chapters 8 and 10 for more detailed discussions).

Adiposity of the Human Neonate and Infant

In comparison with other mammals the human newborn is an exceptionally "obese" neonate.[12] Most mammalian neonates including primates have very little body fat, usually below 3% of body weight. The human newborn on the other hand has a body fat content of about 15%, mainly in subcutaneous regions.[13] The only other mammals known to have comparably high body fat at birth are the harp seal and the guinea pig with about 10% of body weight as fat (reviewed in ref. 12). Although humans are already born relatively plump, they still increase

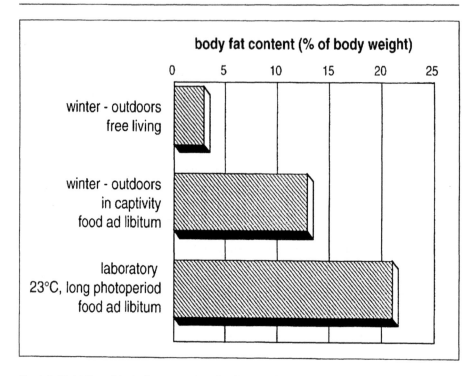

Fig. 1.1. Variability of body fat content in a free living rodent. Body fat content of red backed voles (*Clethrionomys glareolus*) caught in the wild varies depending on the environmental conditions.

body fat in early infancy reaching an adiposity peak of about 25% at 6-9 months of age after which it declines to leaner childhood values. About 40-65% of body weight gain during the first 6 months is accounted for by body fat deposition.[14]

Possible reasons for the "adiposity" of the human neonate and infant were recently discussed exhaustively by C. Kuzawa.[12] A common explanation is that the abundant subcutaneous fat of neonates is a compensation for the lack of insulating fur, thus serving as a protection against heat loss. However, according to Kuzawa there is no empirical support for this hypothesis (see also below). He rather brings forward two related hypotheses which both stress the importance of fat as an energy buffer during human infancy. According to the first hypothesis, the large human brain requires greater energy backups. In the human neonate the brain contributes 50-60% to total energy expenditure (in the adult human it declines to about 20% of resting energy expenditure).[15] The big fat i.e., energy reservoir of infants might thus represent an adaptation to the increased energy requirements associated with hominid encephalization.

The second hypothesis formulated by Kuzawa states that developmental changes in adiposity of the infant parallel the likelihood of nutritional disruption.[12] Prenatal fat deposition seems to be important in species which experience starvation or a negative energy balance in the early postnatal period. Seal pups for instance are subjected to a prolonged postweaning fast and guinea pigs have to mobilize fat after birth because their breast milk is very low in energy content. Human newborns are also forced to mobilize fat at birth to compensate for the disruption in flow of maternal nutritional support at parturition before lactation is established. Weaning is another critical period of major nutritional transition which is furthermore associated with an increased risk of infections as the infant is switched from sterile mother's milk to

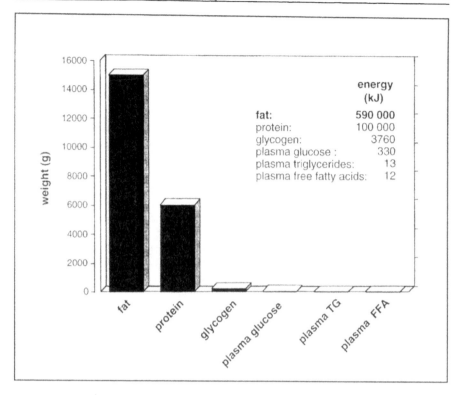

Fig. 1.2. Energy stores in human. Weight of main energy stores of a normal weight male of 70 kg body mass. The inset lists the corresponding energy content. Data are from ref. 10.

solid food. Infections have high nutritional costs, and epidemiological data show clearly that well-nourished infants and children have less severe infections and a lower mortality risk. Peak fat reserves of well-nourished infants are indeed acquired at an age when prevalence of malnutrition and infectious diseases are highest in more unfavorable conditions. Taken together, most evidence points to an importance of early infant body fat as an energy buffer, ensuring survival at critical periods.[12] This does not exclude however, that subcutaneous fat also may contribute to thermal insulation as discussed below.

Additional Functions of Fat and Adipose Tissue

Insulation

The thermal conductivity of fat, i.e., fatty tissues, is only about half that of muscle tissue (reviewed in ref. 16). A subcutaneous fat layer should therefore provide an increased protective thermal insulation against cold by reducing the rate of heat loss. Several human studies have shown a direct correlation of abundance of subcutaneous fat depots with resistance to heat loss in a cold environment. Obese subjects were also shown to have a reduced metabolic response to cold stress compared to lean subjects (for detailed overview see ref. 16). An increased subcutaneous fat deposition is recognized as one tactic available for homeotherms to reduce the rate of heat loss and thus conserve energy for thermogenesis.[17]

The insulator properties of fat are especially important for aquatic mammals, like seals or whales who swim in near freezing water.[18] The thermal conductivity of water is much higher than that of air and taking convection into account, the rate of heat loss in water can be 50 to 100 times higher than in air. Surely everybody has made the experience that even in relatively tepid water one cools out rather rapidly. Aquatic mammals like seals and whales therefore have thick layers of subcutaneous fat or blubber as thick as 50 mm providing effective insulation and thus minimizing heat loss (the skin temperature of seals is indeed very close to environmental temperature whereas inside the blubber layer it is close to core body temperature).[18]

However, the insulator function of fat in terrestrial animals and mammals is controversial. Reviewing literature, several authors concluded that there is very poor experimental evidence supporting this theory; and especially in humans little or no evidence points to fat tissue as an important means of adaptation to cold environments or cold stress.[12,19-22] Epidemiological studies showed that populations from arctic regions, like Eskimos, do not have increased subcutaneous fat; adults but also infants and children have a similar or even reduced skinfold thickness compared to populations from moderate climates.[22,23] It seems that in human adaptation to cold the capability to increase metabolic rate, i.e., heat generation, is more important than the conservation of energy by improving insulation.[21] Although subcutaneous fat certainly contributes to thermal insulation, it seems that at least in humans this is rather a by-product but not the main physiological function of adipose tissue.

Often overlooked is the function of adipose tissue in mechanical insulation, i.e., cushioning against mechanical forces.[4] At many sites where soft tissue lies in contact with bone, fat serves as a cushion. Orbital fat for example serves as a mechanical protection for the eyeballs. Ample subcutaneous fat could also provides some cushioning of externally inflicted forces during a fall for instance.

Source of Metabolic Water

Oxidation of macronutrients leads to the production of CO_2 and H_2O, the so-called metabolic water. Some animals can rely solely on metabolic water for maintenance of their water balance. The kangaroo rat—a desert animal—for example does not need to drink, even on a diet of relatively dry seeds it survives without free water.[24] Oxidation of fat leads to a higher yield in metabolic water than carbohydrates and protein (109 g per 100 g of fat as compared to 60 and 44 g per 100 g of carbohydrates and protein, respectively). Fat is therefore not only an ideal energy store, but also a potential water source. Newborn whales ingest only milk and beside the water content of milk, milk fat is their only source of water for building up the body substance. Animals for which fatty acid oxidation is an important water source include desert animals and marine mammals.[25]

It is not known if the production of metabolic water by fat oxidation has any relevance in humans. An interesting hypothesis has recently been brought forward by J. Stookey who speculates that there might be a link between hydration status and human obesity. According to this theory, poor hydration status leading to a hypertonic environment might stimulate a preferential intake of fat as an oxidative fuel.[26] It is well established that dietary fat intake is related to the development of obesity. However, so far there is no experimental evidence available to support this hypothesis.

Heat Production

Brown adipose tissue (BAT) is a specialized adipose tissue present only in mammals whose main function is the generation of heat for maintenance of body temperature, arousal of hibernation and dissipation of excess food energy. Essential for this thermogenic function is the presence of the uncoupling protein 1 (UCP1). Although BAT shares many metabolic characteristics with WAT, its function is essentially an opposite one to the classical white fat, i.e., energy dissipation in contrast to energy storage. From another point of view, the two types of adipose tissues represent counter actors in energy partitioning, channeling lipid energy either versus accumulation in white fat or oxidation, i.e., dissipation in brown fat. Although brown and white adipocytes represent different cell types (see also Chapter 4 and 5), there is no strict topological separation of the two fat types. Even within typical white fat depots some brown adipocytes can be detected under appropriate conditions. Adipocytes within brown fat depots on the other hand can acquire typical white adipocyte morphology together with loss of thermogenic function e.g., in the state of obesity (see Chapter 2 for further information).

The biochemical mechanisms of heat production in BAT as well as its physiological regulation will be discussed in detail in Chapter 4. Chapter 5 will give insights into differentiation of brown adipocytes, control of UCP1 gene expression and the role of BAT in energy metabolism. For further insight into brown fat ultrastructure and innervation, the reader is referred to Chapters 2 and 6, respectively.

Mammary Adipose Tissue

Two other types of adipose tissue with rather specialized function will be discussed here briefly, that is mammary adipose tissue and bone marrow fat.

Mammary adipose tissue—also called the mammary fat pad—represents a discrete category of white adipose tissue. The ultimate function of the mammary gland is of course the synthesis of milk to nourish the offspring. An important constituent of the mammary gland is the stroma which forms a matrix of connective and adipose tissue termed the "mammary fat pad" (see ref. 27 for review). In recent years it has become more and more apparent that this mammary fat pad plays an important mediatory and supportive role in mammary gland function. Mammary fat provides lipids and energy for milk production but it is also a site for the production of different factors. Mounting evidence is currently arising that the action of several steroid and peptide hormones on mammary gland function is mediated by factors synthesized by constituents of the mammary fat pad such as growth factors, proteases and components of the extracellular matrix (ECM). For more detailed insights the reader is referred to a recent review by R. Hovey and coworkers.[27]

Bone Marrow Fat

Adipocytes are the most abundant stromal cells in adult human bone marrow where apparently they play an important role in health and disease. Although in the human newborn hardly any adipocytes can be found in the marrow, number and size increases with age, and fat eventually occupies over 50% of the human marrow cavity (see ref. 28 for review). Bone marrow stroma contains multipotential mesenchymal stem cells that can differentiate into adipogenic, osteogenic, fibroblastic as well as hematopoietic and myelo-supportive cells (see ref. 28, 29 and references within). Several cell lines have been derived from bone marrow, even a preadipocyte cell line expressing UCP1, i.e., cells capable of differentiation into brown adipocytes.[30] If there are any brown adipocytes present in vivo, i.e., if thermogenesis is a potential function of bone marrow fat is not known today. An immortalized clone from bone marrow of an adult mouse was demonstrated to be quadripotential, able to differentiate into adipose and cartilage and support osteoclast and bone formation.[31] However, not only are there different stem cell types present, but there seems to exist a certain plasticity among the stromal cell phenotypes and a complex interrelationship between bone marrow adipocytes and other bone marrow derived lineages.[28] Clonal adipogenic cells from human bone marrow were recently shown to be able to dedifferentiate and subsequently redifferentiate into osteogenic and adipogenic cells.[29] Plasticity, i.e., the ability of differentiated stromal cells to shift phenotype is apparently a characteristic of bone marrow stroma (see ref. 32 for review). Although adipocytes within the bone marrow certainly function as a localized energy reservoir, other functions in hematopoiesis and osteogenesis, possibly also thermogenesis can not be excluded. Certainly the plasticity of bone marrow stromal cells including adipocytes even in the adult individual warrants further research attention.

Endocrine Function

It is now generally accepted that adipose tissue is not only an inert energy reservoir but an active participator in energy regulation. Notably since the discovery of leptin in the '90s (see ref. 33 for review), it became evident that adipose tissue is a source of factors which have not only paracrine but also endocrine function. According to our current state of knowledge, adipose tissue lies at the heart of a network of autocrine, paracrine and endocrine signals (see ref. 34 and Chapter 9 of this volume for reviews). Factors synthesized and released by adipose tissue include cytokines (e.g., leptin, TNFα, interleukins), components of the complement pathway (e.g., factor D = adipsin, C3adesArg = acylation stimulating protein) and other factors including angiotensinogen and plasminogen activator inhibitor-1 (PAI-1). This list is by no means complete and the exact relevance of the adipose tissue derived portion of many of these factors is still not clear. However, the last few years have witnessed tremendous progresses in our understanding of adipose tissue action in many physiological networks. A detailed overview on adipose tissue as a secretory and endocrine tissue will therefore be presented in Chapter 9 of this volume.

Fat in Tissue Engineering

Only in recent years research effort has been directed towards tissue engineering of fat (for review see ref. 35). Fat could potentially provide soft tissue equivalents important for reconstructive surgery in congenital malformations, posttraumatic repair, cancer rehabilitation, and other soft tissue defects, but also in cosmetic surgery, e.g., breast augmentation, filling out of wrinkles etc. Although autologous transplantation of fat has already been attempted over 100 years ago, transplantation of free fat did not prove to be very successful, mainly because of the gradual absorption of the transplanted fat (see e.g., ref. 35, 36 for review). The advancing discipline of tissue engineering however, has recently become interested in fat, i.e., preadipocytes or adipocytes as a source for implants. The aim of tissue engineering is to provide techniques either allowing the regeneration of tissues ex vivo or to induce site specific regeneration in situ e.g., by injection of potent bioactive factors.

Preadipocytes isolated from adipose tissue should be the ideal basis for tissue engineering of fat equivalents for autologous implantation because they possess proliferative, angiogenic and adipogenic potential. Adipose tissue is furthermore very abundant in humans and can easily be obtained, for instance by the liposuction technique. This relatively mild invasive technique was shown to provide significant numbers of viable preadipocytes and even mature adipocytes.[37]

Recently Patrick et al have used biocompatible polymer scaffolds into which they seeded primary preadipocytes isolated from rat adipose tissue. They could demonstrate that these preadipocytes were able to attach, proliferate, and differentiate into mature adipocytes within the scaffolds, both in vitro and in vivo.[38] An alternative or complementary approach for cell support could involve the use of extracellular matrix (ECM) proteins. It is well established that ECM components are crucial for various cell functions including growth, migration, differentiation and metabolism.[35] The aim of all these strategies is ultimately the development of clinically translatable adipose tissue equivalents.

Today engineering of adipose tissue is still in its infancy. However, as pointed out by Katz et al, autologous cell-based approaches for the engineering of fat equivalents are beginning to emerge, following our rapidly progressing insights into the molecular mechanisms of adipocyte development and differentiation. Final strategies will probably involve autologous preadipocytes, biocompatible polymer scaffolds, EMCs and bioactive factors resulting ultimately in the ex vivo fabrication of adipose tissue implants.[35]

Conclusions and Outlook

Adipose tissue is certainly not a homogenous, inert fat reservoir, but it rather represents a multitude of different cell and tissue types playing active and quite diverse functions, not only in energy metabolism. Many of these functions (e.g., secretion of leptin and its role in body weight homeostasis and reproduction) have only been discovered in recent years and are far from completely understood. According to location and environment, adipocytes can display very different metabolic properties. Different adipose depots might contain different precursor cells and possibly different molecular mechanisms are governing the development and differentiation of adipocytes in different regions.

The elucidation of adipose tissue and adipocyte function in physiological networks as well as the molecular mechanisms of adipocyte development and differentiation is still far from complete although considerable progress has been made in the past few years. The future will certainly bring deeper insights which will be beneficial not only in the context of obesity research but also for the development of tissue engineered fat which will be useful as a soft tissue equivalent in reconstructive and cosmetic surgery.

References

1. Seip M, Trygstad O. Generalized lipodystrophy, congenital and acquired (lipoatrophy). Acta Peaediatr (Suppl) 1996; 413:2-28.
2. McKnight SL. WAT-free mice: Diabetes without obesity. Genes Develop 1998; 12:3145-3148.
3. Cinti S. The adipose organ. Milano: Editrice Kurtis, 1999.
4. Gimble JM, Robinson CE, Clarke SL et al. Nuclear hormone receptors and adipogenesis. Crit Rev Euk Gene Expr 1998; 8(2):141-168.
5. Klaus S. Die jahreszeitliche Akklimatisation der Wärmebildung freilebender Apodemus flavicollis, Apodemus sylvaticus und Clethrionomys glareolus. (Dissertation) In: Intemann C, Intemann A, eds. Wissenschaftliche Forschungsbeiträge Biologie/ Biochemie/ Chemie, Volume 14. München: Intemann, 1988.
6. Bibikov A. Die Murmeltiere. In: Die neue Brehm Bücherei, Wittenberg Lutherstadt: Ziemsen Verlag, 1968:80-84.
7. Dawson WR, Marsh RL, Yacoe ME. Metabolic adjustments of small passerine birds for migration and cold. Am J Physiol 1983; 245(6):R755-67.
8. Marquie G, Duhault J, Jacotot B. Diabetes mellitus in sand rats (*Psammomys obesus*). Metabolic pattern during development of the diabetic syndrome. Diabetes 1984; 33(5):438-443.
9. Collier GR, De Silva A, Sanigorski A et al. Development of obesity and insulin resistance in the Israeli sand rat (*Psammomys obesus*). Does leptin play a role? Ann N Y Acad Sci 1997; 827:50-63.
10. Cahill GF. Starvation in man. N Engl J Med 1970; 282: 668-675.
11. Bray GA, Bouchard C, James WPT (1998) Definitions and proposed current classification of obesity. In: Bray GA, Bouchard C, James WPT, eds. Handbook of Obesity. New York: Marcel Dekker Inc, 1998:31-40.
12. Kuzawa CW. Adipose tissue in human infancy and childhood: An evolutionary perspective. Yrbk Phys Anthropol 1998; 41:177-209.
13. McChance R, Widdowson E. Fat. Pediatr Res 1977; 11:1081-1083.

14. Fomon SJ, Haschke F, Ziegler EE et al. Body composition of reference children from birth to age 10 years. Am J Clin Nutr 1982; 35:1169-1175.
15. Holliday M. Body composition and energy needs during growth. In: Falkner F, Tanner JM, eds. Human growth: A Comprehensive Treatise. New York: Plenum Press, 1986:117-139.
16. Anderson GS. Human morphology and temperature regulation. Int J Biometeorol. 1999; 43(3):99-109.
17. Blix A, Steen J. Temperature regulation in newborn polar homeotherms. Physiol Rev 1979; 59:285-304.
18. Schmidt-Nielsen K. Animal Physiology: Adaptation and Environment. Cambridge: Cambridge University Press. 3rd ed. 1983:262-284.
19. Pond C. Morphological aspects and the ecological and mechanical consequences of fat deposition in wild vertebrates. Annu Rev Ecol Syst 1978; 9:519-570.
20. Pond C. The biological origins of adipose tissue in human. In: Morbeck M, Galloway A, Zihlman A, eds. The Evolving Female: A Life History Perspective. Princeton: Princeton University Press, 1997:147-162.
21. Stini W. Body composition and nutrient reserves in evolutionary perspective. World Rev Nutr Diet 1981; 37:55-83
22. Eveleth P, Tanner J. Worldwide Variation in Human Growth. Cambridge: Cambridge University Press, 1976.
23. Shepard R. Body Composition in Biological Anthropology. Cambridge: Cambridge University Press, 1991.
24. Schmidt-Nielsen K. Desert Animals: Physiological Problems of Heat and Water. Oxford: Clarendon Press, 1964.
25. Garrett RH, Grisham CM. Biochemistry. Fort Worth: Saunders College Publishing, 2nd ed. 1999:790.
26. Stookey JD. Another look at: fuel + O2 —> CO2 + H2O. Developing a water-oriented perspective. Med Hypotheses 1999; 52(4):285-290.
27. Hovey RC, McFadden TB, Akers RM. Regulation of mammary gland growth and morphogenesis by the mammary fat pad: A species comparison. J Mammary Gland Biol Neoplasia 1999; 4(1):53-68.
28. Gimble JM, Robinson CE, Wu X et al. The function of adipocytes in the bone marrow stroma: an update. Bone 1996; 19:421-428.
29. Park SR, Oreffo RO, Triffitt JT. Interconversion potential of cloned human marrow adipocytes in vitro. Bone 1999; 24(6):549-554.
30. Marko O, Cascieri MA, Ayad N et al. Isolation of a preadipocyte cell line from rat bone marrow and differentiation to adipocytes. Endocrinology 1995; 136(10):4582-4588.
31. Dennis JE, Merriam A, Awadallah A et al. A quadripotential mesenchymal progenitor cell isolated from the marrow of an adult mouse. J Bone Miner Res 1999; 14(5):700-709.
32. Bianco P, Riminucci M, Kuznetsov S et al. Multipotential cells in the bone marrow stroma: Regulation in the context of organ physiology. Crit Rev Eukaryot Gene Expr 1999; 9(2):159-173.
33. Friedman JM, Halaas JL. Leptin and the regulation of body weight in mammals. Nature 1998; 395:763-770.
34. Mohamed-Ali V, Pinkney JH, Coppack SW. Adipose tissue as an endocrine and paracrine organ. Int J Obes 1998; 22:1145-1158.
35. Katz AJ, Llull R, Hedrick MH et al. Emerging approaches to the tissue engineering of fat. Clin Plast Surg 1999; 26(4):587-603.
36. Coleman SR. Long-term survival of fat transplants: Controlled demonstrations. Aesthetic Plast Surg 1995; 19:421-425.
37. Moore JH Jr, Kolaczynski JW, Morales LM, et al. Viability of fat obtained by syringe suction lipectomy: Effects of local anesthesia with lidocaine. Aesthetic Plast Surg 1995; 19:335-339.
38. Patrick CW, Chauvin PB, Hobley J et al. Preadipocyte seeded PLGA scaffolds for adipose tissue engineering. Tissue Eng 1999; 5(2):139-151.

CHAPTER 2

Morphology of the Adipose Organ

Saverio Cinti

The adipose organ (Fig. 2.1) of mammals is composed of two different tissues: white and brown adipose tissues.[1-3] In mammals these tissues are organized in distinct depots or are diffuse around or within other organs.

The depots are subcutaneous (anterior and posterior in rats and mice) or visceral (mediastinal, mesenteric, omental, perirenal, perigonadal and retroperitoneal).

Adipocytes also "infiltrate" many organs: skin, synovia, parathyroid, parotid, lymph nodes, bone marrow, pancreas, thymus.

All the subcutaneous and visceral depots contain both tissues (white and brown) and the relative amount of the two types of adipose tissue depends on age, strain and environmental conditions to which the animal is exposed.

Most adipocytes "infiltrating" other organs (see above) show the morphology of white adipocytes.

White Adipose Tissue

General Aspects

The white adipose tissue is composed of spherical cells with a diameter ranging from 15 to 150 μm (Fig. 2.2) in aldehyde-fixed and paraffin or resin embedded specimens.

This enormous variability in size is due to the ability of the cell to accumulate different amounts of lipids (triglycerides) that form a single vacuole in the jaloplasm. Electron microscopy of aldehyde-fixed tissue reveals unstructured, poorly dense material in the lipid droplets (Fig. 2.3). Their surface in contact with the jaloplasm is devoid of any unitary membrane. Mitochondria are elongated and show short and randomly oriented cristae. Rough endoplasmic reticulum is organized in short cisternae, but sometimes it forms stacks of variable size. Smooth endoplasmic reticulum is always well visible. Pinocytotic vesicles are present at the level of the plasma membrane. On the external side of the plasma membrane a distinct external lamina is always visible. The amount of the organelles as well as their size and extension are variable in relationship to the cell's functional and developmental stages.[3]

White Adipocyte Morphology During Lipolysis

During the process of delipidation the size of the lipid droplets reduces progressively. Usually a main lipid droplet (i.e., a lipid droplet that contains about 80-90% of the total lipid

Adipose Tissues, edited by Susanne Klaus. ©2001 Eurekah.com.

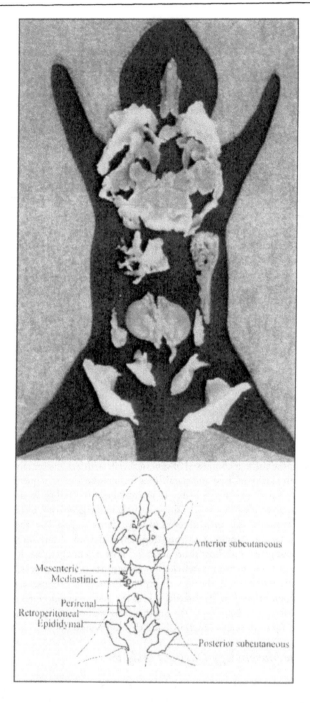

Fig. 2.1. Gross anatomy of the dissected adipose organ of a 12 days old rat. Anterior and posterior subcutaneous depots as well as mesenteric, mediastinal, perirenal, retroperitoneal and epididymal depots are
visible. Note that most of the anterior subcutaneous depot, the mediastinal and the perirenal depots appear
brown (here darker gray scales). 1.3x.

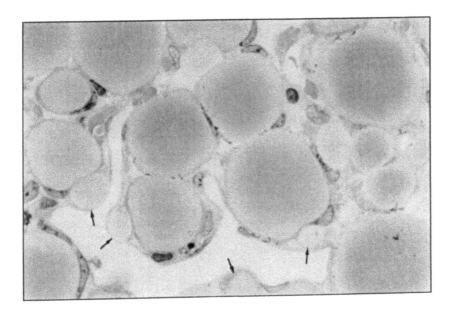

Fig. 2.2. Light microscopy of epididymal adipose tissue of a 12 days old rat. Many unilocular adipocytes are visible. Some unilocular small adipocytes are visible (arrows) along the walls of the capillary visible in the bottom. Toluidine blue-resin embedded tissue. 1,075x.

content of the cell) is visible until the latest stages of delipidation, but some minor lipid droplets can also be seen during the process of delipidation. Smooth endoplasmic reticulum often forms cisternae closely apposed to the lipid droplets.

Adipocytes display cytoplasmic processes, apparently directed towards capillaries, during the first stages of delipidation. These cytoplasmic processes show characteristic invaginations rich in pinocytotic vesicles. During the following steps of delipidation the adipocytes elongate and become progressively thinner. The cytoplasmic processes can be numerous and always in close contact with capillaries. The invaginations with the pinocytotic vesicles can be diffuse along most of the cellular surface.[4]

The tannic acid method demonstrates membranous structures that have been interpreted as free fatty acids.[5] During the delipidation process these membranous structures can be seen at different levels within the adipocytes (Fig. 2.4):

1. at the lipid droplet surface;
2. in close contact with cytoplasmic organelles;
3. apparently free in the jaloplasm;
4. in close association with the invaginations and related pinocytotic vesicles.

The same structures can also be seen outside the adipocytes:

1. in close association with the outer surface of the adipocyte plasma membrane;
2. in the interstitial tissue surrounded by the cytoplasmic invaginations;
3. free in the interstitial tissue;
4. in the wall and lumen of capillaries.

The topographic position of the membranous structures in slimming adipocytes suggests that the cytoplasmic invaginations described above play a role in the process of fatty acid transport from adipocytes into the blood.

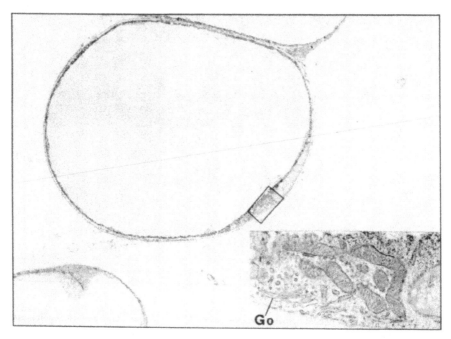

Fig. 2.3. Electron microscopy of inguinal adipose tissue (part of subcutaneous posterior depot) of a 6 days old rat. Unilocular adipocytes are visible. Inset: enlargement of the framed area. Elongated mitochondria with randomly oriented cristae are visible. Go: Golgi complex. 3,300x; inset 22,500x.

White Adipocyte Morphology During Development

In rats, white adipocytes appear in the perinatal period. In newborns, the biggest white depot is the posterior subcutaneous depot (composed by the inguinal, dorsolumbar and gluteal parts).

The first stages of development appear in the inguinal region during the last days of intrauterine life and the histological appearance of this area is that of a well-vascularized mesenchymal tissue.

Glycogen granules and small lipid droplets in poorly differentiated cells of white adipose tissue depots are indicative of adipocytic differentiation.[6] These signs in the developing adipose organ are visible in pericytes, i.e., in poorly differentiated cells of the capillary wall and in splittings of the endothelial external lamina (Figs. 2.5, 2.6).[5,6]

A progressive accumulation of lipids is the major modification in the subsequent steps ending in the formation of mature unilocular spherical cells. This process of lipid accumulation can pass through a transient multilocular stage, but the unilocular stage is reached very early in many adipocytes, giving rise to a unique vacuole even in small unilocular cells. Pinocytotic vesicles and all other organelles described for the mature adipocytes are present in the precursors since the very early steps of development. A distinct external lamina is visible during all the steps of development.

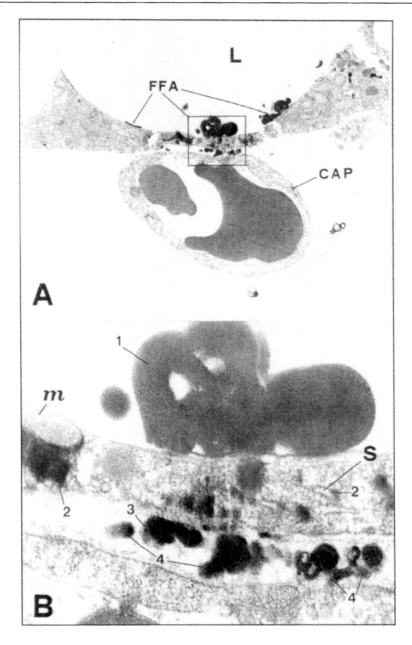

Fig. 2.4. A) Electron microscopy of retroperitoneal adipose tissue of an adult rat treated for 7 days with a beta3-adrenergic agonist. Tannic acid method (see text). The periphery of a unilocular adipocyte close to a capillary (CAP) containing two erythrocytes is visible. At the periphery of the lipid droplet (L) facing the cytoplasm close to the capillary, different amounts of myelinic figures, probably representing the fatty acids (FFA), are visible. 11,200x. B) Enlargement of the framed area in A. Myelinic figures are here visible: at the lipid droplets surface (1), in close contact with organelles (2), in close association with the outer surface of the adipocyte plasma membrane (3), in the interstitial tissue (4). m: mitochondrium; S: smooth endoplasmic reticulum. 80,500x.

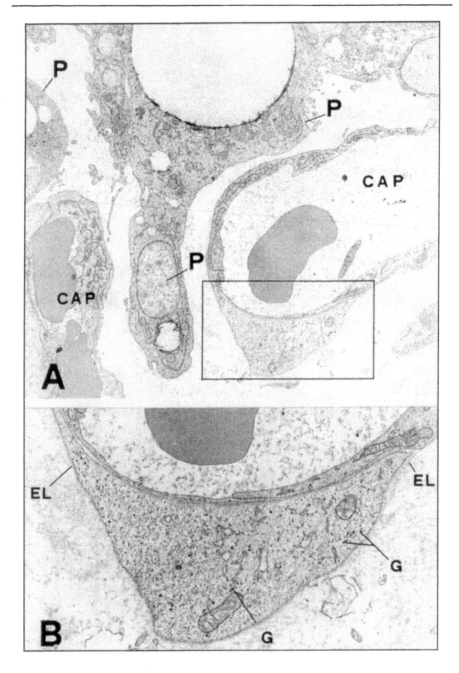

Fig. 2.5. A) Electron microscopy of epididymal adipose tissue of a 12 days old rat. In close association with two capillaries (CAP), adipocyte precursors (P) at various stages of differentiation are visible. 6,300x. B) Enlargement of the framed area in A. This element in pericytic position shows the morphologic character- istics of a blast cell (scarce organelles and cytoplasm rich of ribosomes and polyribosomes). Some glycogen particles (G) are also visible. EL: external lamina. 16,100x.

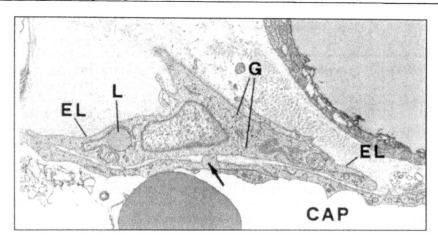

Fig. 2.6. Electron microscopy of epididymal adipose tissue of a 12 days old rat. This poorly differentiated cell is in tight connection (arrow) with the endothelial cell of the capillary (CAP), it is surrounded by a distinct external lamina (EL), and shows early signs for adipocytic differentiation: lipid droplet (L) and glycogen particles (G). 8,700x.

Leptin and S-100 Expression in White Adipocytes

Leptin is the satiety hormone secreted by the adipose organ.[7-10] Immunohistochemistry revealed detectable levels of leptin in white adipocytes of different sizes, even in small and multilocular cells of fed animals.[11] By confocal microscopy, it has been proposed that leptin is present in the endoplasmic reticulum of white adipocytes.[12] White adipocytes of fasting animals show less but still detectable staining of the adipocyte cytoplasm.[11]

S-100 protein is a calcium binding protein with important implications in the cytoskeleton rearrangement in different cell types.[13] Mature white adipocytes and precursors since the early steps of differentiation express S-100 protein.[14] The functional role of S-100 protein in the different cell types in which it is expressed is not yet well elucidated but some data suggest that this protein is more expressed when the cytoskeleton has to be less structured, i.e., when the jaloplasm must be more fluid.[13] Therefore a possible role for S-100 in white adipocytes could be that of a permissive role for lipid accumulation into a single vacuole deforming the cell very early in its ontogenesis. Of interest, classical multilocular brown adipocytes (see below) such as those found in interscapular brown adipose tissue of cold-exposed animals do not express immunoreactive leptin or S-100 protein.[11,14,15]

Other Cell Types in White Adipose Tissue

Besides the cells of vascular and neural structures, a few other cell types are always present in white depots: fibroblasts, histiocytes and mast cells are the most commonly observed. They are present in the tissue since the early phases of development.

Fibroblasts are the constitutive cells of the connective tissue.[16,17] Their main function is to produce the connective matrix. Their morphology shows a very elongated fusiform cell with thin cytoplasmic projections starting at the two cellular poles or from lateral projections (Fig. 2.7). These cytoplasmic projections are always in close connection with collagen fibrils and often they surround groups of collagen fibers. The most characteristic organelle of fibroblasts is the rough endoplasmic reticulum. This organelle occupies most of the fibroblast cytoplasmic volume and it is often dilated and rich with proteinaceous substance that is secreted into the interstitial space via the secretory apparatus: Golgi complex and vesicles.

Fibroblasts and fibroblast-like cells have been claimed to correspond to adipocyte precursors and as delipidized adipocytes.[18,19] The morphology of fibroblasts is highly characteristic as described above and shown in Fig. 2.7 and differs from that of adipocyte precursors as well as from that of delipidized adipocytes in several aspects of which the most important are: absence of a distinct external lamina, presence of thin cytoplasmic projections in close connection with collagen fibrils, abundance of dilated rough endoplasmic reticulum, absence of relevant amounts of glycogen and lipids, absence of S-100 protein expression.[20]

Histiocytes are the tissue counterpart of monocytes.[16,17] Their main role is to phagocyte all tissue debris that must be removed from tissues. In white adipose tissue they probably play a role in the remodeling of the tissue that is necessary for the adaptations to the different amounts of energy (lipids) that have to be stored or removed. When these processes happen suddenly (excess of food intake or fasting), the need for a rapid remodelling of the tissue increases and their function becomes indispensable. Histiocyte morphology shows cells with several irregular cytoplasmic projections and numerous primary and secondary lysosomes in the cytoplasm. Usually the Golgi complex is well developed. During phagocytosis the cytoplasmic projections surround and incorporate the debris that have to be removed. Subsequently the numerous lysosomes of histiocytes digest the phagocyted debris in phagolysosomes. During fasting many histiocytes surround delipidizing adipocytes and many lipid droplets are visible in the phagolysosomes of histiocytes. Probably the abundance of fatty acids present in the tissue is a strong signal for a need of tissue remodelling and this causes histiocyte activation. This would explain also the abundance of histiocytes during the development of white depots in the adipose organ. In fact, in this condition free fatty acids are abundant in the tissue and a continuous remodelling of the tissue is necessary to allow the recruitment and maturation of adipocytes.

Mast cells are the functional counterpart of blood basophils.[16,17] They play a role in the immune system but their significance in white adipose tissue is unknown. Their morphology is highly characteristic mainly because of big dense cytoplasmic granules rich in histamine and leukotrienes.

Brown Adipose Tissue

General Aspects

Brown adipocytes are usually a spheres with a diameter ranging from 10 to 25 μm. The nucleus is spherical and often located in the central part of the cell. The cytoplasm is rich in big mitochondria with numerous cristae. The lipids are organized in many small vacuoles. This multilocular arrangement of triglycerides should be energy demanding for the cell because the physicochemical properties of neutral lipids in an aqueous ambience tend to favor the coalescence and formationof a unique vacuole. Therefore a mechanism should be activated in this cell to allow the multilocularity of the lipid depot. This mechanism is yet unknown but might involve a protein called perilipin which has been localized on the surface of the lipid droplets in brown adipocytes.[21]

The ultrastructural appearance of the lipids is identical to that described for the lipids in white adipocytes including the interface with the jaloplasm. Smooth endoplasmic reticulum is abundant and, when lipolysis is prevalent, long cisternae are closely associated with the lipid droplets. Rough endoplasmic reticulum and Golgi complex are usually poorly developed.

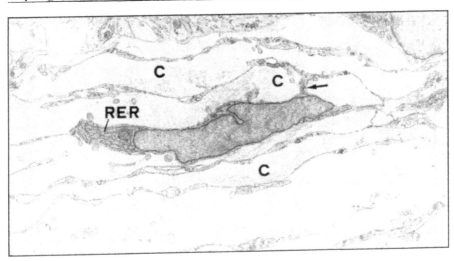

Fig. 2.7. Electron microscopy of mesenteric adipose tissue of a 15 days old rat. A fibroblast with thin cytoplasmic projections is visible. Note that most of the cytoplasm is occupied by rough endoplasmic reticulum (RER). The cytoplasmic projections, whose connection with the body of the cell is sometimes visible (arrow), form parallel structures separating laminae of extracellular matrix rich of collagen fibrils (C). 8,700x.

Glycogen is present and abundant when liposynthesis is prevalent. Many pinocytotic vesicles are present at the plasmalemma level. A distinct external lamina is visible at the external side of the plasma membrane. Brown adipocytes are joined by gap junctions.

Brown Adipocytes in Cold-Exposed Animals

The exposure of mammals to temperatures under that of thermoneutrality causes thermogenesis in brown adipose tissue.[22] This corresponds to morphological modifications. A few hours of cold (4-8°C) exposure in adult mice or rats causes an almost complete disappearance of the lipid droplets from the cytoplasm of most of the brown adipocytes.

In these cells big mitochondria with well-developed cristae occupy most of the cytoplasmic volume. These adipocytes are intensely immunoreactive for UCP1.[23-25]

After chronic cold exposure many lipid droplets are present in the cytoplasm of brown adipocytes (Fig. 2.8).

During the first days of cold exposure many mitoses are visible and electron microscopy revealed that most are due to the proliferation of brown adipocyte precursors and endothelial cells. The gap junctions between brown adipocytes in cold exposed animals enlarge .[26-28]

Both in acute and chronic cold exposure conditions, brown adipocytes do not express detectable levels of leptin and S-100 protein by immunohistochemistry.

Brown Adipocytes in Warm-Exposed Animals

The functional activity of brown adipocytes is reduced in warm environments and, by definition, thermoneutrality is the temperature at which brown adipocytes do not produce heat. Thermoneutrality varies with species and, e.g., it is 28°C for rats and 34°C for mice.[22,29]

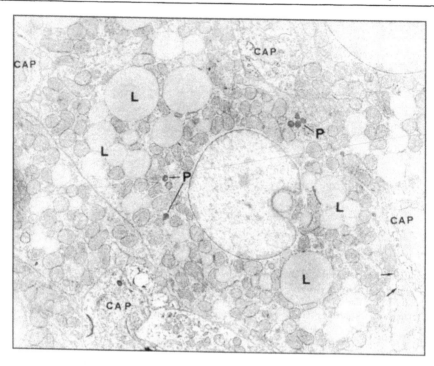

Fig. 2.8. Electron microscopy of mediastinic brown adipose tissue of an adult rat exposed for several weeks to cold (about 5°C). Numerous lipid droplets (L) and typical mitochondria. Small axons (arrows) and capillaries (CAP) are visible. P: peroxisomes. 8,700x.

After few days at thermoneutrality the morphology of brown adipocytes changes: in the interscapular brown adipose tissue (IBAT), most of the cells in mice and many of the cells in rats become unilocular. Therefore, these cells are similar to white adipocytes with a progressive transformation of the organelles (including mitochondria; Fig. 2.9). Furthermore, these brown adipocytes transformed into unilocular white like-cells express proteins usually found in white adipocytes and not found in classical multilocular adipocytes by immunohistochemistry: leptin and S-100 protein.

Brown Adipocytes Morphology During Development

In rats, the anatomical area where the first brown adipose tissue anlage develops is a specific distinct bilateral area, in the dorsal part of the trunk, adjacent to the spinal cord, between the superficial and deep dorsal skeletal muscles. In this area at the embryonal day 15 (E15) only mesenchymal tissue and very large capillaries are visible by light microscopy.

Studying this area by electron microscope in the perinatal period, we observed all the morphological developmental stages of brown adipocytes. The first step of differentiation (adipoblast stage 1) is represented by blast-like cells with numerous pre-typical mitochondria (Fig. 2.10). These cells progressively increase the number of mitochondria whose morphology approach step by step that of typical brown mitochondria (typical mitochondria expressing UCP1 are visible at E19 in brown adipocytes). In the second step of differentiation, glycogen (a small amount of glycogen is usually present since the first step of differentiation) and lipids

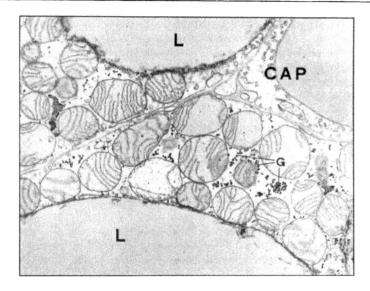

Fig. 2.9. Electron microscopy of brown adipocytes of interscapular brown adipose tissue of an adult rat maintained two weeks at 25°C. Mitochondria are small and their cristae are less numerous then that present in mitochondria of cold exposed rats (compare with Fig. 2.8). L: lipid droplets; G: glycogen; CAP: capillary. 26,400x.

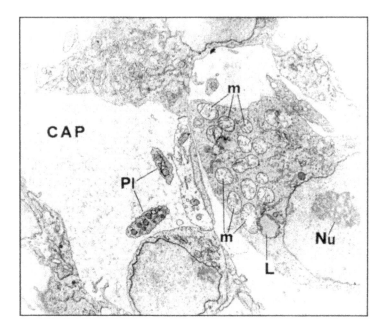

Fig. 2.10. Electron microscopy of brown adipose tissue anlage of a 19 days old fetus. An adipoblast in close association with a capillary (CAP) is visible. The blast-like characteristics of this cell are: the big nucleolus with well visible nucleolonema (Nu) and the cytoplasm rich of ribosomes and polyribosomes. Note the numerous "pre-typical" mitochondria. L: lipid droplet; Pl: platlets. 11,200x.

appear in the cytoplasm of poorly differentiated cells. We distinguish two stages in this step of differentiation: stage 2 (adipoblast stage 2) in which the lipids are in a unilocular form and stage 3 (adipocyte precursors) in which lipids are disposed in a multilocular form. Adipoblasts and precursors are visible in the anlage during all intrauterine life and in IBAT in the first weeks of development after birth. Adipoblasts are always tightly connected with the capillary wall, and many pericytes with the morphological characteristics of adipoblasts are visible in the capillary walls during the very early stages of anlage development (E15-E17). At E18 most of the anlage is occupied by adipoblasts in stage 1 and 2, and in the period E19-E21 all the developmental stages are present: adipoblasts in stages 1 and 2, precursors and brown adipocytes. Not all brown adipocytes show the same morphology in this period and many show abundant glycogen that is barely visible after birth.

Recruitment of Brown Adipocytes in Adult Animals

IBAT of adult rats and mice exposed to cold for few days shows cells with the morphological characteristics of adipoblasts and brown adipocyte precursors as described above. Also in adult animals, adipoblasts are always in tight connection with the capillary wall. This process of proliferation and differentiation of brown adipocytes after cold exposure is also maintained in old rats. In fact, two year old rats exposed for four weeks to 4 °C almost triplicated the number of brown adipocytes.[30]

Other Cell Types in Brown Adipose Tissue

Besides brown adipocytes and the structures of vessels and nerves, brown adipose tissue contains other cell types: fibroblasts, fibroblast-like interstitial cells and mast cells.

The morphology of fibroblasts and mast cells is identical to that described for the same cell types in white adipose tissue. The fibroblast-like interstitial cell is a cell type with a morphology similar to that of fibroblasts, but different in some aspects. These aspects include the presence of glycogen and small lipid droplets. These two characteristics are also present in brown adipoblasts (see above) and therefore these cells have been interpreted as adipoblasts,[31,32] but they differ from adipoblasts because they lack the general aspect of a blast cell (cytoplasm poor of organelles and rich in ribosomes) and they show mitochondria clearly different from the typical and pre-typical mitochondria found, respectively, in brown adipocytes and brown adipoblasts (Fig. 2.11). These cells also differ from fibroblasts because they are tightly connected with endothelial cells and show a distinct external lamina. We think that they represent a distinct cell type with yet unknown function.

Vascular Supply

Many depots of the adipose organ have its own vascular supply, but some use the vascular structures of the organ to which they are tightly connected.

In rats and mice we observed specific vascular structures for the anterior and posterior subcutaneous depots. In the anterior depot arteries and veins are mainly connected with the axillary ones. A single large median vein drains the venous blood into the azygous vein from the interscapular area (mainly composed of brown adipose tissue) of the anterior subcutaneous depot. The femoral vessels supply collaterals to the posterior subcutaneous depots.

Some white visceral depots, i.e., mesenteric, epididymal, and perirenal use the vascular supply of intestine, testis and kidney, respectively. The retroperitoneal depot uses the parietal vascular supply.

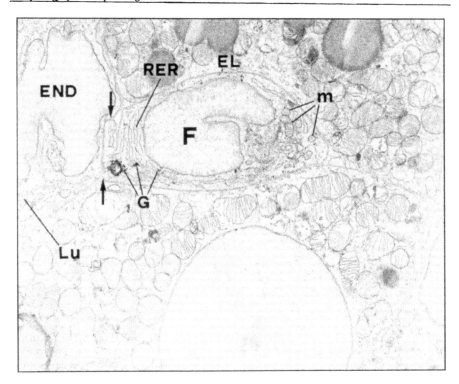

Fig. 2.11. Electron microscopy of interscapular brown adipose tissue of a 3 days old rat. A fibroblast-like interstitial cell (F) is visible among two adipocytes and an endothelial cell (END). Note the abundant dilated rough endoplasmic reticulum (RER), the tight connection with the endothelial cell (arrows), the small mitochondria (m) with a morphology different from that of typical brown adipocytes (compare with Fig. 2.8) and from that of pre-typical mitochondria of adipoblasts (compare with Fig. 2.10), and the external lamina (EL) surrounding the cell. G: glycogen; LU: restricted lumen of the capillary. 10,100x.

The capillaries are formed by endothelial cells of the continuum type, but numerous fenestrae appear during lipolysis. During the development of the organ, many pericytes are present in splittings of their external lamina. In adult animals the endothelium of many capillaries is hypertrophic in well-developed depots (i.e., epididymal or retroperitoneum) and the lumen is restricted or not visible.

The capillary bed in areas of the organ containing brown adipocytes is more developed than in areas formed exclusively by white adipocytes (Fig. 2.12).

Innervation

The general configuration of the nerve supply to the organ is similar to that described for the vascular supply: specific for the anterior and posterior subcutaneous depots and related to other organ innervation in the visceral depots. The retroperitoneal depot receives collaterals from parietal nerves.

The anterior subcutaneous depot receives nerves bilaterally from the brachial plexus and from the intercostal nerves. Nervous fibers penetrate the parenchyma via direct parenchymal nerves or via periarterial plexa. Parenchymal fibers in brown adipose tissue form a network in

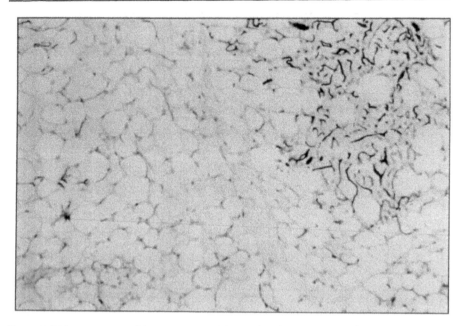

Fig. 2.12. Light microscopy of the retroperitoneal depot of a cold-acclimated adult rat. The animal was perfused with black ink and the paraffin section was counterstained with hematoxylin. The capillary network is marked by the black ink. Note that the area containing multilocular cells (right upper corner) is more vascularized. 215x.

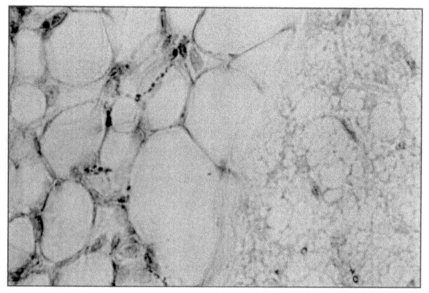

Fig. 2.13. Light microscopy (immunohistochemistry) of the anterior subcutaneous depot at the periphery of the interscapular brown adipose tissue of an adult rat. The unilocular adipocytes surrounding the brown adipose tissue are also visible. Tyrosine hydroxylase immunoreactive fibers (noradrenergic nerves) are visible in dark gray in both tissues (white and brown). The animal was stressed by daily chronic administration of an oral dose of saline. 1,075x.

which single fibers contact both capillaries and adipocytes. These fibers are immunoreactive for tyrosine hydroxylase (TH, noradrenergic fibers) and calcitonin gene-related peptide (CGRP),[33] a neuropeptide usually found in sensory nerves.[34] The extension of the network increases in adult animals after chronic cold exposure with a real branching of the noradrenergic fibers.[35] Electron microscopy shows that parenchymal fibers are mainly unmyelinated fibers contacting the adipocytes and capillaries via synapses "en passant." Brown adipocytes are electrically coupled by gap junctions.[26,27] The gap junctions are also present in old rats and react to cold exposure in the same way as in young rats: increasing their area.[28] The posterior subcutaneous depot receives collaterals from the femoral nerves. We have found parenchymal nerves in all depots of the organ. They are more numerous in the brown areas. The parenchymal innervation of the white adipose tissue surrounding the brown interscapular part of the anterior subcutaneous depot, seems to be different from that of other white depots. Here, in fact, in stressed animals, every single adipocyte appears to be surrounded by TH immunoreactive axons (Fig. 2.13).

In white areas, we found parenchymal TH and CGRP immunoreactive nerves. TH and CGRP parenchymal fibers in contact with adipocytes and capillaries were also found in the periovarian adipose tissue of cold exposed adult rats.[36] Capsaicin treatment reduces considerably the number of parenchymal CGRP fibers in this depot and impairs the brown adipocyte recruitment due to cold exposure although it does not affect the noradrenergic fibers.[37]

References

1. Klaus S. Functional differentiation of white and brown adipocytes. Bioessays 1997; 19(3):215-223.
2. Himms-Hagen J, Ricquier D. Brown Adipose Tissue. In: Bray GA, Bouchard C, James WPT, eds. Handbook of Obesity. New York: Marcel Dekker, 1998:415-441.
3. Cinti S. The Adipose Organ. Milano: Editrice Kurtis, 1999.
4. Carpentier JL, Perrelet A and Orci L. Morphological changes of the adipose cell plasma membrane during lipolysis. J Cell Biol 1997; 72(2):104-117.
5. Blanchette-Mackie EJ and Scow RO. Membrane continuities within cells and intercellular contacts in white adipose tissue of young rats. J Ultrastruct Res 1981; 77(3):277-294.
6. Cinti S, Cigolini M, Bosello O et al. A morphological study of the adipocyte precursor. J Submicrosc Cytol 1984; 16(2):243-251.
7. Zhang Y, Proenca R, Maffei M et al. Positional cloning of the mouse obese gene and its human homologue. Nature 1994; 372(6505):425-432.
8. Tartaglia LA, Dembski M, Weng X et al. Identification and expression cloning of a leptin receptor, OB-R. Cell 1995; 83(7):1263-1271.
9. De Matteis R, Cinti S. Ultrastructural immunolocalization of leptin receptor in mouse brain. Neuroendocrinology 1998; 68(6):412-419.
10. De Matteis R, Dashtipour K, Ognibene A et al. Localization of leptin receptor splice variants in mouse peripheral tissues by immunohistochemistry. Proc Nutr Soc 1998; 57(3):441-448.
11. Cinti S, Frederich RC, Zingaretti MC et al. Immunohistochemical localization of leptin and uncoupling protein in white and brown adipose tissue. Endocrinology 1997; 138(2):797-804.
12. Barr VA, Malide D, Zarnowski MJ et al. Insulin stimulates both leptin secretion and production by rat white adipose tissue. Endocrinology 1997; 138(10):4463-4472.
13. Donato R. Functional roles of S-100 proteins, calcium-binding proteins of the EF-hand type. Biochim Biophys Acta 1999; 1450(3):191-231.
14. Cinti S, Cigolini M, Morroni M et al. S-100 protein in white preadipocytes: An immuno-electronmicroscopic study. Anat Rec 1989; 224(4):466-472.
15. Cancello R, Zingaretti MC, Sarzani R et al. Leptin and UCP1 genes are reciprocally regulated in brown adipose tissue. Endocrinology 1998; 139(11):4747-4750.
16. Fawcett DW. A Textbook of Histology. 12th ed. New York: Chapman & Hall, 1994.
17. Ghadially FN. Ultrastructural Pathology of the Cell and Matrix. 4th ed. Boston: Butterworth-Heinemann, 1997.
18. Napolitano L. The differentiation of white adipose cells. J Cell Biol 1963; 663-679.

19. Slavin BG. Fine structural studies on white adipocyte differentiation. Anat Rec 1979; 195(1):63-72.

20. Barbatelli G, Morroni M, Vinesi P et al. S-100 protein in rat brown adipose tissue under different functional conditions: a morphological, immunocytochemical, and immunochemical study. Exp Cell Res 1993; 208(1):226-231.

21. Greenberg AS, Egan JJ, Wek SA et al. Perilipin, a major hormonally regulated adipocyte-specific phosphoprotein associated with the periphery of lipid storage droplets. J Biol Chem 1991; 266(17):11341-11346.

22. Himms-Hagen J. Brown adipose tissue and cold-acclimation. In: Trayhurn P, Nicholls DG, eds. Brown Adipose Tissue. London: Edward Arnold 1986; 214-268.

23. Cadrin M, Tolszczuk M, Guy J et al. Immunohistochemical identification of the uncoupling protein in rat brown adipose tissue. J Histochem Cytochem 1985; 33(2):150-154.

24. Cinti S, Zancanaro C, Sbarbati A et al. Immunoelectron microscopical identification of the uncoupling protein in brown adipose tissue mitochondria. Biol Cell 1989; 67(3):359-362.

25. Klaus S, Casteilla L, Bouillaud F et al. The uncoupling protein UCP: A membraneous mitochondrial ion carrier exclusively expressed in brown adipose tissue. Int J Biochem 1991; 23(9):791-801.

26. Schneider-Picard G, Carpentier JL and Orci L. Quantitative evaluation of gap junctions during development of the brown adipose tissue. J Lipid Res 1980; 21(5):600-607.

27. Schneider-Picard G, Carpentier JL, Girardier L. Quantitative evaluation of gap junctions in rat brown adipose tissue after cold acclimation. J Membr Biol 1984; 78(2):85-89.

28. Barbatelli G, Heinzelmann M, Ferrara P et al. Quantitative evaluations of gap junctions in old rat brown adipose tissue after cold acclimation: a freeze-fracture and ultra-structural study. Tissue Cell 1994: 26:667-676.

29. Cinti S. Morphological and functional aspects of brown adipose tissue. In: Giorgi PL, Suskind RM, Catassi C, eds. The Obese Child. Basel: Karger, 1992.

30. Morroni M, Barbatelli G, Zingaretti MC et al. Immunohistochemical, ultrastructural and morphometric evidence for brown adipose tissue recruitment due to cold acclimation in old rats. Int J Obes Relat Metab Disord 1995; 19(2):126-131.

31. Bukowiecki L, Collet AJ, Follea N et al. Brown adipose tissue hyperplasia: A fundamental mechanism of adaptation to cold and hyperphagia. Am J Physiol 1982; 242(6):E353-E359.

32. Bukowiecki L, Geloen A and Collet A. Proliferation and differentiation of brown adipocytes from interstitial cells during cold acclimation. Am J Physiol 1986; 250(6 Pt 1):C880-C887.

33. Norman D, Mukherjee S, Symons D et al. Neuropeptides in interscapular and perirenal brown adipose tissue in the rat: A plurality of innervation. J Neurocytol 1988; 17(3):305-311.

34. Hokfelt T. Neuropeptides in perspective: The last ten years. Neuron 1991; 7(12):867-879.

35. De Matteis R, Ricquier D and Cinti S. TH-, NPY-, SP-, and CGRP-immunoreactive nerves in interscapular brown adipose tissue of adult rats acclimated at different temperatures: an immuno-histochemical study. J Neurocytol 1998; 27(12):877-886.

36. Giordano A, Morroni M, Santone G et al. Tyrosine hydroxylase, neuropeptide Y, substance P, calcitonin gene-related peptide and vasoactive intestinal peptide in nerves of rat periovarian adipose tissue: an immunohistochemical and ultrastructural investigation. J Neurocytol 1996; 25(2):125-136.

37. Giordano A, Morroni M, Carle F et al. Sensory nerves affect the recruitment and differentiation of rat periovarian brown adipocytes during cold acclimation. J Cell Sci 1998; 111(17):2587-2594.

CHAPTER 3

Development of White Adipose Tissue and Adipocyte Differentiation

Gérard Ailhaud

In humans, the development of white adipose tissue (WAT) occurs to a large extent post-natally and continues throughout life, in contrast to the development of brown adipose tissue (BAT) which takes place mainly before birth and disappears thereafter. The acquisition of fat cells appears to be an irreversible process, as apoptosis has not been shown to be significant under physiological conditions. At the cellular level, this phenomenon raises the question of the characteristics of the cells constituting the adipose tissue organ. This leads in turn to the question of the nature of the factors that regulate the formation of new fat cells from dormant adipose precursor cells. Once adipose tissue is formed, adipocytes represent between one-third and two-thirds of the total number of cells. The remaining cells are blood cells, endothelial cells, pericytes, adipose precursor cells of varying degree of differentiation, and, most likely, fibroblasts. However, when studying WAT, the most provocative question which comes to mind when one considers the physiopathological consequences of overweight and obesity epidemic in developed and developing countries is "do we need WAT for normal life?" Part of the answer comes from evolution as, among invertebrates, adipose tissue represents an important organ in insects whereas its quantitative importance decreases in arachnids, crustaceans, and mollusks in which liver appears as a new organ. Among vertebrates, adipose tissue develops extensively in homeotherms, although its proportion of body weight can vary greatly between species (up to 40% of body weight in cetaceans) or within a species as it is the case in migrating birds and hibernating animals. But the most straightforward answer comes from the recent generation of transgenic mice largely devoid of WAT (vide infra: see below). The KO mice exhibit anatomical and physiological properties very similar to patients suffering from congenital generalized lipodystrophy which is characterized by severely decreased fat mass, hypertriglyceridemia and non-ketotic diabetes. It is now realized that WAT is an endocrine organ which secretes leptin in proportion to its mass. As leptin is known among many effects to stimulate the gonadal axis and to promote reproductive functions WAT, which represents the main energy storage organ in the body, has become a major player in connecting reproduction and energy requirements.

When and Where Does White Adipose Tissue Develop in the Body?

White adipose tissue has been long known to behave like an organ located in various depots which differ according to gender. Like all tissues, organogenesis of WAT takes place in early life during gestation.[1] In humans, abdominal fat mass is predominant in the male whereas the subcutaneous fat mass is more important in the female. Sex steroid hormones are likely to

Adipose Tissues, edited by Susanne Klaus. ©2001 Eurekah.com.

play a major role in these regional differences, but their causal and differential effects remain uncertain in relation to growth of the various WAT depots. Gaining insights in WAT development is related to a fair estimate of WAT cellularity, i.e., the number and size of adipocytes. Cell number relies on the enumeration of triglyceride-containing cells whereas cell size measurements rely on the fact that the whole population of adipocytes has to remain intact during tissue dissociation. In most instances some underestimation occurs but, more importantly, precursor cells and triglyceride-poor, immature (early differentiating) fat cells are not counted (Fig. 3.1). Regarding the adipocyte phenotype, in addition to triglyceride accumulation which is predominant but not unique to this cell type, there is no single marker but rather a combination of markers predominantly but not exclusively expressed in adipocytes which appears characteristic of this cell type (Fig. 3.2), i.e., the occurrence of lipoprotein lipase (LPL) and the α chain 2 of type VI collagen (A2COL6)/pOb24 among early markers, associated to that of adipocyte lipid-binding protein (ALBP), glycerol-3-phosphate dehydrogenase (GPDH), hormone-sensitive lipase (HSL), adipsin and leptin among late markers. The population of adipose precursor cells in vivo corresponds presumably to those cells previously defined in developing and adult rodents as "interstitial cells", "non-lipid-filled mesenchymal cells" or "other mesenchymal cells", which includes both "undifferentiated" and "poorly differentiated" mesenchymal cells (Fig. 3.1).[2] At present, it cannot be assessed to which extent various morphological descriptions represent only different stages of the cell lineage leading to the characteristic adipocyte phenotype.[3] From the biological perspective, hundreds of different cell types present in the whole body arise from a single fertilized egg. With development, stem cells become increasingly committed to specific lineages. In that respect, the establishment of embryonic stem (ES) cell lines has opened new experimental approaches as ES cells are able to generate various lineages under appropriate culture conditions, including the adipose lineage.[4] The latter originates from mesenchymal multipotent stem cells that develop into unipotent adipoblasts by largely unknown mechanisms. So far, most preadipocyte clonal lines have been considered as unipotent cells termed adipoblasts (Table 3.1) (see § "Cellular models for the study of WAT development and cell plasticity").[2,5-9] Commitment of adipoblasts gives rise in vitro to preadipose cells (usually termed preadipocytes), i.e., cells that have expressed early markers and that have not yet accumulated triacylglycerol stores. It remains unclear whether adipoblasts, which are formed during embryonic development, are still present postnatally or whether only preadipocytes are present in vivo. Because there are no specific markers that can be used to this date to identify adipoblasts and preadipocytes in vivo, very little is known about the ontogeny of adipose tissue.

During the postnatal period, numerous studies using ^{3}H-thymidine have been performed in order to distinguish between cell proliferation and the lipid-filling process. The proliferative activity is highest in preadipocytes, which suggests that committed cells are able to proliferate, whereas fully differentiated cells lose the capacity to divide.[10] Moreover the highest labeling index is observed in cells negative for GPDH, a late marker of adipose cell differentiation. A dramatic decrease of the labeling index precedes the rise of GPDH activity, which is detected subsequently in all triacylglycerol-containing cells.[11] Thus it appears that early marker-expressing cells undergo mitoses before terminal differentiation and lipid filling take place. However a late hyperplastic development of WAT still remains possible. Sprague-Dawley or Osborne-Mendel rats are able to increase the fat cell number in most of their adipose depots (retroperitoneal, perirenal) in response to a high-carbohydrate or a high-fat diet.[12] The life-long potential to make new fat cells has been clearly illustrated in rodents. For instance, the perirenal fat depots of very old mice from both sexes contain large amounts of early markers of differentiation, i.e., A2COL6/pOb24, LPL, and IGF-1 mRNAs, indicating that preadipocytes are indeed present in these depots (vide infra).[13] A similar conclusion can be indirectly drawn in humans, since a significant proportion of stromal-vascular cells from the subcutaneous fat tissue of elderly men

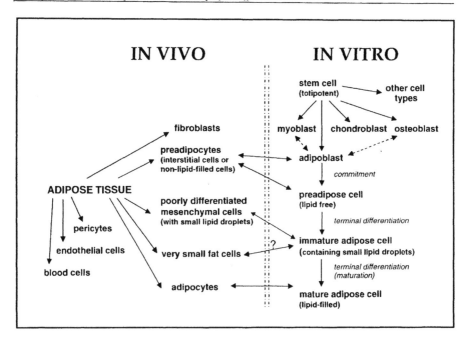

Figure 3.1. Relationships between morphological types in vivo and stages of cell differentiation in vitro. The adipocyte fraction corresponds to adipocytes and some very small fat cells. The stromal-vascular fraction corresponds to a mixture of the other cell types.

and women is able to differentiate into adipose cells.[14] This finding indicates that these adipose precursor cells should be responsible for the formation of new fat cells known to continue to take place at the adult age in severely obese patients.[15]

Sex- and site-related differences in body fat distribution are well known both in human and various animal species. Possibly, part of the explanation regarding the differential hyperplastic growth of adipose tissue lies at the cell level (see § "Cellular models for the study of WAT development and cell plasticity"). In any event, the ability of adult rodents and humans to increase the number of adipocytes, depending on the localization of the adipose depot, the nature of the diet, and the environmental conditions to which the animals were exposed, has been long known.[16] Some controversy exists, however, as to whether the formation of new fat cells takes place during refeeding after a prolonged period of food deprivation. A regimen causes loss and recovery of endothelial and non-lipid filled mesenchymal cells only and no loss or gain of fat cells occurs.[17] However, a similar nutritional study in adult mice that uses improved ^{3}H-thymidine austoradiography shows that poorly differentiated mesenchymal cells can replicate.[18] Some loss of adipocytes appears possible under severe pathological conditions: a reduction in the number of fat cells from parametrial and retroperitoneal sites has been reported in streptozotocin-treated diabetic rats,[19] but the turnover of adipocytes clearly is a slow process.[20] This conclusion goes against the current tide of apoptosis of white adipocytes. In vitro it is clear that human preadipocytes and adipocytes undergo apoptosis in response to serum deprivation or tumor necrosis factor-α (TNF-α), whereas depot specific differences are observed ex vivo in rates of preadipocyte apoptosis.[21,22] Tumor necrosis fator-α blocks clonal expansion of 3T3-L1 preadipocytes and favors apoptosis.[23] Apoptosis occurs also in serum-starved primary fetal brown adipocytes[24] or immortalized brown adipocytes.[25] In vivo, brain administration of leptin causes cell loss by apoptosis in retroperitoneal and parametrial WAT

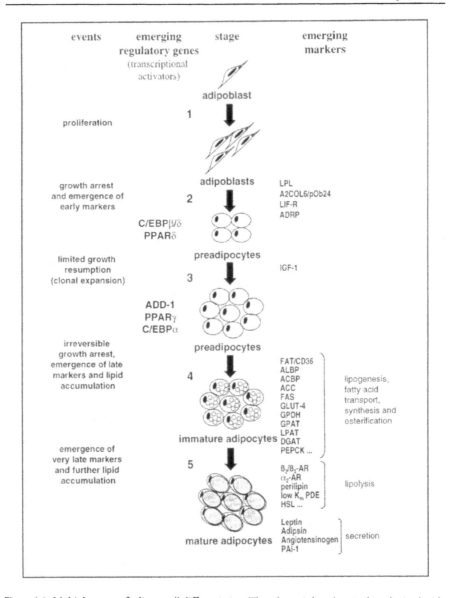

Figure 3.2. Multiple stages of adipose cell differentiation. The scheme is based upon data obtained with 3T3-L1, 3T3-F442A, and Ob17 cells as well as with rodent adipose precursor cells. The abbreviations used are: LPL, lipoprotein lipase; A2COL6/pOb24, α2-chain of collagen VI; LIF-R, leukemia inhibitory factor-receptor; ADRP, adipocyte-differentiation related protein; C/EBP, CCAAT/enhancer binding protein; PPAR, peroxisome proliferator-activated receptor; IGF-1, insulin-like growth factor-1; ADD-1, adipocyte determination and differentiation factor-1; FAT/CD36, fatty acid translocase; ALBP, adipocyte lipid-binding protein (aP2); ACBP, acyl CoA binding protein; ACC, acetyl-CoA carboxylase; FAS, fatty acid synthase; GLUT4, insulin sensitive glucose transporter-4; GPDH, glycerol-3-phosphate dehydrogenase; GPAT, glycerophosphate acyltransferase; LPAT, lysophosphatidate acyltransferase; DGAT, diglyceride acyltransferase; PEPCK, phosphoenolpyruvate carboxykinase; β_2-AR, β_2-adrenoreceptor; β_3-AR, β_3-adrenoreceptor; low K_m PDE, low K_m phosphodiesterase; HSL, hormone-sensitive lipase; PAI-1, plasminogen activator inhibitor-1

depots whereas apoptosis is observed in pair-fed rats.[26] A decrease in the number of large size adipocytes has been reported in both retroperitoneal and subcutaneous WAT depots of obese Zucker rats treated with troglitazone, a PPARγ ligand, to relieve insulin-resistance.[27]

Contrary to these observations, no apoptosis occurs in the epididymal white fat pad when mice previously exposed to cold are then transferred to warmth. Under these conditions of cessation of the sympathetic stimulus, important apoptotic events are in contrast observed in BAT.[28] The capacity of white adipocytes to develop resistance to apoptosis is supported by (I) the differentiation-dependent dramatic rise in 3T3-L1 cells of the neural apoptosis inhibitory protein (NAIP) which attain levels similar to those found in mature adipocytes isolated from WAT and (II) the lack of apoptosis induced by growth factor deprivation in 3T3-L1 adipocytes, in contrast to preadipocytes.[29] In aggregate, apoptosis does occur to some extent in WAT under drastic conditions including pharmacological treatment but its quantitative importance under physiological conditions remains questionable.

White Adipose Tissue Requires a Few Hormones to Develop

A limited number of hormones can affect both the adipose tissue mass and possibly its distribution pattern. This observation is consistent with the hormonal requirements observed in vitro for the differentiation of adipose precursor cells (vide infra).

From a general point of view, it is most important to distinguish the effect of an hormone in the adipose differentiation process, i.e., in differentiating preadipocytes, from its effect in fully differentiated cells, i.e., adipocytes per se. In vivo during development as well as in vitro, both events may take place and make data difficult to interpret. Regarding growth hormone (GH), it has long been known that GH deficiency is associated with increased body fat mass. Children with GH deficiency have enlarged fat cells but a reduced number compared to healthy children.[30] On the other hand, GH exhibits intrinsic lipolytic activity, which can be demonstrated in fat cells from GH-deficient adults and is even enhanced after long-term GH administration.[31] As many biological actions of GH are mediated via the induction of IGF-1 synthesis, attention has been paid to the question of whether the metabolic effects of GH on adipose tissue are exerted by IGF-1 (see § "Adipogenic factors that trigger adipocyte differentiation"). Hypothyroidism in rats induces a transient hypoplasia, whereas hyperthyroidism induces a transient hyperplasia of retroperitoneal and epididymal fat tissues,[32] suggesting a precocious formation of mature fat cells in response to triiodothyronine (T_3). The role of insulin in adipose cell differentiation has been investigated in streptozotocin-diabetic rats, taking advantage of the fact that after one month of diabetes the majority of adipocytes

1. have lost the bulk of their triacylglycerol,
2. show a very small diameter, and
3. contain several tiny triacylglycerol droplets (termed pauciadipose cells).

A rapid and dramatic hypertrophic effect of insulin on pauciadipose cells transformed into adipocytes is observed, followed by a slower but potent effect of insulin on the proliferation of interstitial (non-lipid-filled) cells, which gives rise to neo-formed pauciadipose cells.[19]

Glucocorticoids have been long known to increase adipose tissue mass via their hypertrophic effect. In Cushing's syndrome, hypercortisolism, which leads to centrally localized adipose tissue as in abdominal obesity, is likely accompanied by cell hyperplasia.[33] The role of glucocorticoids and the importance of the hypothalamic-pituitary-adrenal axis in the excessive development of adipose tissue have been documented in two models of transgenic mice. In a first model,[34] a reduction of the glucocorticoid receptor in the brain and also in liver and kidneys was obtained with an antisense mRNA, whereas in a second model,[35] overexpression of corticotropin-releasing factor (CRF) was achieved. In both cases, the observed increase in corticotropin and corticosterone levels was accompanied by increase in adipose tissue mass,

Table 3.1 Main cellular models of adipocyte differentiation

I Cell Lines

Cell lines	Origin	Phenotypes	Ref.
Totipotent			
Embryonic Stem cells	Mouse blastocysts	All	4
Multipotent			
10 T1/2	Mouse embryo	Muscle cells, white adipocytes, chondrocytes, osteocytes	2,5
Unipotent			
3T3-L1 3T3-F442A	Swiss mouse embryo (3T3 cells)	White adipocytes	2, 5
Ob17	Epididymal fat pad of ob/ob mice	White adipocytes	2, 5
BFC-1	Mouse BAT	White adipocytes	2, 5
TA1	Sub-clone of 10 T1/2 treated with 5-azacytidine		2, 5
HIB1B	Mouse hibernoma	Brown adipocytes	6
T37i	Mouse hibernoma	Brown adipocytes	7
BIP	Bovine intramuscular WAT	White adipocytes	8
PAZ6	BAT of human fetuses (immortalization with SV40T antigen)	Brown adipocytes	9

continued on next page

with the development of Cushing's syndrome in CRF-overexpressing mice. Finally, glucocorticoids are known to be potent promoters of the adipose differentiation process, a mechanism that may also contribute to adipose tissue expansion during glucocorticoid excess (see § "Adipogenic factors that trigger adipocyte differentiation").

WAT is recognized as an important site of estrogen biosynthesis and steroid hormone storage but limited information on the receptor equipment of animal and human WAT is available. In humans, receptors for progesterone (PR) and androgen (AR) are present in stromal-vascular cells of WAT (containing preadipocytes). PR at very low levels and AR remain present in adipocytes in which estrogen receptor (ER) is detectable.[36-38] Similarly, in the rat,

Table 3.1.—Continued

II Primary cultures

Primary cultures	Origin	Phenotypes	Ref.
Subcutaneous, epididymal, retroperitoneal WAT of newborn and adult	Rat	White adipocytes	2, 5
Interscapular BAT	Rat	White adipocytes	2, 5
Perirenal WAT	Rabbit	White adipocytes	2, 5
Subcutaneous and perirenal WAT of foetus and newborn	Pig	White adipocytes	2, 5
Subcutaneous and abdominal WAT of newborn and adult	Human	White adipocytes	2, 5

the presence of PR and AR in stromal-vascular cells and adipocytes has also been reported as well as that of ER in adipocytes (Table 3.1). After castration, in vitro differentiation of rat preadipocytes from the perirenal fat depot is increased whereas that of the epididymal fat pad is severely decreased. Both effects were partially corrected by prior testosterone treatment of the rats.[39] After ovariectomy, the capacity of preadipocytes from the perirenal fat depot is also increased and fully reversed by prior progesterone and estradiol treatment of female rats whereas that of preadipocytes from parametrial and subcutaneous fat depots remains unaltered.[40] Thus, although these studies provide evidence of site-specific differences between fat depots, the evidence of a direct endocrine effect and/or an intracrine effect of sex steroids in WAT development remains to be shown. In addition to hormones, normal glucose homeostasis appears to be essential for adipose tissue development. Transgenic mice overexpressing the glucose transporter GLUT4 selectively in adipose tissue have enhanced glucose disposal in vivo and develop increased adiposity due to adipocyte hyperplasia.[41] In contrast, mice deficient in GLUT4 by disrupting the gene exhibit growth retardation accompanied by a severe reduction in adipose depots despite a nearly normal glycemia.[42]

Dietary Fat Intake and Development of White Adipose Tissue

High dietary fat intake is now widely recognized to be associated with gain of fat and body mass in adult animals and humans.[43-50] In humans, under normal dietary conditions and in contrast to rodents, adipose tissue relies on the exogenous supply of fatty acids (FA) from chylomicrons for triglyceride synthesis, although de novo fatty acid synthesis is likely taking place in WAT of LPL-deficient patients[51] as reported in WAT of LPL-knockout mice.[52] In vivo, the effects of fatty acid flux on adipose tissue development are remarkably illustrated in transgenic mice overexpressing human LPL in skeletal and cardiac muscle. Increased fatty acid entry in both tissues is associated with weight loss due to some loss of muscle mass and, quite strikingly, to the virtually complete disappearance of adipose tissue.[53] Both amount and dietary fat type are important in regulating fat pad weight through hyperplasia and/or hypertrophy. It is now known that fatty acids and some of their metabolites are recognized by molecular sensors, i.e., nuclear receptors of the peroxisome proliferator-activated receptor (PPAR) family, which are implicated in the differentiation of precursor cells into adipocytes. The properties and mechanisms of action of PPARs have provided an important link not only between fatty acids and adipocyte differentiation in vitro but also between high-fat intake and augmented development of adipose tissue in vivo.[54] Regarding the relationships between the amount/type of dietary fats and WAT development, for a similar caloric intake, hypertrophy of perirenal and epididymal adipose tissues is lower with high-fat diets enriched in mono- and polyunsaturated fatty acids than with high-fat diets enriched with saturated fatty acids.[55] In another study, hyperplasia appears as the main phenomenon.[56] More recently,[57] mother rats have been fed diets containing either coconut oil resembling milk fat (high in saturated fatty acids, COD rats) or safflower oil (high in mono- and diunsaturated fatty acids, SOD rats) and have been mated to provide Fa/Fa and Fa/fa lean offspring as well as fa/fa obese offspring. Suckling rats were analyzed at 17 days of age whereas additional 28-day-old male rats were maintained after weaning on the same diet as their mothers until 12 weeks of age. Interestingly, although inguinal fat pad weights are similar in the 17-day-old pups, hyperplasia is observed in the inguinal fat pad of the three genotypes in SOD pups whereas adipocyte hypertrophy is observed in COD pups. In the 12-week-old rats, when analyzing inguinal, retroperitoneal and epididymal fat pads, fat cell number is nearly 2-fold higher in SOD rats than in COD rats of the three genotypes whereas fat cell size is significantly higher in COD rats in fa/fa rats only. Thus, regardless of the genotype, long-chain polyunsaturated fatty acids favor hyperplasia through presumably enhanced differentiation of adipose precursor cells whereas long-chain saturated fatty acids favor hypertrophy through enhanced triacylglycerol accumulation in differentiated cells. The mechanisms for such differences remain unclear but should implicate, at least in part, differences in the amount of arachidonic acid and some of its metabolites acting as signal transducing molecules and triggering adipogenesis, i.e., the differentiation of precursor cells into adipocytes (see § "Adipogenic factors that trigger adipocyte differentiation").

Cellular Models for the Study of Adipocyte Differentiation and Cell Plasticity

The establishment of clonal preadipocyte cell lines has greatly facilitated the study of the different steps of adipocyte differentiation which may be implicated in WAT development. Several cell lines are used for in vitro study of the process of differentiation (Table 3.1) and basically display the same sequence of events.[2,5] Adipocyte differentiation is a multistep process characterized by a sequence of events during which adipoblasts divide until confluence (Fig. 3.2). Adipoblasts are fibroblast-like cells and no specific markers of that stage have been identified so far. Growth arrest leads to the emergence of early markers of differentiation and to the formation of preadipose cells that have not yet accumulated neutral lipids. Terminal differentiation of preadipocytes into adipocytes leads to the emergence of a very large set of markers which are enzymes and proteins involved in lipid accumulation and mobilization, resulting in the formation of triacylglycerol-containing cells. Mature adipose cells become able to secrete many proteins. Primary and secondary cultures of adipose precursor cells isolated from the stromal-vascular fraction of WAT from various species, including humans, have been also extensively used. In contrast to cells from clonal lines which are aneuploid, these cells remain diploid but have a limited life span. Although the presence of adipoblasts cannot be excluded, the differentiation process of adipose precursor cells isolated from fat tissue corresponds primarily to the sequence: preadipose cell (preadipocyte) immature adipose cell mature adipose cell (adipocyte). When injected into animals at the undifferentiated state, cells from established lines (3T3-F442A, Ob17) (clonal preadipocytes) or rat stromal-vascular cells (primary preadipocytes) develop into mature fat cells.[2] When seeded at clonal densities, stromal-vascular cells from rat perirenal and epididymal fat depots show varying capacities for replication and differentiation, irrespective of donor age. At any given age, stromal-vascular cells from perirenal fat tissue showed a greater proportion of clones with a high frequency of differentiation than was found in epididymal fat tissue.[2] In humans, stromal-vascular cells from abdominal fat tissue show a higher capacity for differentiation than those of the femoral depot,[58] in agreement with clinical observations regarding the differential adipose tissue development observed for both sites. Both in rat for the perirenal and epididymal adipose tissues,[59] and in human for the subcutaneous adipose tissue,[14] aging is associated with a decrease in the proportion of cells undergoing differentiation.

One of the most exciting findings in the recent years is the possibility on the one hand to direct adipocyte differentiation from pluripotent cells and on the other hand to induce transdifferentiation into adipocytes from supposedly unipotent cells. The ability of bone marrow stromal vascular cells to give rise to adipocytes, osteoblasts, chondroblasts, and fibroblasts suggests that these cells are derived from a multipotential stem cell. Depending upon the differentiation conditions, rat bone marrow cells in primary cultures exhibit an inverse relationship in their capacity to differentiate into adipose and osteogenic cells.[60] Cultured human osteosarcoma SaOS-2B10 and MG-63 cells as well as murine stromal BMS2 cells can convert to adipose cells.[61,62] Cells of human marrow stromal cells immortalized by retroviral transduction as well as osteogenic cells derived from explants of adult human trabecular bone give rise to bipotential precursor cells able according to culture conditions to differentiate into either oesteoblast or adipocyte.[63,64] Even more striking, stromal-vascular cells from human WAT can be directed to undergo either adipogenesis or osteoblastogenesis, strongly suggesting that, in contrast to previous belief, adipose precursor cells are indeed (at least) bipotential cells. Transdifferentiation appears also possible as both myoblasts of C2C12 clonal line and muscle satellite cells are able to convert significantly to adipose cells upon exposure of the cells to fatty acids or PPARγ ligand agonists.[65] Collectively, these observations emphasize that cell plasticity is more important than anticipated and that unipotentiality should be better and more cautiously defined in vitro as the only pathway that a precursor cell can follow in response to a given hormonal milieu. The permissive stage at which the multipotential precursor cells respond to this milieu remains unclear, particularly in relation to the irreversible state of growth arrest (vide infra).

Sequential Events of Adipogenesis

General Considerations

Morphologically, after reaching confluence, fibroblast-like cells become round, enlarge, and accumulate later triacylglycerol droplets in their cytoplasm. Once differentiated, mature adipose cells show most biochemical characteristics and hormonal responses, if not all, of adipocytes. The main events are summarized in Figure 3.2. Growth arrest at the G_1/S stage of the cell cycle (stage 2), rather than contact among arrested cells, is necessary to trigger the process of cell commitment.[66] This commitment is associated with the induction of genes such as A2COL6/pOb24, clone 5, ADRP, LPL mRNAs and LPL activity. The regulation of expression of these early genes takes place primarily at a transcriptional level and does not require the various hormones which, in contrast, are required for subsequent steps. The expression of late (ALBP) and very late (adipsin, leptin) genes is associated with limited growth resumption of these committed, early marker-expressing cells. At least one round of cell division has been consistently observed and this process of clonal amplification of preadipocytes (defined as post-confluent mitoses or clonal expansion) is limited both in magnitude and duration. Terminal differentiation (defined by the emergence of GPDH activity and triacylglycerol accumulation) then takes place, providing that cells are exposed to the appropriate hormonal milieu. These observations made in vitro are consistent with those made in vivo concerning the relationships in rodent adipose tissue between cell proliferation and differentiation.[2]

During adipocyte differentiation, the dramatic changes observed in cell morphology are associated with alterations in extracellular matrix components. Type I and type III procollagens and fibronectin decrease dramatically during differentiation of 3T3-L1 and 3T3-F442A cells, respectively, whereas secretion of type IV collagen and entactin increases.[5] Expression of $\alpha 2$-collagen type IV occurs at growth arrest and is transiently increased in preadipocytes.[2] Soluble and cell-associated chondroitin-sulfate proteoglycan-I (versican) is also increased in differentiating 3T3-L1 cells.[5] So far, three families of transcriptional activators are known to be induced during adipogenesis. Among members of the first family, PPARδ (= PPARβ = NUC-1 = FAAR) emerges rapidly at confluence followed during clonal expansion by the emergence of PPARγ. PPARs belong to the superfamily of nuclear hormone receptors. They form heterodimers with retinoic X receptors (RXRs) which are present at the adipoblast stage and thereafter.[54,68] PPARα, γ and β recognize as ligands a variety of fatty acids and fatty acid metabolites as well as different pharmaceutical agonists and antagonists.[69-76] Among members of the second family of basic-helix-loop-helix (bHLH) leucine zipper transcription factors, active as unliganded homodimer, has been characterized the adipocyte determination and differentiation factor-sterol regulatory element binding protein-1c (ADD-1/SREBP-1c), which emerges also quite early.[77] CCAAT/enhancer binding proteins (C/EBPs) are part of the third large family of transcription factors possessing both leucine zipper (bZip) and basic and acidic domains and which are also active as unliganded homodimers. C/EBPβ and C/EBPδ are expressed at the onset of the differentiation program and are able to induce the late expression of C/EBPα through regulatory elements identified in the proximal promoter of this gene.[78] Co-expression of C/EBPα and PPARγ induces marked synergy for the maintenance of the adipocyte phenotype.[79]

Growth Arrest and Adipogenesis

After exposure to adipogenic agents, density-arrested (confluent) cells reenter the cell cycle for a limited proliferation commonly referred to as clonal expansion (stage 3).[*] (see bottom of p. 37) Post-mitotic growth arrest is followed by terminal differentiation of preadipocytes to adipocytes (stages 4 and 5), which are states of irreversible growth arrest as mature fat cells do not divide.[80] Recent data have brought new insights into the molecular events which take place at growth arrest and allow permanent exit from the cell cycle at the onset of terminal differentiation.

PPARδ appears to initiate the effects of fatty acids on post-confluent mitoses, independently of the other adipogenic factors. Ectopic expression of PPARδ renders confluent 3T3-C2 fibroblasts capable to respond to long-chain fatty acids in a dose-dependent manner by one to two rounds of mitoses. A mutated dominant-negative form of PPARδ, which does not exhibit any transcriptional activity, appears unable to initiate in 3T3-C2 stable transfectants a fatty acid-induced proliferation (P. Grimaldi et al, personal communication). As PPARδ is implicated in the activation of a limited set of genes, including PPARγ,[83] and as PPARγ induces cell cycle withdrawal and irreversible growth arrest (vide infra), it would appear that PPARδ and PPARγ are playing both opposite and complementary role.

In addition to fatty acids, upon adipogenic stimulation by a mixture of dexamethasone and MIX, synchronous activation within 16 h of cyclin genes characteristic of the G_1 phase (cyclins D1, E and A) is accompanied by activation of C/EBPβ and C/EBPδ genes, the role of which appears critical for subsequent differentiation events. At that point C/EBPα and PPARγ are not detectable.[84] Most likely but not proven, PPARδ, LPL and A2COL6/pOb24 may have also been expressed. The onset of C/EBPα and PPARγ within 48 h coincides with the switch in the disappearance of the expression of cyclin genes and in that of histone 2B (a marker of S phase), strongly suggesting the mutually exclusive nature of growth and terminal differentiation. Cyclin-dependent kinase inhibitors (CKIs) are known to prevent the phosphorylating activity of cyclin/cyclin-dependent kinase complexes. The fine regulation of p18, p21 and p27 CKIs has allowed to define distinct states of growth arrest during clonal expansion and terminal differentiation. In PPARγ ectopically expressing fibroblasts, ligand activated PPARγ results in adipocyte differentiation associated with p18 and p21 CKI expression.[84] Collectively, these observations indicate that PPARγ induces cell cycle withdrawal and terminal differentiation accompanied by irreversible growth arrest. Interestingly, mutant forms of PPARγ which are unable to bind to cis-acting DR-1 sequences (DR-1, direct repeat sequences separated by one nucleotide) or mutants of PPARγ which lack the carboxy-end activation domain have no effect on growth arrest, demonstrating that the activity of PPARγ as a DNA-binding protein and transcription activator is required to promote this phenomenon.[84] A mechanism distinct of CKIs has also been reported by which ligand-induced PPARγ activation decreases the expression of the protein phosphatase PP2Ac, leading in turn to an enhanced steady-state of phosphorylation of the DP-1 transcription factor of the E2F-DP complex implicated in promoting S phase entry, leading to a decrease of its DNA-binding activity.[85] The role of retinoblastoma protein (pRB) in the control of clonal expansion and terminal differentiation remains unclear as, in pRB -/- mouse embryo fibroblasts, ligand-induced activation of PPARγ is sufficient to initiate terminal differentiation accompanied by cell cycle withdrawal.[86] Phosphorylation inactivates pRB and allows the release of E2F1 from the E2F-DP complex[87] and, when adipogenesis goes to completion, pRB shifts to its hypo-phosphorylated state.[88] It is of interest that C/EBPβ and δ bind in vitro to the hypophosphorylated form of the pRB that predominates in the G1 phase of the cell cycle and enhances as a chaperone the binding of C/EBPs to cognate DNA sequences.[89] It is also of interest to note that C/EBPβ and δ exhibit a delayed acquisition of DNA binding activity, between 12 and 16 h after induction of differentiation. This acquisition depends probably upon their phosphorylation, as it occurs immediately before phosphorylation of pRB and entry into the S phase of the cell cycle. Importantly, the localization of C/EBPβ and δ to centromeres occurs through C/EBP binding sites in centromeric satellite DNA and might control mitotic clonal expansion.[90] Last but not least, C/EBPα implicated in growth arrest may stabilize p21 CKI at the protein level as it does in hepatocytes.[91]

Taken together, the above observations indicate that, upon stimulation by adipogenic factors, activated transcription factors that regulate adipogenesis also regulate the expression of cell cycle inhibitors. These molecular events which demonstrate a switch between preadipocyte proliferation and terminal differentiation are associated with a shift in the composition of cell surface receptors, nuclear receptors and trans-acting factors during the differentiation process.

---

* Post-confluent mitoses are observed in primary cultures of rat preadipocytes[81] but not of human preadipocytes.[82] However, it cannot be excluded that amplification of human committed cells has already occurred in vivo.

Receptor Equipment and Adipogenesis

Table 3.2 summarizes the receptor composition of adipoblasts, preadipocytes and adipocytes. Not unexpectedly, preadipocytes express the cognate receptors of adipogenic hormones required for subsequent events to occur. The alterations in hormone responsiveness as a function of differentiation correspond to a critical "time-window" of action in preadipocytes during which adipogenic hormones induce the terminal differentiation process.

Growing adipoblasts express cell surface receptors of the major prostaglandins produced by these cells [prostacyclin (PGI$_2$), prostaglandins F$_{2\alpha}$ (PGF$_{2\alpha}$) and prostaglandin E$_2$ (PGE$_2$)] [92] as well as nuclear receptors recognizing glucocorticoids, triiodothyronine (T$_3$) and retinoic acid (RA).[2.5.68] Pref-1 is a EGF repeat-containing transmembrane protein that belongs to the EGF-repeat family. The soluble ectodomain of pref-1 inhibits the expression of key transcription factors of adipogenesis, suggesting the existence of pref-1 mediated intracellular signals and emphasizing the need to identify the interacting protein(s).[93] Evidence for the role of this protein is reminiscent of that played by a member of this protein family, i.e., the transmembrane protein Notch active via its ligands Delta and Serrate. In that case, however, Notch-1 appears required for adipogenesis as its impaired expression by antisense Notch-1 prevents PPARδ and PPARγ expression as well as adipocyte differentiation.[94] Commitment is characterized by the emergence of the cell surface receptor of Leukemia Inhibitory Factor (LIF) as well as that of the nuclear receptor PPARδ and the transcription factors C/EBPβ and δ.[95] The adenosine receptor A2 subtype,[96] the prostacyclin IP receptor and the PGE$_2$ receptor (EP4 subtype),[92] all positively coupled to adenylate cyclase, are present in preadipocytes and then become expressed at low levels, if any, in adipocytes, consistent with the importance of the PKA pathway in preadipocytes to trigger terminal differentiation of preadipocytes. These observations are also consistent with those made on the comparative study of prostaglandin sensitivity of adenylate cyclase of rat primary preadipocytes and isolated adipocytes which show that G$_i$ activity is reduced or absent in preadipocytes.[97] This may also account for the potent stimulatory effect of PGE$_1$ and 6-keto-PGE$_1$ (ligands of IP receptor) and the weak stimulatory effect of catecholamines (ligands of β-adrenoreceptors) on the adenylate cyclase activity of preadipocyte membranes.[98] Expression of IP receptor mRNA decreases[92] and responses to carbacyclin (rise in cAMP and Ca^{2+} levels) are abolished.[99] The EP$_3$ subtype (α, β and γ isoforms) of the PGE$_2$ receptor and the adenosine receptor A$_1$ subtype,[92.100,101] coupled negatively to adenylate cyclase, then become detectable in adipocytes, consistent with the well-known antilipolytic effects of PGE$_2$ and adenosine observed in mature fat cells whereas the transmembrane pref-1 protein disappears rapidly.[102]

Adipogenic Factors that Trigger Adipocyte Differentiation

The responsiveness of preadipose cells to external signals may vary according to differences in the stage of adipose lineage at which clonal lines have been established and that at which cells of a given line are exposed to various agents. Moreover, the requirement for adipogenic hormones shows differences between clonal and primary preadipocytes. Last but not least, serum-supplemented media are (too) often used. As many serum components are uncharacterized and thus not controllable, the use of serum-free chemically-defined media appears more appropriate. This renders rather difficult comparisons between various studies. Nevertheless, in serum-free conditions, the most common requirements of primary preadipocytes appear to be glucocorticoids, insulin-like growth hormone factor I (IGF-1) and insulin. Clonal preadipocytes also require growth hormone and triiodothyronine and/or retinoids. A salient feature is that the required hormones have to be present simultaneously within a physiological range above threshold concentrations to trigger terminal differentiation. Glucocorticoids, including the synthetic glucocorticoid dexamethasone, stimulate the differentiation of clonal preadipocytes and that of primary preadipocytes of various species including human.[2] They increase in Ob1771 preadipocytes the production of prostacyclin, a potent adipogenic hormone. This enhanced production is likely due to increased cycloxygenase activity as COX-1 and COX-2 mRNA levels are enhanced under these conditions.[103] Glucocorticoids also repress quite dramatically at a transcriptional level the expression of pref-1, thus allowing the differentiation process to proceed.[102]

IGF-1 has been shown to stimulate adipogenesis of clonal mouse 3T3-L1 preadipocytes and primary preadipocytes (rat, rabbit, porcine). In addition to IGF-1, clonal and primary preadipocytes also secrete insulin-like growth factor binding proteins (IGFBPs) as a function of differentiation. The IGF-1/IGFBPs system is assumed to act in a paracrine/autocrine manner upon GH stimulation, although conflicting results have been reported with respect to the adipogenic activity of GH in vitro. Growth hormone triggers IGF-1 gene transcription, IGF-1 secretion and adipogenesis of 3T3-F442A and Ob1771 preadipocytes.[104,105] However GH appears to have an additional effect besides stimulating mitoses through IGF-1 production. Growth hormone promotes diacylglycerol production from phosphatidylcholine breakdown and thus activates protein kinase C activity (PKC). Protein kinase C activators can substitute for GH in modulating cell proliferation in the presence of IGF-1.[106] However the most critical role of GH is related to cell differentiation. It appears that the JAK-STAT signaling pathway (via STAT5) is implicated in transducing GH-dependent signals during the initiation of 3T3-F442A preadipocyte differentiation.[107] This conclusion is in agreement with the generation by homologous recombination of GH-receptor deficient mice or STAT5a- and STAT5b deficient mice in which fat deposition was reduced.[108]

Despite all these observations, the role of GH in adipogenesis remains somewhat confusing. In rat primary preadipocytes, although GH increases IGF-1 production and although IGF-1 promotes cell proliferation, GH still exhibits anti-adipogenic properties independent of IGF-1 mitogenic activity.[109] GH inhibits the differentiation process at a late stage, i.e., at a step downstream of ADD-1/SREBP-1c expression. This inhibition appears to be due to the maintenance of high levels of pref-1.[110] In order to reconcile conflicting data in GH action, it is assumed that clonal and primary preadipocytes represent early and late stages of the differentiation process, respectively, as primary preadipocytes should have been previously exposed to GH and «primed» in vivo. This hypothesis is consistent with the lipolytic effect of GH observed in more differentiated fat cells, and also consistent with the dramatic shift in receptor/signaling transduction pathways when preadipocytes undergo terminal differentiation into adipocytes.

In addition to its well-known effects on lipogenesis, insulin can now be considered as triggering adipogenesis. Targeted invalidation of the insulin receptor in 3T3-L1 preadipocytes impairs their ability to differentiate into adipocytes.[111] When a chimeric receptor is constructed in which the ligand binding domain of the CSF-1 receptor is fused to the cytosolic tyrosine kinase of the insulin receptor, stable transfectants of 3T3-L1 preadipocytes are able to differentiate upon CSF stimulation. This indicates that insulin-derived signals are sufficient to induce this process.[112] Triiodothyronine appears to be essential for the differentiation of Ob17 preadipocytes[113] and partially required for that of 3T3-F442A preadipocytes.[114] This requirement remains questionable as

1. carbacyclin can substitute for T_3 in the adipocyte differentiation of Ob1771 cells in serum-free conditions[115] and
2. no obvious requirement for T_3 is observed in primary preadipocytes isolated from pig, rat, rabbit and human, suggesting like for GH that prior exposure to T_3 in vivo has "primed" the cells.[2]

Table 3.2. Cell surface and nuclear receptor equipment of adipose precursor cells and adipocytes

Receptors[b]	Adipoblasts[a]	Preadipocytes[a]	Adipocytes[a]	Ref.
Cell surface receptors				
LIF-R	·	+	+	95
IP-R	++	+	(+)	95
FP-R	++	+	·	135
EP$_1$-R	++	+	(+)	92
EP$_4$-R	++	+	(+)	92
EP$_3$-R	·	·	+	92
(α, β, γ isoforms)				
A1-R	N.D.	·	+	96,100
A2-R	N.D.	+	(+)	96,100
GH-R	N.D.	+	+	25
IR	+	++	++	25
pref-1	+	+	·	93
Nuclear receptors				
GR	+	+	+	2,5
AR	N.D.	(+)	+	37
ER	N.D.	(+)	+	36,37
PR	N.D.	+	(+)	37,38
RARα and RARγ	+	++	+++	68
RXRα and RXRβ	+	++	+++	68
TR	+	+	++	114,115
PPARδ	·	+	++	67
PPARγ	·	(+)	++	54
AhR	+	(+)	·	158-161
Arnt	+	(+)	·	158,159

continued on next page

The ability of retinoic acid (RA) and synthetic retinoids to promote or inhibit differentiation processes has been recognized for several years. At supraphysiological concentrations, RA inhibits adipogenesis of clonal and primary preadipocytes via RA receptors (RARs). Inhibition appears to take place at an early stage, i.e., through C/EBPβ repression.[116] In contrast, at concentrations close to the K_d values of RARs, RA and retinoids behave as potent adipogenic hormones in triggering the differentiation of Ob1771 preadipocytes and rat primary preadipocytes.[67]

Fatty acids have been shown to act as hormones as the main adipogenic component of serum that has been identified is arachidonic acid (AA). Both the long-term adipogenic effect as assayed by GPDH activity and the short-term intracellular production of cAMP induced by AA are blocked by cyclooxygenase inhibitors suggesting that prostanoid(s) are involved in these responses.[117] Prostacyclin, under the form of its stable analogue carbacyclin (cPGI$_2$), induces short- and long-term effect of AA,[118] and this stimulatory effect can be extended to the differentiation of rat and human primary preadipocytes.[99]

Table 3.2.—Continued

Transcription factors				
CREB	+	+	+	144
C/EBPβ and C/EBPδ	(+)	+	(+)	95,139,140
ADD1/SREBP-1c	-	+	+	77
C/EBPα	-	(+)	++	78,90
AP-2α	+	(+)	-	162
AEBP-1	+	(+)	-	164
Id2	+	(+)	-	154,155
Id3	+	(+)	-	154,155

Based upon data obtained in cells of clonal lines and primary cultures, [a]The signs refer to comparative levels of each mRNA or protein between adipoblasts, preadipocytes and adipocytes, i.e. the absence— or the presence at very low levels (+), low levels (+), or higher levels (++) and (+++). N.D., not determined. [b]Abbreviations: IP-R, prostacyclin receptor; FP-R, PGF2a receptor; EP-R, PGE2 receptor; A1-R and A2-R, adenosine (sub-types 1 and 2) receptor; GH-R, growth hormone receptor; IR, insulin receptor; GR, glucocorticoid receptor; AR, androgen receptor; ER, estrogen receptor; PR, progesterone receptor; RAR, retinoic acid receptor; RXR, retinoic X receptor; TR, triiodothyronine receptor; CREB, cAMP response element-binding protein; AEBP-1, AE binding protein-1; Id, inhibitory protein of DNA binding; AhR, arylhydrocarbon receptor; Arnt, AhR nuclear translocator. The other abbreviations are given in the legend of Fig. 3.2.

Antibodies direct against PGI_2 strongly diminish the adipogenic effect induced by AA. Thus, PGI_2 secreted into the culture medium of Ob1771 preadipocytes upon exposure to AA, acts as autocrine/paracrine effector of adipose cell differentiation.[119] PGI_2 and $cPGI_2$ are able, in addition and concomitantly to a rise in cAMP levels, to induce intracellular mobilization of calcium which appears to be independent of inositol triphosphate generation.[98] Consistent with the importance of this dual intracellular signaling pathway for adipogenesis, a synergistic effect between a stable analogue of cAMP (8-bromo cAMP) and a calcium ionophore (ionomycin) clearly takes place in serum-free medium.[99]

Several additional observations demonstrate that PGI_2 is a potent adipogenic hormone.

1. PGI_2 is active within a critical time-window during preadipocyte differentiation, consistent with the decrease in the expression of the cell surface IP receptor,[99,101]
2. through activation of its cell surface receptor, PGI_2 rapidly up-regulates the expression of critical early transcription factors, C/EBPβ and C/EBPδ,[120]
3. like $cPGI_2$,[74] PGI_2 might also bind to PPARδ and act at the gene level as $cPGI_2$ up-regulates the expression of several genes including a-FABP, angiotensinogen and UCP2 whereas this effect is not mimicked by PGE_1, 6-keto-PGE_2 or BMY45778 despite the fact that all are ligands of cell surface IP receptor and that all alter cAMP and calcium levels.[121,122]

Clearly, it cannot be ruled out that the natural ligand(s) of PPARδ differ from PGI_2 but this prostaglandin, along with PGE_2 and trace levels of $PGF_{2\alpha}$, is the only metabolite of AA so far characterized in preadipose cells.[123,124] In agreement with a role in early events of adipogenesis, evidence of the ability of PGI_2 to recruit rat adipose precursor cells to differentiate ex vivo as well as in vivo has been recently obtained using co-culture of preadipocytes and adipocytes[125] as well as explants of adipose tissue and in situ microdialysis of periepididymal fat pads of anesthetized rats (C. Darimont et al, unpublished results).

The critical role of PGI$_2$ as a locally-produced hormone raising cAMP and Ca^{2+} levels cannot be dissociated from the developmental stage of the cells at which the effectors trigger transducing pathways. First, cyclic AMP promotes the initiation of differentiation of 3T3-F442A preadipocytes but inhibits the late stage(s) of this process.[126] Second, the Ca^{2+}-calmodulin -sensitive protein kinase II, which is required for the differentiation of 3T3-L1 preadipocytes, exhibits at an early stage within a few hours two temporally distinct phases of activity after stimulation of adipogenesis.[127] It is likely that the time frame of exposure of clonal preadipocytes to calcium ionophores, i.e., the developmental stage at which the cells respond, explains the inhibitory[128] or stimulatory[99] effect previously reported.

In addition to AA, the evidence that other fatty acids are important regulators of adipo-genesis relies on the fact that saturated and unsaturated long-chain fatty acids (LCFA) act as transcriptional activators of lipid-related genes, as first reported for the gene encoding ALBP [also termed adipocyte-fatty acid binding protein (a-FABP) or aP2].[129] Metabolism of LCFA is not required as α-bromopalmitate, which is not metabolized by preadipocytes and has no effect on cAMP and Ca^{2+} levels, is more potent than natural LCFA in activating the ALBP gene and enhancing adipogenesis.[130] When compared to control conditions, a brief exposure of confluent preadipocytes (1-3 days) to LCFA appears sufficient in the presence of adipogenic hormones to bring a maximal differentiating effect and leads within 1-2 weeks in culture both to hyperplasia, i.e., an increase in the number of differentiated cells and to hypertrophy, i.e., an increase in triglyceride accumulation accompanied by enhanced overexpression of terminal differentiation-related genes.[131] Similar observations are made in rat and human primary preadipocytes.[132]

Antiadipogenic Factors

Various factors that inhibit or abolish differentiation of adipose precursor cells have been reported, including platelet-derived growth factor (PDGF) and TGF-β. The most intriguing results are those obtained in preadipose cells with TGF-β since, except for pig preadipocytes,[2] terminal differentiation appears irreversibly blocked after its removal, in contrast to various other factors where partial or complete recovery of the differentiated phenotype is observed.[133] The situation remains unclear with regard to fibroblast growth factor (FGF) and epidermal growth factor (EGF), which have been claimed to be either inhibitory or without effect, de-pending on the origin of target cells. In vivo, subcutaneous administration of EGF to newborn rats results in a large decrease in the weight of inguinal fat pads, which suggests the delayed formation of adipocytes from preadipocytes.[2] Tumor necrosis factor-α, which is secreted by adipocytes, inhibits the expression of several differentiation-specific genes and induces a phe-notypic "dedifferentiation" on a long-term basis. The inhibitory effect of TNF-α on adipocyte differentiation is only observed if it is administered at an early stage. TNF-α appears to dis-rupt the normal pattern of expression of p107 and p130 proteins, two members of the retino-blastoma family of proteins, and leads to complete block in clonal expansion.[23] Many fat-soluble vitamins, as well as dehydroepiandrosterone and some structural analogs, are known to abolish terminal differentiation of 3T3-L1 preadipose cells.[2,134] Retinoids have been reported also to inhibit differentiation, but the supraphysiological concentrations used in those studies raise serious doubt about their physiological relevance. So far, the most interesting antiadipogenic agent appears to be PGF$_{2α}$. It inhibits differentiation of clonal and primary preadipocytes, most likely by means of the specific FP prostanoid receptors.[135] Inhibition of adipogenesis by PGF$_{2α}$ has been reported to occur through MAP kinase-mediated phosphorylation of PPARγ, thus inhibiting its transcriptional activity.[136] Nevertheless this inhibition is rather unlikely to take place during the differentiation process as

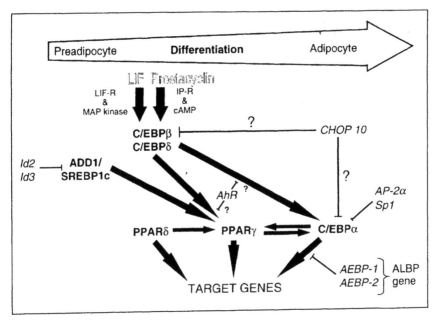

Figure 3.3. Transcriptional activators (bold face) and transcriptional repressors (italics) of adipogenesis. (see text for comments and abbreviations)

1. only trace levels of $PGF_{2\alpha}$ are produced in preadipocytes compared to the production of PGI_2 active as a potent adipogenic hormone,[123]
2. a full reversibility of $PGF_{2\alpha}$ inhibitory action occurs upon subsequent exposure to (carba)prostacyclin[133] and
3. secreted prostacyclin overcomes the inhibitory effect of secreted $PGF_{2\alpha}$, if any, when preadipocytes are differentiating in the presence of co-cultured adipocytes stimulated by angiotensin II.[125]

Transcriptional Activators of Adipogenesis

Recent evidence indicates that PPARγ is essential for placental, cardiac and adipose tissue development during mouse embryogenesis.[137,138] Various adipogenic signaling pathways are converging to up-regulate PPARγ both as a common and essential regulator of terminal events of adipogenesis and as a regulator of adipocyte hypertrophy. Altogether, current data indicate that several transcription factors are implicated in the expression of PPARγ (Fig. 3.3). These redundant pathways in up-regulating PPARγ emphasize the importance of adipose tissue development to ensure physiological functions.

Elegant studies over the last years have shown that forced ectopic expression of C/EBPβ in cells of various clonal lines of fibroblastic origin (NIH-3T3, BALB/c-3T3, and others), which normally express very low levels of C/EBPα, C/EBPβ and PPARγ, induces adipogenesis, with C/EBPδ acting in synergy with C/EBPβ.[140] The forced expression of ADD1/SREBP-1c or ADD-1 403, a super-active truncated form of ADD-1, up-regulates the expression of PPARγ in 3T3-L1 cells.[141] In most studies, significant adipogenesis requires the simultaneous presence of PPAR ligands, in contrast to authentic preadipocyte target cells (3T3-L1, 3T3-F442A, Ob1771), suggesting that authentic differentiating cells are able to express PPARγ and simultaneously to synthesize PPARγ ligand(s). Although ADD-1 has been shown to favor the production of endogenous ligand(s) of PPARγ, a full characterization of these ligands remains to be done.[142] PGD_2, PGJ_2 and derivatives are naturally-occurring ligands of PPARγ but unlikely candidates since neither endogenous production nor any adipogenic effect of these prostaglandins are observed in differentiating cells, in contrast to prostacylin.[143] Not surprisingly, intracellular pathways which up-regulate the expression of C/EBPβ and C/EBPδ should promote adipogenesis. This is the case for preadipocytes which have the unique capacity both to synthesize and secrete LIF and prostacyclin and to express at the same time their cognate cell surface receptors LIF-R and IP-R. Binding to LIF-R and IP-R triggers the MAP kinase and PKA pathways, respectively, which in turn activate the critical expression of C/EBPβ and C/EBPδ and promote adipocyte differentiation.[95] Recently the CRE-binding protein (CREB) has been implicated in the regulation of adipocyte-related genes whose promoters contain cAMP response elements (CRE). Like C/EBPs, CREB exhibits a bZip DNA binding/dimerization domain. Overexpression of CREB enhances adipogenesis of 3T3-L1 cells whereas a dominant-negative form of CREB blocks this process.[144] It is no surprise that signal transduction proteins involved in the up-regulation of transcriptional activators may also regulate adipogenesis. This has been reported for ectopic expression of the differentiation-enhancing factor-1 (DEF-1) in NIH-3T3 and BALB/c-3T3 fibroblastic cells. DEF-1 contains a Src homology 3 (SH3) domain found in many signal transduction proteins, including the Src family of protein tyrosine kinase. DEF-1 NIH-3T3 cells exhibit enhanced levels of PPARγ and, in the presence of pioglitazone as PPARγ ligand, terminate differentiation.[145] Regarding ADD-1, which up-regulates the expression of fatty acid synthase and leptin genes in adipose cells, its expression appears controlled positively by insulin thus establishing a link between this transcription factor and carbohydrate availability.[146] Stable transfectants able to express constitutively various components of the insulin signaling pathways have also been shown to trigger adipogenesis, i.e., activated Ras mutants, Raf, PI3-kinase, p70S6 kinase and more recently PKB (c-Akt) which appears to act downstream of PI3-kinase but upstream of p70S6 kinase.[147] However the relationship between the up-regulation of ADD-1 gene by insulin and these signal transduction proteins, if any, remain to be investigated.

Taken together, these studies emphasize the wide panoply of transcription factors and their link with nutritional status, i.e., first PPARs and fatty acid availability and second ADD-1 and carbohydrate availability. In vivo, it cannot be excluded that alterations in the activity of PPARγ mutants could lead to significant changes in adipose tissue development. However the polymorphism in codon 12 (Pro 12 Ala) does not appear associated with a higher prevalence of early-onset obesity. In the German population, the reported Pro 115 Gln variant next to the phosphorylation site of Ser 112 modulating PPARγ activity should be a rare event as either association[148] or no association[149] have been reported. Physiologically, more important is the comparative response to low- and high-fat diet of heterozygous PPARγ +/- deficient and homozygous +/+ wild-type mice. No difference is observed when mice are fed a low-fat diet but, upon high fat feeding, heterozygous mice in which PPARγ levels are presumably 50% of WT mice appear, at least in part, protected from increased fat mass. The size of adipocytes is significantly smaller in heterozygous mice whereas the number of adipocytes remains similar in both genotypes.[150] Thus, as compared to mutational events, it is believed that subtle alterations between individuals in the activity level of any of the components of the pathways regulating expression and/or activity of PPARγ may be sufficient and quite important to bring significant differences in fat tissue mass. If it were so, minor changes would be clearly difficult to estimate but would appear consistent with the polygenic nature of overweight and obesity and also consistent with the slow pace at which the formation of fat cells occurs in vivo. In contrast, invalidation of key genes regulating PPARγ expression or invalidation of PPARγ gene lead as predicted to dramatic alterations in adipose tissue development.

Fat Tissue Development in Targeted Gene Knockout Mice

Gene knockout mice lacking C/EBPβ, C/EBPδ or both are severely impaired in adipose tissue development.[151] Despite a 70% reduction in weight of epididymal fat pad, the levels of C/EBPα and PPARγ were expressed to an extent similar to those of wild-type mice, fully consistent with the existence of other pathways, i.e., PPARδ and ADD-1 up-regulating PPARγ expression which in turn cross-regulates C/EBPα expression (Fig. 3.3). C/EBPα -/- mice develop poorly adipose tissue and adipocytes are relatively devoid of triglycerides, also consistent with the postulated role of this transcriptional activator in adipocyte hypertrophy.[152] PPARγ-deficient mice embryos and PPARγ -/- ' PPARγ +/+ chimeric mice have been recently generated.[137] Both studies provide direct evidence that PPARγ is indispensable for adipocyte formation and adipose tissue development but the data do not exclude that early marker-expressing, lipid-free preadipocytes have been formed. Clearly, these studies establish definitely that C/EBPs and PPARγ can be considered as true "adipogenic" transcription factors but the multiple regulatory pathways leading to PPARγ expression would imply that adipocyte formation remains potentially a permanent process as preadipocytes remain present even at a very late age in animals and humans.[14] This permanent acquisition of fat cells is not likely in humans as two peaks of accelerated tissue growth have been reported, one after birth and another in preadolescent years,[153] suggesting in the meantime the role of additional players in slowing down or inhibiting adipogenesis.

Transcriptional Repressors Implicated in Adipogenesis

Different mechanisms of transcriptional repression have been proposed. The first involves a passive repression in which the negative regulator binds to activators and forms inactive heterodimers, thus playing a dominant-negative role. The second mechanism involves active repression in which the repressor binds to cis-acting sequences of a gene promoter and interferes with the transcriptional machinery. Both types of repression take place in preadipocytes (Fig. 3.3). Passive repression by Id2 and Id3 proteins (inhibitors of DNA binding) has been recently shown.[154] Both proteins do not possess the DNA-basic region directly adjacent to the HLH domain and act as dominant-negative protein. Id2 and Id3 decrease dramatically during adipogenesis and forced expression of Id3 protein blocks this process. Direct interaction of Id2 and Id3 with ADD-1 takes place and this postulated heterodimerization decreases the functional capacity of ADD-1 to up-regulate the expression of fatty acid synthase gene and most likely that of several other lipid-related genes controlled transcriptionally by ADD-1.[155] CHOP-10/GADD153 is a member of the C/EBP family and acts also as a dominant-negative inhibitor of transcription. It is able to bind strongly with C/EBPβ and C/EBPα but its role remains uncertain during adipogenesis as it is only induced by cellular stresses, including glucose starvation.[156,157] Another member of the bHLH proteins which plays some undefined inhibitory role is the arylhydrocarbon receptor (AhR). Although a physiological ligand of AhR has yet to be identified, it binds 2,3,7,8-tetrachlorodibenzo-p-dioxin (TCDD) and related compounds. Upon binding to the ligand, AhR dissociates from a cytosolic complex and translocates to the nucleus where it heterodimerizes with the AhR nuclear translocator (Arnt). This heterodimer then binds to xenobiotic response element (XRE) present within the promoter region of various genes. Both AhR and Arnt proteins are present in adipoblasts and disappear rapidly in differentiating preadipocytes. TCDD appears only effective when added at the adipoblast or at the early preadipocyte stage and decreases C/EBPα and PPARγ2 but not C/EBPβ and C/EBPδ levels.[158] Although TCDD clearly inhibits adipogenesis through the AhR pathway as shown in AhR -/- mouse embryo fibroblasts,[159] the role of AhR remains unclear. It appears that TCDD increases the phosphorylation of the repressor COUP-TF, thus enhancing its capacity to bind to the response element DR-1 recognized by the PPARγ/RXR heterodimer.[160] In addition, Arnt may play a role independent of its binding to AhR as it heterodimerizes with HIF-1α which is also induced at early stage of adipogenesis.[161] So far, the various nuclear factors which inhibit adipogenesis act indirectly, i.e., by passive repression of transcription. However active repressors which bind to regulator sequences of promoters of several key genes of adipogenesis have also been reported, i.e., promoters of the genes encoding C/EBPα and ALBP. Two repressor binding sites designated CUP-1/AP-2 and CUP-2/AP-2 are present in the C/EBPα promoter and are recognized by an isoform of the transcription factor AP-2α, a member of the bHLH family. Providing that dual CUP sites are present, AP-2α inhibits the transcriptional activation of the C/EBPα gene and thus may be implicated in the maintenance of this gene in a repressed state as well as in the maintenance of the preadipocyte stage.[162] A member of the Sp family, notably Sp1, binds to a consensus Sp binding site of C/EBPα unique among the promoters of adipocyte genes. By doing so, it prevents the trans-activation of C/EBPα gene by the C/EBPs. Raising cAMP levels causes down-regulation of Sp1, consistent with the initial PKA-mediated events of adipogenesis.[163] Within the ALBP proximal promoter region of the ALBP gene, the AE-1 sequence is recognized by a transcriptional repressor active through its carboxypeptidase activity (termed AEBP-1)[164] and by another one recently reported as a zinc finger protein (termed AEBP-2).[165] AEBP-1 may play a more specific role in adipogenesis than AEBP-2 as its expression, which is abolished in differentiating preadipocytes, is confined in humans to adipose tissue and osteoblast-like cells.[166]

Although the detailed mechanisms of transcriptional inhibition of adipogenesis remain largely unknown, it seems that some rationale is emerging by which broad and passive mechanisms, which are functional at early stage(s), i.e., at the time of C/EBPß and ADD-1 expression, are reinforced by specific and active mechanisms. Both mechanisms then become non-functional at the onset of terminal differentiation, i.e., at the time of C/EBPα and PPARγ expression, leading to the adipocyte phenotype.

A Speculative and Integrated View of Adipogenesis

A current hypothetical view of adipogenesis based upon collective data obtained with various cellular models could be summarized as follows. Growing adipoblasts express constitutively CREB as a transcriptional activator as well as various transcriptional repressors (Id2, Id3, AhR and Arnt) and the family of retinoblastoma proteins (pRB, p107, p130). At that time nuclear receptors of glucocorticoids, T_3 and RA are also expressed as well as pref-1 and cell surface receptors positively coupled to adenylate cyclase. Growth arrest at the G_1/S phase of the cell cycle triggers the expression of PPARδ and that of the LIF receptor. As production of PGI_2 (enhanced by glucocorticoids) takes place at that stage, the LIF/LIF-R and PGI_2/IP prostanoid receptor systems can operate and trigger the MAP kinase and protein kinase A pathways, respectively, up-regulating the expression of C/EBPβ and C/EBPδ genes. The rise in cAMP levels induced by the couple PGI_2/IP leads first to CREB activation through its phosphorylation by protein kinase A and second to a decrease in Sp1 levels, allowing C/EBPβ and C/EBPδ to gain access to the C/EBP binding site present in the C/EBPα gene promoter and favoring the subsequent expression of C/EBPα. Although p38 MAP kinase activated by mitogen-activated protein kinase 6 (MKK6) appears necessary for the differentiation of 3T3-L1 cells, and although C/EBPβ has been identified as one of its substrates, the role of this kinase is difficult to estimate as no hormonal effector and no transducing machinery upstream of MKK6 are known.[167,168] The rise in Ca^{2+} levels (concomitant with that of cAMP levels) leads to activation of Ca^{2+}-calmodulin-sensitive protein kinase II leading in turn to the phosphorylation of C/EBPβ and possibly to that of C/EBPδ which are likely required for their full DNA-binding activity. In the meantime, PPARδ undergoes activation after binding of exogenous fatty acids and/or binding of endogenous prostacyclin which have entered the nucleus. At that point C/EBPβ and C/EBPδ are active in promoting the expression of both C/EBPα and PPARγ whereas active C/EBPβ/δ, activated PPARδ and ADD-1/SREBP-1c cooperate for the critical expression of PPARγ. The adipogenic hormones, which are required at that stage are:

1. glucocorticoids which enhance prostacyclin production and promote the inhibition of pref-1 gene transcription,
2. GH, which triggers the JAK/STAT pathway, promotes IGF-1 production and triggers the inhibition of Id3 gene expression and thus enhances the transcriptional activity of ADD-1/SREPB-1c and
3. insulin which enhances the expression of ADD-1/SREBP-1c.

The role of the adipogenic hormones, if any, in down-regulating the expression of transcriptional repressors other than Id3 by GH and Sp1 by cAMP-elevating agents, i.e., AhR, Arnt, AP-2α, AEBP-1 and AEBP-2, is unknown and remains to be investigated. Once expressed, the levels of PPARγ and C/EBPα are maintained through cross-regulation, thereby sustaining the adipocyte phenotype. In contrast to C/EBPβ/δ and PPARδ, C/EBPα and PPARγ are anti-mitotic. PPARγ up-regulates the differentiation-dependent cascade expression CKIs thus providing a molecular coupling between irreversible growth arrest and terminal differentiation. Meanwhile PPARγ up-regulates the expression of various lipid-related genes encoding proteins and enzymes leading to the terminal phenotype characterized by triacylglycerol accumulation.

Differentiation of Embryonic Stem Cells and the Adipose Lineage

So far, most studies on adipogenesis have relied on the use of supposedly unipotent adipoblastic cells and rarely on that of multipotent cells of mesodermal origin (see Table 3.1). However the gene(s) that commits progression from the multipotent mesodermal stem cell to the adipoblast stage of development (see Fig. 3.1) has not been yet identified. Nor have the factors involved in self-renewal of adipoblast stem cell precursors and specific adipoblast markers.

Embryonic stem cells (ES cells) present in the inner mass of blastocysts (3.5 day-post-coïtum) contribute to all cell lineages of developing mouse. ES cells, after maintenance in a totipotent state in vitro, can differentiate spontaneously into various lineages. In this case, the number of ES cell-derived adipocytes is low because spontaneous commitment into the adipocyte lineage, i.e., adipoblast formation, is rare.

Recently conditions to induce from ES cells the adipocyte lineage and adipocyte formation at a high frequency have been described.[4] Embryonic stem cell-derived embryoid bodies treated with RA or specific retinoids for a precise period of time (between day 2 and day 5) leads to the appearance of large clusters of mature fat cells in embryoid body outgrowths. Between the 2nd and 5th day, C/EBPβ at low levels and PPARδ but not PPARγ are expressed. Adipogenic hormones and/or PPARγ activators/ligands could not substitute for RA. Thus two phases can be distinguished: the first phase, between the 2nd and the 5th day, which corresponds to a permissive period for the commitment to adipoblast, and a second phase which corresponds to the permissive period for differentiation to adipocytes and which requires adipogenic hormones. The model system of ES cells should provide a means to identify novel regulatory genes implicated in early determination events of adipogenesis.

Conclusions and Perspectives

A limited number of signaling pathways regulated by adipogenic hormones cooperate to regulate the expression of PPARγ which is the critical transcriptional activator of terminal differentiation and adipose tissue development. In preadipocytes, initial events which modulate the subsequent expression of PPARγ rely on the cooperation between secreted factors produced endogenously which up-regulate C/EBPβ and C/EBPδ expression/activity (LIF and prostacyclin) and exogenous factors arising from triacylglycerol hydrolysis (fatty acids including arachidonic acid) which activate PPARδ. Initial events rely also on circulating hormones such as insulin required for ADD-1/SREBP-1c expression/activity. Many questions still remain unanswered regarding adipogenesis which may prove important pharmacologically:

1. the identification of the natural ligand(s) of PPARγ synthesized by differentiating, authentic target cells,
2. the gene structure and the hormonal regulation of the expression of transcriptional repressors,
3. the characterization of new repressors and their mechanisms of action.

Above all, however, the question of intrinsic differences between precursor cells from different adipose depots remains critical in order to gain some insights in humans on the factors that control the differential development of WAT, including in patients suffering from HIV infection and who exhibit lipodystrophy after multiple drug therapy. Whether important differences do exist between adipocytes from subcutaneous and intra-abdominal depots in the pathways that regulate the synthesis and secretion of proteins (angiotensinogen, PAI-1, adiponectin, etc.), it would also help to understand the increased risk to develop cardiovascular diseases in overweight and obese patients compared to lean individuals.

Acknowledgements

The author wishes to thank the colleagues and the members of the technical staff who have contributed efficiently all over the years to the study of adipose tissue development. Thanks are due to Drs. E. Amri, C. Dani and P. Grimaldi for critical reading of the manuscript. Special thanks are due to Prof. R. Négrel and for outstanding secretarial assistance to Mrs. G. Oillaux.

References

1. Poissonnet CM, Burdi AR, Garn SM. The chronology of adipose tissue appearance and distribution in human fetus. Early Hum Dev 1984; 10:1-11.
2. Ailhaud G, Grimaldi P, Négrel R. Cellular and molecular aspects of adipose tissue development. Annu Rev Nutr 1992; 12:207-233.
3. Cinti S. The Adipose Organ. Milano: Editrice Kurtis, 1999.
4. Dani C, Smith A, Dessolin S et al. Differentiation of embryonic stem cells into adipocytes in vitro. J Cell Sci 1997; 110:1279-1285.
5. Grégoire F, Smas C, Sul HS. Understanding adipocyte differentiation. Physiol Rev 1998:783-809.
6. Ross SR, Choy L, Graves RA et al. Hibernoma formation in transgenic mice and isolation of a brown adipocyte cell line expressing the uncoupling protein gene. Proc Natl Acad Sci USA 1992; 89:7561-7565.
7. Zennaro MC, Viengehareun, Le Menuet D et al. Establisement of a novel brown adipose cell line by targeted oncogenesis: Involvement of aldosterone in adipocyte differentiation. Abstract—Satellite Symposium of the 9th EASO Congress, Ancona; 1999.
8. Aso H, Abe H, Nakajima I et al. A preadipocyte clonal line from bovine intramuscular adipose tissue: Nonexpression of GLUT-4 protein during adipocyte differentiation. Biochem Biophys Res Commun 1995; 213:369-375.
9. Zilberfarb W, Piétri-Rouxel F, Jockers R et al. Human immortilized brown adipocytes express functional β3-adrenoreceptor coupled to lipolysis. J Cell Science 1997; 110:801-807.
10. Pilgrim C. DNA synthesis and differentiation in developing white adipose tissue. Dev Biol 1971; 26:69-76.
11. Cook JR, Kozak LP. sn-Glycerol-3-phosphate dehydrogenase gene expression during mouse adipocyte development in vivo. Dev Biol 1982; 92:440-448.

12. Faust IM, Johnson PR, Stern JS et al. Diet-induced adipocyte number increase in adult rats : A new model of obesity. Am J Physiol 1978; 235:E279-E286.

13. Ailhaud G, Amri E, Bertrand B et al. Cellular and molecular aspects of adipose tissue growth. In: Bray G, Ricquier D, Spiegelman, BM, eds. Obesity: Towards a Molecular Approach. New York: Alan R. Liss Inc., 1990; 133:219-236.

14. Hauner H, Entenmann G, Wabitsch M et al. Promoting effect of glucocorticoids on the differentiation of human adipocyte precursor cells cultured in a chemically defined medium. J Clin Invest 1989; 84:1663-1670.

15. Salans LB, Cushman SW, Weismann RE. Studies of human adipose tissue. Adipose cell size and number in nonobese and obese patients. J Clin Invest 1973; 52:929-941.

16. Faust IM, Miller Jr WH. Hyperplastic growth of adipose tissue in obesity. In: Angel A, Hollenberg CH, Roncari DAK, eds. The Adipocyte and Obesity: Cellular and Molecular Mechanisms. New York: Raven Press, 1983:41-51.

17. Miller Jr WH, Faust IM, Golberger AC et al. Effects of severe long-term food deprivation and refeeding on adipose tissue cells in the rat. Am J Physiol 1983; 245:E74-E80.

18. Ochi M, Yoshioka H, Sawada T et al. New adipocyte formation in mice during refeeding after long-term deprivation. Am J Physiol 1991; 260:R468-R474.

19. Geloen A, Roy PE, Bukowecki LJ. Regression of white adipose tissue in diabetic rats. Am J Physiol 1989; 257:E547-E553.

20. Klyde BJ, Hirsch J. Increased cellular proliferation in adipose tissue of adult rats fed a high-fat diet. J Lipid Res 1979; 20:705-715.

21. Prins JB, Niesler CU, Winterford CM et al. Tumor necrosis factor-α induces apoptosis of human adipose cells. Diabetes 1997; 46:1939-1944.

22. Niesler CU, Siddle K, Prins JB. Human preadipocytes display a depot-specific susceptibility to apoptosis. Diabetes 1998; 74:1365-1368.

23. Lyle RE, Richon VM, McGehee Jr RE. TNF α disrupts mitotic clonal expansion and regulation of retinoblastoma proteins p130 and p107 during 3T3-L1 adipocyte differentiation. Biochem Biophys Res Commun 1998; 274:373-378.

24. Navarro P, Valverde AM, Conejo R et al. Inhibition of caspases rescues brown adipocytes from apoptosis downregulating BCL-XS and upregulating BCL-2 gene expression. Exp Cell Res 1999; 246:301-307.

25. Navarro P, Valverde AM, Benito M et al. Insulin/IGF-1 rescues immortalized brown adipocytes from apoptosis down-regulating Bcl-xS expression, in a PI 3-kinase- and Map kinase-dependent manner. Exp Cell Res 1998; 243:213-221.

26. Qian H, Azain MJ, Compton MM et al. Brain administration of leptin causes deletion of adipocytes by apoptosis. Endocrinology 1998; 139:791-794.

27. Okuno A, Tamemoto H, Tobe K et al. Troglitazone increases the number of small adipocytes without the change of white adipose tissue mass in obese Zucker rats. J Clin Invest 1998; 101:1354-1361.

28. Lindquist JM, Rehnmark S. Ambient temperature regulation of apoptosis in brown adipose tissue. Erk1/2 promotes norepinephrine-dependent cell survival. J Biol Chem 1998; 273:30147-30156.

29. Magun R, Gagnon AM, Yaraghi Z et al. Expression and regulation of neuronal apoptosis inhibitory protein during adipocyte differentiation. Diabetes 1998; 47:1948-1952.

30. Bonnet FP, Rocour-Brumioul D. Fat cells in leanness, growth retardation, and adipose tissue dystrophic syndromes. In: Bonnet FP, ed. Adipose Tissue in Childhood. Boca Raton: CRC Press, 1981:155-163.

31. Carter-Su C; Schwartz J, Smit LS. Molecular mechanism of growth hormone action. Annu Rev Physiol 1996; 58:197-207.

32. Levacher C, Sztalryd C, Kinebanyan M et al. Effects of thyroid hormones on adipose tissue development in Sherman and Zucker rats. Am J Physiol 1984; 246:C-C56.

33. Rebuffé-Scrive M, Krotkiewski M, Elfverson J et al. Muscle and adipose tissue morphology and metabolism in Cushing's syndrome. J Clin Endocrinol Metab 1988; 67:1122-1128.

34. Pepin MC, Pothier F, Barden N et al. Impaired type II glucocorticoid-receptor function in mice bearing antisense RNA transgene. Nature 1992; 355:725-728.

35. Stenzel-Poore MP, Cameron VA, Vaughan J et al. Development of Cushing's syndrome in corticotropin-releasing factor transgenic mice. Endocrinology 1992; 130:3378-3386.

36. Mizutani T, Nishikawa Y, Adachi H et al. Identification of estrogen receptor in human adipose tissue and adipocytes. J Clin Endocrinol Metab 1994; 78:950-954.

37. Pedersen SB, Fuglsig S, Sjögren P et al. Identification of steroid receptors in human adipose tissue. Eur J Clin Invest 1996; 26:1051-1056.

38. O'Brien SN, Welter BH, Mantzke KA et al. Identification of progesterone receptor in human subcutaneous adipose tissue. J Clin Endocrinol Metab 1988; 83:509-513.

39. Lacasa D, Garcia E, Henriot D et al. Site-related specificities of the control by androgenic status of adipogenesis and mitogen-activated protein kinase cascade/c-fos signaling pathway in rat preadipocytes. Endocrinology 1997; 138:3181-3186.

40. Lacasa D, Garcia E, Agli B et al. Control of rat preadipocyte adipose conversion by ovarian status: regional specificity and possible involvement of the mitogen-activated protein kinase-dependent and c-fos signaling pathway. Endocrinology 1997; 138:2729-2734.

41. Shepherd PR, Gnudi L, Tozzo E et al. Adipose cell hyperplasia and enhanced glucose disposal in transgenic mice overexpressing GLUT4 selectively in adipose tissue. J Biol Chem 1993; 268:22243-22246.

42. Katz EB, Stenbit AE, Hatton K et al. Cardiac and adipose tissue abnormalities but not diabetes in mice deficient in GLUT4. Nature 1995; 377:151-155.

43. Oscai LB, Brown NM, Miller WC. Effect of dietary fat on food intake, growth and body composition in rats. Growth 1984; 48:415-424.

44. Romieu I, Willett WC, Stampfer MJ et al. Energy intake and other determinants of relative weight. Am J Clin Nutr 1988; 74:406-412.

45. Schutz Y, Flatt JP, Jequier E. Failure of dietary fat intake to promote fat oxidation: A factor favoring the development of obesity. Am J Clin Nutr 1989; 50:307-314.

46. Klesges RC, Klesges LM, Haddock CK et al. A longitudinal analysis of the impact of dietary intake and physical activity on weight change in adults. Am J Clin Nutr 1992; 55:818-822.

47. Tucker LA, Kano MJ. Dietary fat and body fat: A multivariate study of 205 adult females. Am J Clin Nutr 1992; 56:616-612.

48. Klesges RC, Klesges LM, Eck LH et al. A longitudinal analysis of accelerated weight gain in preschool children. Pediatrics 1995; 95:126-130.

49. Blundell JE, Cotton JR, Delargy H et al. The fat paradox: Fat-induced satiety signals versus high fat over-consumption. Int J Obes 1995; 19:832-835.

50. Nguyen VT, Larson DE, Johnson RK et al. Fat intake and adiposity in children of lean and obese parents. Am J Clin Nutr 1996; 63:507-513.

51. Brun LD, Gagne C, Julien P et al. Familial lipoprotein lipase-activity deficiency : study of total body fatness and subcutaneous fat tissue distribution. Metabolism 1989; 38:1005-1010.

52. Weinstock PH, Levak-Frank S, Hudgins LC et al. Lipoprotein lipase controls fatty acid entry into adipose tissue, but fat mass is preserved by endogenous synthesis in mice deficient in adipose tissue lipoprotein lipase. Proc Natl Acad Sci USA 1997; 94:10261-10266.

53. Levak-Frank S, Radner H, Walsh AM et al. Muscle-specific over-expression of lipoprotein lipase causes a severe myopathy characterized by proliferation of mitochondria and peroxisomes in transgenic mice. J Clin Invest 1995; 96:976-986.

54. Lemberger T, Desvergne B, Wahli W. Peroxisome proliferator-activated receptors: A nuclear receptor signaling pathway in lipid physiology. Annu Rev Dev Biol 1996; 12:335-363.

55. Parrish CC, Pathy DA, Angel A. Dietary fish oil limits adipose tissue hypertrophy in rats. Metabolism 1990; 39:217-219.

56. Shillabeer G, Lau DCW. Regulation of new fat cell formation in rats: The role of dietary fats. J Lipid Res 1994; 35:592-600.

57. Cleary MP, Phillips FC, Morton AA. Genotype and diet effects in lean and obese Zucker rats fed either safflower or coconut oil diets. P Soc Exp Biol Med 1999; 220:153-161.

58. Hauner H, Entenmann G. Regional variation of adipose differentiation in cultured stromal-vascular cells from the abdominal and femoral adipose tissue of obese women. Int J Obes 1991; 15:121-126.

59. Kirkland JL, Hollenberg CH, Gillon WS. Age, anatomic site, and the replication and differentiation of adipocyte precursors. Am J Physiol 1990; 258:C206-C210.

60. Beresford JN, Bennett JH, Devlin C et al. Evidence for an inverse relationship between the differentiation of adipocytic and osteogenic cells in rat marrow stromal cell cultures. J Cell Sci 1992; 102:341-351.
61. Diascro Jr DD, Vogel RL, Johnson TE et al. High fatty acid content in rabbit serum is responsible for the differentiation of osteoblasts into adipocyte-like cells. J Bone Miner Res 1998; 13:96-106.
62. Dorheim MA, Sullivan M, Vandapani V et al. Osteoblastic gene expression during adipogenesis in hematopoietic supporting murine bone marrow stromal cells. J Cell Physiol 1993; 154:317-328.
63. Houghton A, Oyajobi BO, Foster GA et al. Immortalization of human marrow stromal cells by retroviral transduction with a temperature sensitive oncogene: Identification of bipotential precursor cells capable of directed differentiation to either an osteoblast or adipocyte phenotype. Bone 1998; 22:7-16.
64. Nuttall ME, Patton AJ, Olivera DL et al. Human trabecular bone cells are able to express both osteoblastic and adipocytic phenotype: implications for osteopenic disorders. J Bone Miner Res 1998; 13:271-382.
65. Teboul L, Gaillard D, Staccini L et al. Thiazolidinediones and fatty acids convert myogenic cells into adipose-like cells. J Biol Chem 1995; 276:28183-28187.
66. Amri E, Dani C, Doglio A et al. Coupling of growth arrest and expression of early markers during adipose conversion of preadipocyte cell lines. Biochem Biophys Res Commun 1986; 137:903-910.
67. Amri EZ, Bonino F, Ailhaud G et al. Cloning of a protein that mediates transcriptional effects of fatty acids in preadipocytes. Homology to peroxisome proliferator-activated receptors. J Biol Chem 1995; 270:2367-2371.
68. Safonova I, Darimont C, Amri E et al. Retinoids are positive effectors of adipose cell differentiation. Mol Cell Endocrinol 1994; 104:201-211.
69. Forman BM, Tontonoz P, Chen J et al. 15-deoxy-$\delta^{12,14}$-prostaglandin J$_2$ is a ligand for the adipocyte determination factor PPARγ. Cell 1995; 83:803-812.
70. Krey G, Braissant O, L'Horset F, et al. Fatty acids, eicosanoids, and hypolipidemic agents identified as ligands of peroxisome proliferator-activated receptors by coactivator-dependent receptor ligand assay. Mol Endocrinol 1997; 11:779-791.
71. Kliewer SA, Sundseth SS, Jones SA et al. Fatty acids and eicosanoids regulate gene expression through direct interactions with peroxisome proliferator-activated receptors α and γ. Proc Natl Acad Sci USA 1997; 94:4318-4323.
72. Forman BM, Vhen J, Evan RM. Hypolipidemic drugs, polyunsaturated fatty acids, and eicosanoids are ligands for peroxisome proliferator-activated receptors α and δ. Proc Natl Acad Sci USA 1997; 94:4312-4317.
73. Nolte RT, Wisely GB, Westin S et al. Ligand binding and co-activator assembly of the peroxisome proliferator-activated receptor γ. Nature 1998; 395:137-143.
74. Xu HE, Lambert MH, Montana VJ et al. Molecular recognition of fatty acids by peroxisome proliferator-activated receptors. Molecular Cell 1999; 3:397-403.
75. Berger JB, Leibowitz MD, Doebber TW et al. Novel peroxisome proliferator-activated receptor (PPAR)γ and PPARδ ligands produce distinct biological effects. J Biol Chem 1999; 274:6718-6725.
76. Wright HM, Clish CB, Mikami T et al. A synthetic antagonist for the peroxisome proliferator-activated receptor γ inhibits adipocyte differentiation. J Biol Chem 2000; 275:1873-1877.
77. Kim JB, Spiegelman BM. ADD1/SREBP1 promotes adipocyte differentiation and gene expression linked to fatty acid metabolism. Genes Dev 1996; 10:1096-1107.
78. Mandrup S, Lane MD. Regulating adipogenesis. J Biol Chem 1997; 272:5367-5370.
79. Wu Z, Rosen ED, Brun R et al. Cross-regulation of C/EBPα and PPARγ controls the transcriptional pathway of adipogenesis and insulin sensitivity. Molecular Cell 1999; 3:151-158.
80. Négrel R. Fat cells cannot divide. In: Angel A, Anderson H, Bouchard C, Lau D, Leiter L, Roncari DAK, Mendelson R, eds. The Adipocyte and Obesity: Cellular and Molecular Mechanisms. Vol. 7. John Libbey & Company Ltd., 1996:121-125.
81. Deslex S, Négrel R, Ailhaud G. Development of a chemically defined serum-free medium for complete differentiation of rat adipose precursor cells. Exp Cell Res 1987; 168:15-30.
82. Entenmann G, Hauner H. Relationship between replication and differentiation in cultured human adipocyte precursor cells. Am J Physiol 1996; 270:C1011-C1016.

83. Bastié C, Holst D, Gaillard D et al. Expression of peroxisome proliferator-activated receptor δ promotes induction of PPARγ and adipocyte differentiation in 3T3C2 fibroblasts. J Biol Chem 1999; 274:21920-21925.

84. Morrison RF, Farmer SR. Role of PPARγ in regulating a cascade expression of cyclin-dependent kinase inhibitors, p18(INK4c) and p21(Waf1/Cip1), during adipogenesis. J Biol Chem 1999; 274:17088-17097.

85. Altiok S, Xu M, Spiegelman BM. PPARγ induces cell cycle withdrawal: Inhibition of E2F/DP DNA-binding activity via down-regulation of PP2A. Genes Dev 1997; 11:1987-1998.

86. Hansen JB, Petersen RK, Larsen BM. Activation of peroxisome proliferator-activated receptor γ bypasses the function of the retinoblastoma protein in adipocyte differentiation. J Biol Chem 1999; 274:2386-2393.

87. Nevins JR. A link between the Rb tumor suppressor protein and vital oncoproteins. Science 1992; 258:424-429.

88. Shao D, Lazar MA. Peroxisome proliferator-activated receptor γ, CCAAT/enhancer-binding protein α, and cell cycle status regulate the commitment to adipocyte differentiation. J Biol Chem 1997; 272:21473-21478.

89. Chen PL, Riley DJ, Chen Y et al. Retinoblastoma protein positively regulates terminal adipocyte differentiation through direct interaction with C/EBPs. Genes Dev 1996; 10:2794-2804.

90. Lane MD, Tang QQ, Jiang MS. Role of CCAAT enhancer binding proteins (C/EBPs) in adipocyte differentiation. Biochem Biophys Res Commun 1999; 266:677-683.

91. Timchenko NA, Harris TE, Wilde M et al. CCAAT/enhancer binding protein a regulates p21 protein and hepatocyte proliferation in newborn mice. Mol Cell Biol 1997; 17:7353-7361.

92. Börglum JD, Pedersen SB, Ailhaud G et al. Differential expression of prostaglandin receptor mRNAs during adipose cell differentiation. Prostaglandins Other Lipid Mediat 1999; 57:305-317.

93. Smas CM, Chen L, Sul HS. Cleavage of membrane-associated pref-1 generates a soluble inhibitor of adipocyte differentiation. Mol Cell Biol 1997; 17:977-988.

94. Garcès C, Ruiz-Hidalgo MJ, Font de Moras J. Notch-1 controls the expression of fatty acid-activated transcription factors and is required for adipogenesis. J Biol Chem 1997; 272:29729-29734.

95. Aubert J, Dessolin S, Bemonte N. Leukemia inhibitory factor and its receptor promote adipocyte differentiation via the mitogen-activated protein kinase cascade. J Biol Chem 1999; 274: 24965-24972.

96. Börglum J, Vassaux G, Richelsen B et al. Changes in adenosine A1- and A2-receptor expression during adipose cell differentiation. Mol Cell Endocrinol 1996; 117:17-25.

97. Lu Z, Pineyro MA, Kirkland JL et al. Prostaglandin-sensitive adenylyl cyclase of cultured preadipocytes and mature adipocytes of the rat: Probable role of Gi in determination of stimulatory or inhibitory action. J Cell Physiol 1988; 136:1-12.

98. Kirkland JL, Pineyro MA, Zhongdin L et al. Hormone-sensitive adenylyl cyclase in preadipocytes cultured from adipose tissue: comparison with 3T3-L1 cells and adipocytes. J Cell Physiol 1987; 133:449-460.

99. Vassaux G, Gaillard D, Ailhaud et al. Prostacyclin is a specific effector of adipose cell differentiation: its dual role as a cAMP- and Ca²⁺-elevating agent. J Biol Chem 1992; 267:11092-11097.

100. Vassaux G, Gaillard D, Mari B et al. Differential expression of A1- and A2-receptors in preadipocytes and adipocytes. Biochem Biophys Res Commu 1993; 193:1123-1130.

101. Vassaux G, Gaillard G, Darimont C et al. Differential response of preadipocytes and adipocytes to PGI2 and PGE2: physiological implications. Endocrinology 1992; 131:2393-2398.

102. Smas CM, Chen L, Zhao L et al. Transcriptional repression of pref-1 by glucocorticoids promotes 3T3-L1 adipocyte differentiation. J Biol Chem 1999; 274:12632-12641.

103. Börglum J, Richelsen B, Darimont C et al. Expression of the two isoforms of prostaglandin endoperoxide synthase (PGHS-1 and PGHS-2) during adipose cell differentiation. Mol Cell Endocrinol 1997; 131:67-77.

104. Doglio A, Dani C, Grimaldi P et al. Growth hormone stimulates c-fos gene expression via protein kinase C without increasing inositol lipid turnover. Proc Natl Acad Sci USA 1989; 86:1148-1152.

105. Kamai Y, Mikawa S, Endo K et al. Regulation of insulin-like growth factor-I expression in mouse preadipocyte Ob1771 cells. J Biol Chem 1996; 271:9883-9886.

106. Catalioto RM, Gaillard D, Ailhaud G et al. Terminal differentiation of mouse preadipocyte cells: the mitogenic-adipogenic role of growth hormone is mediated by the protein kinase C signaling pathway. Growth Factors 1992; 6:255-264.
107. Yarwood SJ, Sale EM, Sale GJ et al. Growth hormone-dependent differentiation of 3T3-F442A preadipocytes requires Janus kinase/signal transducer and activator of transcription but not mitogen-activated protein kinase or p70 S6 kinase signaling. J Biol Chem 1999; 274:8662-8668.
108. Teglund S, McKay C, Schuetz E et al. Stat5a and STAT5b proteins have essential and nonessential or redundant roles in cytokine responses. Cell 1998; 93:841-850.
109. Wabitsch M, Heinze E, Hauner H et al. Biological effects of human growth hormone in rat adipocyte precursor cells and newly differentiated adipocytes in primary culture. Metabolism 1996; 45:34-42.
110. Hansen LH, Madsen B, Teisner B et al. Characterization of the inhibitory effect of growth hormone on primary preadipocyte differentiation. Mol Endocrinol 1998; 12:1140-1149.
111. Accili D, Taylor SI. Targeted inactivation of the insulin receptor gene in mouse 3T3-L1 fibroblasts via homologous recombination. Proc Natl Acad Sci USA 1991; 88:4708-4712.
112. Chaika OV, Chaika N, Volle DJ et al. CSF-1 receptor/insulin receptor chimera permits CSF-1-dependent differentiation of 3T3-L1 preadipocytes. J Biol Chem 1997; 272:11968-11974.
113. Grimaldi P, Djian P, Négrel R et al. Differentiation of Ob17 preadipocytes to adipocytes: requirement of adipose conversion factor(s) for fat cell cluster formation. EMBO J 1982; 1:687-692.
114. Flores-Delgado G, Marsh-Moreno M, Kuri-Harcuch W. Thyroid hormone stimulates adipocyte differentiation of 3T3 cells. Mol Cell Biochem 1987; 76:35-43.
115. Darimont C, Gaillard D, Ailhaud G et al. Terminal differentiation of mouse preadipocyte cells: adipogenic and antimitogenic role of triiodothyronine. Mol Cell Endocrinol 1993; 98:67-73.
116. Xue JC, Schwartz EJ, Chawla A et al. Distinct stages in adipogenesis revealed by retinoid inhibition of differentiation after induction of PPARγ. Mol Cell Biol 1996; 16:1567-1575.
117. Gaillard D, Négrel R, Lagarde M et al. Requirement and role of arachidonic acid in the differentiation of preadipose cells. Biochem J 1989; 257:389-397.
118. Négrel R, Gaillard D, Ailhaud G. Prostacyclin as a potent effector of adipose cell differentiation. Biochem J 1989; 257:399-405.
119. Catalioto RM, Gaillard D, Maclouf J et al. Autocrine control of adipose cell differentiation by prostacyclin and $PGF_{2\alpha}$. Biochim Biophys Acta 1991; 1091:364-369.
120. Aubert J, Saint-Marc P, Belmonte N et al. Prostacyclin IP receptor up-regulates the early expression of C/EBPβ and C/EBPδ in preadipose cells. Mol Cell Endocrinol 2000; 160:149-156.
121. Aubert J, Ailhaud G, Négrel R. Evidence for a novel regulatory pathway activated by (carba)prostacyclin in preadipose and adipose cells. FEBS Lett 1996; 397:117-121.
122. Aubert J, Champigny O, Saint-Marc P et al. Up-regulation of UCP-2 gene expression by PPAR agonists in preadipose and adipose cells. Biochem Biophys Res Commun 1997; 238:606-611.
123. Négrel R, Ailhaud G. Metabolism of arachidonic acid and prostaglandin synthesis in Ob17 preadipocyte cell line. Biochem Biophys Res Commun 1981; 98:768-777.
124. Gaillard D, Wabitsch M, Pipy B et al. Control of terminal differentiation of adipose precursor cells by glucocorticoids. J Lipid Res 1991; 32:569-579.
125. Darimont C, Vassaux G, Ailhaud G et al. Differentiation of preadipose cells : paracrine role of prostacyclin upon stimulation of adipose cells by angiotensin II. Endocrinology 1994; 135:2030-2036.
126. Yarwood SJ, Kilgour E, Anderson NG. Cyclic AMP potentiates growth hormone-dependent differentiation of 3T3-F442A preadipocytes: possible involvement of the transcription factor CREB. Mol Cell Endocrinol 1998; 138:41-50.
127. Wang HY, Goligorsky MS, Malbon CC. Temporal activation of Ca2+-calmodulin-sensitive protein kinase type II is obligate for adipogenesis. J Biol Chem 1997; 272:1817-1821.
128. Ntambi JM, Takova T. Role of Ca^{2+} in the early stages of murine adipocyte differentiation as evidenced by calcium mobilizing agents. Differentiation 1996; 60:151-158.
129. Amri E, Bertrand B, Ailhaud G et al. Regulation of adipose cell differentiation I) fatty acids are inducers of the aP2 gene expression. J Lipid Res 1991; 32:1449-1456.
130. Grimaldi P, Knobel SM, Whitesell RR et al. Induction of the aP2 gene by nonmetabolized long chain fatty acids. Proc Natl Acad Sci USA 1992; 89:10930-10934.

131. Amri E, Ailhaud G, Grimaldi PA. Fatty acids as signal transducing molecules : involvement in the differentiation of preadipose to adipose cells. J Lipid Res 1994; 35:930-937.
132. Ailhaud G, Abumrad N, Amri EZ et al. A new look at fatty acids as signal transducing molecules. World Rev Nutr Diet 1994; 75:35-45.
133. Vassaux G, Négrel R, Ailhaud G et al. Proliferation and differentiation of rat adipose precursor cells in chemically defined medium : Differential action of anti-adipogenic agents. J Cell Physiol 1994; 161:249-256.
134. Kawada T, Aoki N; Kamei Y et al. Comparative investigation of vitamins and their analogues on terminal differentiation, from preadipocytes to adipocytes, of 3T3-L1 cells. Comp Biochem Physiol 1990; 96:323-326.
135. Miller CW, Casimir DA, Ntambi JM. The mechanism of inhibition of 3T3-L1 preadipocyte differentiation by prostaglandin $PGF_{2\alpha}$. Endocrinology 1996; 137:5641-5650.
136. Reginato MJ, Krakow SL, Bailey ST et al. Prostaglandins promote and block adipogenesis through opposing effects on peroxisome proliferator-activated receptor γ. J Biol Chem 1997; 273:1855-1858.
137. Rosen ED, Sarraf P, Troy AE et al. PPARγ is required for the differentiation of adipose tissue in vivo and in vitro. Mol Cell 1999; 4:611-617.
138. Barak Y, Nelson MC, Ong ES et al. PPARγ is required for placental, cardiac and adipose tissue development. Mol Cell 1999; 4:585-595.
140. Wu Z, Xie Y, Bucher NLR et al. Conditional ectopic expression of C/EBPβ in NIH-3T3 cells induces PPARγ and stimulates adipogenesis. Genes Dev 1995; 9:2350-2363.
141. Fajas L, Schoonjans, K, Gelman L et al. Regulation of peroxisome proliferator-activated receptor γ expression by adipocyte differentiation and determination factor 1/sterol regulatory element binding protein 1: Implications for adipocyte differentiation and metabolism. Mol Cell Biol 1999; 19:5495-5503.
142. Kim JB, Wright HM, Wright M et al. ADD1/SREBP1 activates PPARγ through the production of endogenous ligand. Proc Natl Acad Sci USA 1998; 95:4333-4337.
143. Ailhaud G. Cell surface receptors, nuclear receptors and ligands that regulate adipose tissue development. Clin Chim Acta 1999; 286:181-190.
144. Reusch JEB, Colton LA, Klemm DJ. CREB activation induces adipogenesis in 3T3-L1 cells. Mol Cell Biol 2000; 20:1008-1020.
145. King FJ, Hu E, Harris DF et al. DEF-1, a novel Src SH3 binding protein that promotes adipogenesis in fibroblastic cell lines. Mol Cell Biol 1999; 19:2330-2337.
146. Kim JB, Sarraf P, Wright M et al. Nutritional and insulin regulation of fatty acid synthase and leptin gene expression through ADD1/SREBP1. J Clin Invest 1998; 101:1-9.
147. Sorisky A. From preadipocyte to adipocyte: differentiation-directed signals of insulin from the cell surface to the nucleus. Critical Reviews in Clinical Laboratory Sciences 1999; 36:1-34.
148. Ristow M, Müller-Wieland D, Pfeiffer A et al. Obesity associated with a mutation in a genetic regulator of adipocyte differentiation. N Engl J Med 1998; 14:953-959.
149. Hamann A, Münzberg AH, Buttron P et al. Missense variants in the human peroxisome proliferator-activated receptor-γ2 gene in lean and obese subjects. Eur J Endocrinol 1999; 141:90-92.
150. Kubota N, Terauchi Y, Miki H et al. PPARγ mediates high-fat diet-induced adipocyte hypertrophy and insulin resistance. Molecular Cell 1999; 4:597-609.
151. TanakaT, Yoshida N; Kishimoto T et al. Defective adipocyte differentiation in mice lacking the C/EBPβ and/or C/EBPδ gene. EMBO J 1997; 16:7432-7443.
152. Wang ND; Finegold MJ, Bradley CN et al. Impaired energy homeostasis in C/EBPα knockout mice. Science 1995; 269:1108-1112.
153. Ailhaud G, Hauner H. Development of white adipose tissue. In: Bray G, Bouchard C, James PT, eds. Handbood of Obesity. New York: M. Dekker Inc., 1997:359-378.
154. Moldes M; Lasnier F, Fève B et al. Id3 prevents differentiation of preadipose cells. Mol Cell Biol 1997; 17:1796-1804.
155. Moldes M, Boizard M, Le Liepvre X et al. Functional antagonism between inhibitor of DNA binding (Id) and adipocyte determination and differentiation factor 1/sterol regulatory element-binding protein-1c (ADD1/SREBP-1c) trans-factors for the regulation of fatty acid synthase promoter in adipocytes. Biochem J 1999; 344:873-880.

156. Ron D and Habener JF. CHOP, a novel developmentally regulated nuclear protein that dimerizes with transcription factors C/EBP and LAP and functions as a dominant-negative inhibitor of gene transcription. Genes Dev 1992; 6:439-453.

157. Carlson SG, Fawcett TW, Bartlett JD et al. Regulation of the C/EBP-related gene gadd 153 by glucose deprivation. Mol Cell Biol 1993; 13: 4736-4744.

158. Liu PCC, Phillips MA and Matsumura F. Alteration by 2,3,7,8-tetrachlorodibenzo-p-dioxin of CCAAT/enhancer binding protein correlates with suppression of adipocyte differentiation in 3T3-L1 cells. Mol Pharmacol 1996; 49:989-997.

159. Alexander DL, Ganem LG, Fernandez-Salguero P et al. Aryl-hydrocarbon receptor is an inhibitory regulator of lipid synthesis and of commitment to adipogenesis. J Cell Sci 1998; 111:3311-3322.

160. Brodie AE, Manning VA and Hu CY. Inhibitors of preadipocyte differentiation induce COUP-TF binding to a PPAR/RXR binding sequence. Biochem Biophys Res Comm 1996; 228:655-661.

161. Shimba S, Todoroki K, Aoyagi T et al. Depletion of arylhydrocarbon receptor during adipose differentiation in 3T3-L1 cells. Biochem Biophys Res Commun 1998; 249:131-137.

162. Jiang MS, Tang QQ, McLenithan J et al. Derepression of the C/EBPα gene during adipogenesis: identification of AP-2α as a repressor. Proc Natl Acad Sci USA, 1998; 95:3467-3471.

163. Tang QQ, Jiang MS and Lane D. Repressive effect of Sp1 on the C/EBPα gene promoter: role in adipocyte differentiation. Mol Cell Biol 1999; 19:4855-4865.

164. He G, Muise A, Li AW et al. A eukaryotic transcriptional repressor with carboxypeptidase activity. Nature 1995; 378:92-96.

165. He GP, Kim S and Ro HS. Cloning and characterization of a novel zinc finger transcriptional repressor: A direct role of the zinc finger motif in repression. J Biol Chem 1999; 274:14678-14684.

166. Ohno I, Hashimoto J, Shimizu K et al. A cDNA cloning of human AEBP1 from primary cultured osteoblasts and its expression in a differentiating osteoblastic cell line. Biochem Biophys Res Commun, 1996; 228:411-414.

167. Engelman JA, Lisanti MP, Scherer PE. Specific inhibitors of p38 mitogen-activated protein kinase block 3T3-L1 adipogenesis. J Biol Chem 1998; 273:3211-32120.

CHAPTER 4

Brown Adipose Tissue:

Thermogenic Function and Its Physiological Regulation

Susanne Klaus

Brown adipose tissue (BAT) is a specialized form of adipose tissue whose function is opposite to classical white fat function. As a thermogenic tissue it is a site of energy dissipation in contrast to the energy storing white fat. Whereas white, i.e., storage fat is known throughout the whole animal kingdom, BAT seems to be a unique mammalian invention. Heat production in brown fat is very important for the survival of small mammals in a cold environment and also for arousal from hibernation (Fig. 4.1). A protein unique to brown adipocytes, the uncoupling protein (UCP1), is central to uncoupling of brown fat mitochondrial respiratory chain, the mechanism of heat production in this tissue. Research interest in brown fat has risen greatly upon the discovery that in certain animal models increased BAT thermogenesis can be a means of dissipating excess energy and thus preventing obesity.

BAT was first recognized in the 1960s to be a thermogenic tissue and in the following twenty years research interest was mainly focused on the biochemistry of heat production in brown fat. Advances in molecular biology techniques and development of brown adipocyte cell culture systems in the last decade have allowed closer investigation of the molecular mechanisms of gene regulation and differentiation of brown adipocytes. Advancements in transgenic animal technology and the increasing availability of transgenic animal models also greatly contributed to our understanding of brown fat function in various physiological situations. Studies on the UCP1 gene have ultimately led to the discovery of other uncoupling proteins similar to UCP1 expressed in various tissues whose functions are not yet clear but which might play important roles in energy balance.

This chapter will give an overview on where brown adipose tissue is present, what its function is and how brown adipose tissue thermogenesis is physiologically regulated during acute thermogenesis as well as in adaptation to various environmental conditions. The following chapter (Chapter 5) will focus on brown adipocyte differentiation and the function of BAT in energy metabolism.

Nonshivering Thermogenesis: Physiological Function of Brown Fat

In contrast to reptiles, amphibians, insects etc., mammals are homeothermic (also termed endothermic) animals. This means that they are able to regulate their body temperature independently from environmental temperature. For a mammal living in the arctic region this can easily signify a temperature gradient of 50 to 60°C between the environment and the body core. Large mammals have therefore developed very efficient means of insulation to prevent

Adipose Tissues, edited by Susanne Klaus. ©2001 Eurekah.com.

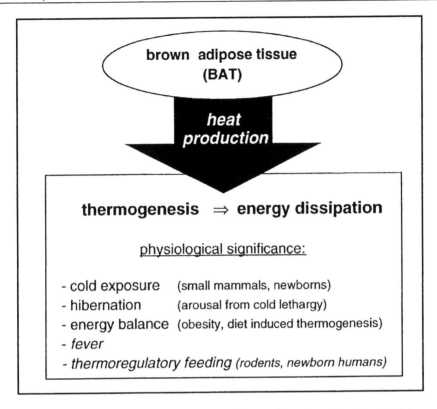

Fig. 4.1. Physiological significance of heat production in brown adipose tissue. The role of BAT thermogenesis during cold exposure, for arousal from hibernation and in energy balance is well established. More speculative is the physiological significance of BAT in the control of fever and as a control organ of thermoregulatory feeding. See text for further explanations and references.

heat loss. Small mammals on the other hand display an unfavorable volume-surface relationship which favors rapid cooling when exposed to cold. A mammal the size of a mouse would thus require a fur so thick that it would not be able to move about any more. Small as well as newborn mammals therefore rely on their capacity for non-shivering thermogenesis (NST) in order to maintain normothermia.[1] This mechanism of facultative thermogenesis is independent of muscle contractions, i.e., shivering. Upon cold exposure NST is activated by the sympathetic nervous system.[2] NST can also be induced experimentally and thus quantified by injection of norepinephrine.[3] The NST capacity is negatively correlated with body size and was calculated to be insignificant above a body weight of approximately 10 kg.[4] Larger mammals can maintain normothermia at low ambient temperatures solely by increasing their fur insulation. Small mammals, however are dependent on their capacity for NST. The bank vole for example—a common rodent in Middle and Northern Europe—is able in winter to double its heat production by activating NST.[5] NST is also the mechanism by which hibernators like marmots, squirrels, dormice etc. regain normothermia after a bout of hibernation during which body temperature drops drastically to values around and even below 5°C.[6]

It is generally accepted today that brown adipose tissue is the effector organ of NST. Using different methods, the contribution of BAT alone to total cold-induced NST was estimated to be between 30 and 70%.[7,8] Using radioactive microspheres to measure blood flow, BAT was calculated to be by far the dominant anatomical site of heat production in cold adapted rats.[7] Similar results were obtained by partial BAT-ectomy in the Siberian dwarf hamster *Phodopus sungorus*.[8] These studies showed that cold adapted animals who have an increased NST-capacity also display the highest contribution of BAT thermogenesis to overall energy expenditure.

NST, or rather brown fat thermogenesis is obviously important for maintenance of normothermia when small animals are exposed to a cold environment and for arousal of hibernators. However, as illustrated in Fig. 4.1, BAT thermogenesis is also thought to play an important role in energy balance (as will be discussed in the next chapter),[9] in fever,[10] and in the control of thermoregulatory feeding of rats and possibly also newborn humans. The latter refers to a hypothesis formulated by J. Himms-Hagen some years ago.[11,12] Putting together experimental data derived from different studies, she proposed the following model: food intake of rats occurs in episodes throughout the activity period which are proposed to occur during an increase in sympathetic nervous activity which stimulates BAT thermogenesis. BAT thermogenesis leads to increased fuel oxidation and a transient dip in blood glucose which then initiates the feeding episode. During feeding BAT thermogenesis continues and leads eventually to an increase in body temperature resulting in decreased sympathetic activity and thus thermogenesis, and termination of feeding. Body temperature gradually declines until the next bout of sympathetic activity. In this model stimulated BAT thermogenesis provides signals for both initiation and termination of the feeding episode. Feeding is thus viewed as the outcome of a thermoregulatory event. The feeding pattern of newborn humans during the first few weeks of life resembles rather closely the pattern displayed by rats. Therefore Himms-Hagen proposes that this theory also applies to the human infant which possesses abundant brown fat and is very sensitive even to mild cold.[12]

For many years BAT has been recognized as an effector organ in fever, its heat production being of importance especially during the onset phase of fever.[13] Recently however, Cannon and coworkers have proposed that BAT might be more than just an effector. They speculate that through the release of interleukins (IL-1 and IL-6) BAT might be actively involved as an endocrine organ in regulation of fever response.[10]

Definition of BAT In Contrast To WAT

The basic and name giving feature of BAT is its brown color, making it macroscopically very distinct from the well known WAT which is of more white or yellowish color. Because of this and its localization in discreet anatomical deposits, it was evident to recognize brown fat as a distinct tissue. However, a precise definition depended on the physiological function which was not known until barely 40 years ago, although brown fat had already been described in the 16th century. Interestingly, in the 19th century brown fat was also called "hibernation gland" because of its good vascularization and its abundance in hibernators.[14] Only in 1961, was BAT first hypothesized to be the source of NST.[15] In the '60s brown fat became an object of intensive research and as early as 1970 a whole book was dedicated to this topic.[16]

The early definition of BAT was based mainly on morphology and histology. Brown adipose tissue shows a high degree of vascularization and brown adipocytes are multilocular in appearance and equipped with numerous well-developed mitochondria.[17] The brown appearance of the tissue is in fact due to the high content of respiratory chain pigments, i.e., cytochromes on one hand and the rich capillary network on the other hand.[19] In Chapter 2 of this volume S. Cinti presents in detail the morphology and ultrastructure of white and brown adipose tissue and adipocytes.

Nowadays the molecular basis of BAT thermogenesis has been deciphered and a unique protein of BAT mitochondria was identified which is crucial for heat generation. This protein

Table 4.1. Main distinguishing features of white and brown adipose tissue

	White adipose tissue (WAT)	Brown adipose tissue (BAT)
Color	white or yellow	brown (light to very dark)
Sympathetic innervation	low	high
Vascularization	low	high
Lipid content	up to 90%	30-40%
Adipocyte morphology	unilocular diameter 15 to 150 μ	multilocular, diameter 10 to 25 μm
Mitochondria	few, poorly developed	many, well developed
Respiratory activity	low	high
Gene expression of UCPs:		
UCP-1	absent	present (unique to brown fat)
UCP 2	present	present
UCP-3	absent or very low	present

was found to uncouple the mitochondrial respiratory chain from ATP production and was thus termed uncoupling protein (UCP, now called UCP1).[20-22] UCP1 is still today the only known qualitative marker for brown adipocytes and the general agreement during the last decade has been to consider an adipocyte expressing UCP1 as a brown adipocyte in contrast to white adipocytes lacking the ability for UCP1 expression. However, there are numerous quantitative differences between brown and white adipose tissue mainly related to the high metabolic capacity of BAT. Table 4.1. summarizes the main features which distinguish WAT and BAT, and white and brown adipocytes, respectively. Although most genes which are expressed in BAT are also expressed in WAT, abundance and physiological regulation of gene products and their activity can vary considerably between BAT and WAT. The responses of the two types of adipose tissue to environmental stress (e.g., cold exposure) or nutritional stress (fasting and refeeding) can be quite opposite.[23] Lipogenesis is decreased in WAT of fasted or cold exposed rats, whereas it is increased in BAT, which shows a high, cold-induced activity of glycerokinase activity and fatty acid synthesis.[23,24] The activity and gene expression of lipoprotein lipase (LPL) is also very differently regulated in both tissues. LPL activity is increased by cold exposure in BAT but not in WAT.[25-27] This differential regulation of lipid metabolism in WAT and BAT reflects the different physiological function of the two tissues. During cold for example, fatty acids are used as a substrate for BAT thermogenesis and therefore fatty acids are channeled from white fat tissue towards brown adipose tissue.

The use of UCP1 as a marker has revealed the presence of UCP1 or its mRNA, i.e., the presence of brown adipocytes also in depots normally considered as typically white (reviewed in 28). This has led to some confusion regarding the definition of brown and white fat, and brown and white adipocytes, respectively. However, it is clear now that the presence of brown preadipocytes is not restricted to typical brown depots. There seems to be a certain plasticity of adipose tissue with cells (preadipocytes or stem cells) present in various fat depots which can undergo the brown adipocyte differentiation program upon appropriate external stimulation. This induction of brown adipocytes within white depots was recently shown to be a highly variable genetic trait in mice.[29] This may account for some of the controversies regarding the nature of specific fat depots. For example it was found that UCP1 expression could be induced in epididymal fat—a typical white fat depot—in A/J mice but not in B6 mice.[29] Considering this variability even within one species, it is not surprising that in different species the distribution of brown and white adipocytes might be considerably different.

The advance of PCR-technology nowadays allows the detection of very low levels of mRNA using reverse transcription PCR (RT-PCR). Using this method UCP1 expression has been reported in different adipose depots and adipocyte cell cultures. For example, when cultured human preadipocytes were treated with thiazolidinedione, a PPARγ ligand (which will be discussed in the next chapter), UCP1 mRNA expression could be detected by RT-PCR in all depots studied, even the subcutaneous one which is considered a typical white depot.[30] This renders problematic the current definition of a brown adipocyte as an adipocyte expressing UCP1. There might be a residual very low expression of the UCP1 gene in many adipocytes not connected to any functionality. This would imply that in addition to UCP1 gene expression other criteria should be fulfilled before an adipocyte can be considered brown, for example high respiratory activity, high mitochondrial density and multilocularity. However, this would mean a reversion to the rather quantitative criteria used for BAT definition before the cloning of UCP1. Several groups are right now employing various techniques of differential hybridization and amplification in order to identify genes exclusively expressed in one but not the other type of adipocyte or adipose tissue.[31,32] So hopefully in the near future more molecular markers will be available, including markers which are expressed in white but not brown adipocytes or preadipocytes. As described in Chapter 2 of this volume, S. Cinti could not detect immunoreactive S-100 protein (a calcium binding protein implicated in cytoskeleton rearrangements) in brown adipocytes, whereas it was present in white adipocytes. If confirmed by other methods, this could represent the first positive marker for white adipocytes in contrast to brown adipocytes.

Occurrence and Abundance

Although many attempts to identify brown fat in non-mammals like birds and marsupials have been made, there is no conclusive evidence for its occurrence outside of the mammalian class. Clearly UCP1 expression could not be detected in non-mammalian species, so the development of brown fat as a thermogenic tissue is considered a mammalian specialty.[33] It is still not clear however if marsupials express UCP1, i.e., if they have developed brown fat. Recently it was shown that *Sminthopsis crassicaudatus*, a small nocturnal marsupial has an interscapular fat pad which structurally resembles brown fat. The presence of UCP1 was also demonstrated by immunoblot.[34] However, antibodies against UCP1 show some cross reactivity with other uncoupling proteins which also occur in other tissues (e.g., UCP2). Physiological evidence—measuring the ability of β-adrenergic agonists to induce NST—does not strongly support the existence of brown fat in marsupials.[35] In a marsupial which shows norepinephrine-induced NST (the Tasmanian bettong) no evidence of UCP1 gene expression could be found using rat cDNA or oligonucleotide probes.[36] This seems to imply that if marsupials have an uncoupling protein it most probably is not very closely related to the UCP1 of other mammals. Only by

Table 4.2. Species of the mammalian orders in which the existence of UCP1 was demonstrated (for review see ref. 33, 41)

	Order	Species
1. Marsupialia	marsupials	*Sminthopsis crassicaudatus* (immunoblotting)
2. Insectivora	insectivores	Etruscan shrew
3. Macroscelidae	elephant shrews	NI[*]
4. Scandentia	tree shrews	NI
5. Primates	primates	humans, different species of monkeys
6. Chiroptera	bats	pipistrelle bat
7. Dermoptera	flying lemurs	NI
8. Xenarthra	xenarthra	NI
9. Pholidota	pangolins	NI
10. Rodentia	rodents	rats, mice, voles, hamsters, lemming, marmots etc.
11. Carnivora	carnivores	dog, cat
12. Lagomorpha	hares and rabbits	rabbit
13. Cetacea	whales	NI
14. Tubulidentata	aardvark	NI
15. Perissodactyla	odd-toed ungulates	NI
16. Hyracoidea	hyraxes	NI
17. Sirenia	dugongs and seecows	NI
18. Proboscidea	elephants	NI
19. Artiodactyla	even-toed ungulates	sheep, goat, bovine, reindeer, red deer

*NI, not investigated

identification and sequencing of cDNAs or genes of possible marsupial uncoupling proteins will we be able to establish if marsupials do possess brown fat or other tissues providing thermogenic function.

Smaller animals are more dependent on NST and thus on BAT thermogenesis than bigger animals because of their unfavorable surface/volume relationship. Accordingly, the relative mass of brown fat is inversely correlated with body mass. The Etruscan shrew (*Suncus etruscus*) is one of the smallest known mammals weighing as little as 1-2 grams. It has the highest known weight specific oxygen consumption and brown fat constitutes up to 10% of its body weight.[37] In the Siberian dwarf hamster *Phodopus sungorus*, which undergoes daily torpor and can withstand cold as low as –50 °C, BAT amounts to about 5% of body weight.[38] Similar relative amounts of brown fat are found in newborn guinea pigs and rabbits, whereas in adult rats it only represents about 1% of body weight.[39] Hibernating mammals have about twice as much brown fat as non-hibernators, and with increasing body mass the relative weight of brown fat decreases parallel to the decrease in NST capacity. By extrapolation of these data it was postulated that mammals with a body weight over 10 kg do not need brown fat thermogenesis any more.[4]

As shown in Table 4.2, the presence of BAT, i.e., UCP1, was confirmed in many mammalian orders including primates, rodents and even ungulates (for review see ref. 40,41). Actual evidence of brown fat was found in almost any mammalian species investigated so far, one exception being the domestic pig.[42] Rodents with body masses rarely ever exceeding a few kilograms have significant BAT depots throughout their life. In larger mammals like sheep, goat and reindeer, BAT is evident in newborns but disappears quite rapidly after some weeks or months.[43-45] The human neonate is no exception and also has ample brown fat in several locations.[46]

Brown Fat in Humans

Newborn humans undoubtedly possess brown adipose tissue in appreciable amounts. It was first suggested by Hatai in 1902 that dorsal and cervical fat pads in human embryos appear to be homologues to the brown fat—then called hibernation gland—of lower mammals.[47] Since that time the localization of BAT in humans has been reported in various anatomical sites in the cervical and axillary region as well as inside the thoracic and abdominal cavities (perirenal, periadrenal, along thoracic and abdominal aorta and intercostal vessels) (for review see 28,44,48,49). BAT is apparently present in the human fetus from 25 gestational week one and increases until birth.[50,51] The amount of brown fat decreases after infancy,[52] but brown fat can still be found in adults where it was shown to be thermogenic functional.[53] More recent studies using UCP1 as a marker of brown fat confirmed the presence of BAT in infants and also in patients suffering from a pheochromocytoma.[51,54,55] There are also rare cases of benign brown fat tumors called hibernomas in which UCP1 can readily be detected.[56,57] But also in adults not suffering from endocrine pathologies, UCP1 or its mRNA can be detected quite frequently.[58,59] Human preadipocytes especially from the perirenal region, which are differentiated in cell culture in vitro, can be induced to increase the expression of UCP1 either by adrenergic receptor agonists,[60] or by thiazolidinediones, potent agonists for the nuclear hormone receptor peroxisome proliferator-activated receptor γ (PPARγ).[30] This shows clearly that even in adult humans BAT is still present and that human brown adipocytes retain the ability to increase UCP1 expression. However, if BAT has a major contribution to thermogenesis in adults except in pathological situations is still not clear. Cunningham and coworkers measured the respiratory capacity of total perinephric fat in adult humans which contains brown fat.[53] They concluded that in the average adult this fat depot could account for less than 0.2% of increased oxygen consumption induced by norepinephrine stimulation. Another study however, showed that UCP1 content of axillary adipose tissue was greater than that of the perirenal depot, in infants as well as in adults.[51] This means that the 0.2% is certainly an underestimation. However, consistent in all human studies is the great inter-individual variation in the amount of BAT and UCP1, respectively, whatever criterion is used for the definition of BAT. The cause for these large variations is not yet clear.

For many years, brown adipose tissue has been implicated in sudden infant death syndrome (SIDS or cot death) (see ref. 61-63 for review). However, only circumstantial evidence is available. There are conflicting hypotheses which attribute SIDS to increased, i.e., hyperactive, BAT thermogenesis on the one hand as well as to decreased BAT thermogenesis on the other hand. In autopsies of SIDS victims, increased amounts of BAT as well as UCP1 immunostaining was observed,[59] whereas others found decreased BAT.[62] Another theory describes SIDS as a magnesium-dependent disease of the transition from chemical to physical thermoregulation which would be compatible with both increased and decreased amounts of BAT postmortem.[64] In any case, there is strong evidence that SIDS is linked to disordered BAT thermogenesis.

Mechanism of Heat Production: Function of UCP1

Heat production of brown adipocytes is due to an uncoupling of the mitochondrial respiratory chain, responsible for which is the uncoupling protein UCP1 (for reviews see ref. 65,66). After BAT was postulated to be a site of thermogenesis, research in the late '60s and early '70s focused on the biochemical basis, and the uncoupling mechanism of brown fat mitochondria was subsequently characterized mainly by D. Nicholls and his coworkers.[65] The bioenergetic model of this mechanism is based on Mitchell's chemiosmotic theory according to which the electron flow in the respiratory chain is tightly coupled to a proton extrusion and the proton reentry through the ATPase is tightly coupled to ATP synthesis. BAT mitochondria possess an alternative pathway, which allows proton reentry without ATP synthesis, bypassing the ATPase complex and thus resulting in a dissipation of energy as heat (Fig. 4.2.). This pathway is normally inhibited by purine nucleotides (ATP, GTP, ADP, GDP). Fatty acids, provided by lipolysis, open the uncoupling pathway leading to elevated respiration rates, i.e., heat production. Fatty acids act thus as both substrate for respiration and cytosolic second messenger for UCP1 function. Very recently is was shown that retinoic acid is also able to activate the proton transport by UCP1.[67] The identification of a protein related to this uncoupling pathway began with the detection of a 32000 Dalton component in SDS-PAGE of rat BAT mitochondria, which was specifically increased after cold acclimation.[68] This protein was called uncoupling protein (UCP) or thermogenin and found to be unique to brown adipocytes. For over 20 years UCP1 was believed to be a unique protein for mammalian brown adipocytes only distantly related to other mitochondrial carriers.[41] Only very recently similar uncoupling proteins have been identified which are expressed not only in other mammalian tissues, but also plants and protozoans.[69-73] Therefore the brown fat uncoupling protein is now called UCP1.

The amino acid sequence of UCP1 of several species was determined by either direct protein sequencing or deduced from the cDNA sequence. Complete or partial sequences are now available for UCP1 from mouse,[74] rat,[75,76] hamster,[77] rabbit,[78] human,[79] bovine,[80] and the etruscan shrew.[81,82] UCP1 (as well as the other UCPs) belongs to a superfamily of transporters of the inner mitochondrial membrane including the ATP/ADP carrier, the phosphate carrier and others.[83-85] Figure 4.3 shows the relationship of UCP1 of different species with other UCPs, as deduced from alignment of amino acid sequences. Interspecies homology of UCP1 is clearly higher than homology to other UCPs.

Like the other members of this family, UCP1 has a tripartite structure consisting of three repeats of about 100 amino acids. Based on the primary structure, a folding model of UCP1

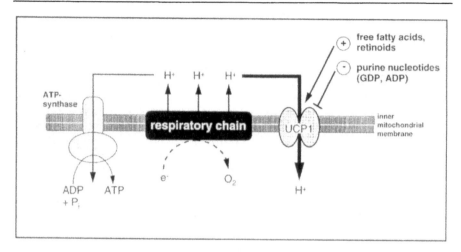

Fig. 4.2. UCP1 as an uncoupler of the respiratory chain. UCP1 functions as a dimer in the inner mitochondrial membrane where it acts as a proton translocator. When activated by fatty acids it dissipates the proton gradient across the inner mitochondrial membrane, allowing maximum activities of the respiratory chain which is uncoupled from ATP-synthesis. Purine nucleotides inhibit the proton translocation. In the normal state, when acute thermogenesis is not required, the intracellular purine nucleotide concentration is high enough to completely block UCP1 activity, and respiratory control is maintained. Adapted from ref. 65, 67.

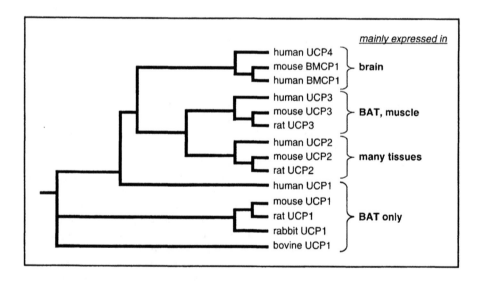

Fig. 4.3. Similarity between uncoupling proteins (UCPs) from different species based on their amino acid sequence (clustal analysis).

was proposed containing six transmembrane α-helices, two in each repeat (see Fig. 4.4.).[77] More detailed information on the UCP1 structure-function relationship was obtained using different experimental approaches. Membrane topology of UCP1 was investigated by epitope analysis using purified antibodies and isolated from intact or sonicated mitochondria.[86] The

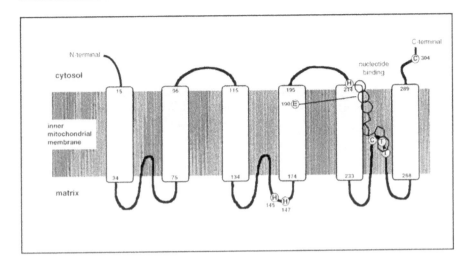

Fig. 4.4. Topology of UCP1 in the inner mitochondrial membrane. It consists of six α-helices (indicated by boxes) spanning the membrane with both N- and C-terminal towards the cytosolic side. Numbers refer to amino acid residues. Indicated is the nucleotide binding site and residues involved in nucleotide binding (E190, H214, C253, T259, T264). Histidine 145 and 147 are important for proton transport activity of UCP1. Cysteine 304 had been identified as a residue involved in fatty acid activation of UCP1; however a truncated protein lacking residues 296 to 306 was apparently still activated by fatty acids. The three matrix loops have been proposed to form a hydrophobic pocket accommodating the purine moiety of the nucleotide. Adapted from ref. 66, 82.

ion-transport properties of UCP1 and their dependence on fatty acids were investigated in detail by reconstitution studies where isolated UCP1 was reconstituted into liposomes or proteoliposomes (for review see ref. 66). Heterologous expression of UCP1 was attempted in several systems, like *Xenopus* oocytes[87] and the mammalian cell line CHO.[88] The most successful heterologous system for UCP1 expression however, proved to be yeast (*Saccaromyces cerevisiae*) (reviewed in 82).[89-91] With these approaches the main characteristics of UCP1, that is transport activity, fatty acid, and nucleotide binding, have been investigated using amino acid modification reagents but also site-directed mutagenesis in order to identify involved epitopes and amino acids.

Although proton transport is generally accepted to be crucial for UCP1 uncoupling activity, there is still a controversy regarding the role of fatty acids in this process.[92,93] According to one theory, UCP1 is an anion transporter catalyzing the transport of fatty acids. In this model fatty acids serve as cycling protonophores, crossing the membrane in a flip-flop mechanism. They would then become protonated and after a flip-flop back they would release the proton on the matrix side (reviewed in 66,93,94). The other hypothesis, termed "local H^+-buffering" proposes that fatty acids serve as prosthetic group for proton transport, delivering protons to a site in UCP1 from which they are transported across the membrane.[95] Rial and coworkers found that proton conductance through UCP1 can occur without fatty acids being present.[93] However, they propose that both types of interactions may occur depending on fatty acid concentrations: normal, nanomolar concentrations would activate UCP1 uncoupling by the "local H^+-buffering" mechanism. Only when fatty acid concentrations increase to several micromolar would they then become transportable substrates to UCP1, acting as cycling protonophores.[93]

Table 4.3. Cold induced adaptive changes in brown fat function

	Increased	Decreased
General features	tissue mass (not all species) cell number (not all species) respiratory capacity blood flow fatty acid synthesis lipogenesis glucose uptake mitochondrial content mitochondrial protein synthesis (116) number of peroxisomes	lipid content
Enzyme acitivities or amount of protein	UCP1 cytochrome c oxidase lipoprotein lipase (25, 105, 106) T4-5' deiodinase glycerokinase (24) hormone sensitive lipase (117) glycogen synthase (118) integral plasma membrane glycoprotein CE9 (120) C/EBPβ (119)	Gα proteins in plasma membrane (121) C/EBPα (119)

Gene expression	UCP1 (122)
	UCP2 (123, 124)
	UCP3 (70, 125, 126)
	(no increase of UCP3 observed in 124)
	lipoprotein lipase (127)
	S14 (128)
	GLUT4 (129, 131)
	GAPDH (129)
	IGF-1 (transient) (130)
	heat shock proteins (132)
	metallothionein (133)
	H-FABP (134)
	angiogenic factors (VEGF, bFGF) (135, 136)
	guanosine monophosphate reductase (137)
	integral plasma membrane
	    glycoprotein CE9 (120)
	FAD-linked GPDH (138)
	Cig30 (139)
	PPARδ (140)
	C/EBPβ (119)
	PGC-1 (141)
		interleukin-1 a (143)
		leptin 144, 145
		β3-adrenergic receptor
		    (146, 147)
		PPARγ and α (initial phase)
		C/EBPα (119)

Where no specific references are given, see ref. 6 for review; Cig30, cold-inducible-glycoprotein; FAD-linked GPDH, FAD-linked glycerol-3-phosphate dehydrogenase; H-FABP, heart type fatty acid binding protein. See text for further abbreviations. References are in parentheses.

The nucleotide binding properties of UCP1 have been a key feature to the first identification and isolation of this protein. Nucleotide binding (especially GDP binding) to isolated brown fat mitochondria has also widely been used to quantitate UCP1 activity before the availability of antibodies. The mechanism of nucleotide binding and the binding site are today the best characterized features of UCP1. Nucleotides bind from the cytosolic side, but the binding site is deep inside the protein (see Fig. 4.4.) because it was shown to involve the loop between the fifth and sixth transmembrane domains located on the matrix side. Apparently this loop projects into the translocation channel area, making it accessible from the cytosol.[66,82,96]

Despite numerous studies there are still gaps in our knowledge about UCP1 structure-function relationships especially regarding proton translocation and its dependence on fatty acids. The recent discovery of closely related proteins expressed in numerous tissue makes this an all the more interesting problem. Although the newly found proteins have also been named uncoupling proteins, it is far from clear if they display any uncoupling function in vivo (in vitro uncoupling has been demonstrated), and what their physiological function is.[97] So far all attempts have been unsuccessful to obtain the three-dimensional structure of UCP1 or any other mitochondrial transporter, which would be the prerequisite of a more detailed insight into the molecular function of these carriers.

Physiological Regulation of Brown Fat Function

Acute Thermogenesis

Brown fat thermogenesis is subject to acute regulation, i.e., turning the heat on and off, but also to long term adaptive regulation of its thermogenic function. The acute activation of BAT thermogenesis is mediated by the sympathetic nervous system and its neurotransmitter norepinephrine (reviewed in 20,65). However, norepinephrine plays a dual role in BAT metabolism. It not only activates acute thermogenesis but also serves as a morphogen by inducing the expression of genes required for the thermogenic function of BAT. Chronic adrenergic stimulation may even result in recruitment of brown adipocytes.[98] Recently it was shown that furthermore norepinephrine has an inhibitory effect on proteolysis in differentiated brown adipocytes.[99]

The acute activation of BAT thermogenesis is illustrated in Fig. 4.5. It is important for overall energy homeostasis that the thermogenic activity of BAT can be tightly regulated and is only manifest when extra heat production is required. As this process has been reviewed in detail elsewhere, (see ref. 20, 28, 33, 65, 100) only a short overview will be given here. Acute cold exposure leads to an activation of the sympathetic nervous system. BAT is very well innervated as discussed in more detail in Chapter 6. Norepinephrine is released from sympathetic nerve endings and interacts with β-adrenergic receptors on adipocytes. The predominant β-adrenoceptor of brown adipocytes is the β_3-adrenergic receptor (β_3-AR). Similar to other β-adrenergic receptors, β_3-AR stimulation leads to an increase in cyclic AMP (cAMP) via G_s-protein activation of adenylate cyclase. cAMP stimulates protein kinase A which subsequently phosphorylates and thereby activates the hormone sensitive lipase (HSL). This leads to increased lipolysis and elevated levels of free fatty acids (FFA). FFAs have a dual function: they serve as a substrate for oxidation, i.e., fuel for thermogenesis, and they also activate the proton conductance function of UCP1 as already described. Because ATP-synthesis is bypassed, the respiratory chain is uncoupled and can function at maximum speed, assuring a high oxidation rate of FFA and release of energy as heat. This lasts as long as lipolysis is activated and FFA levels are high.

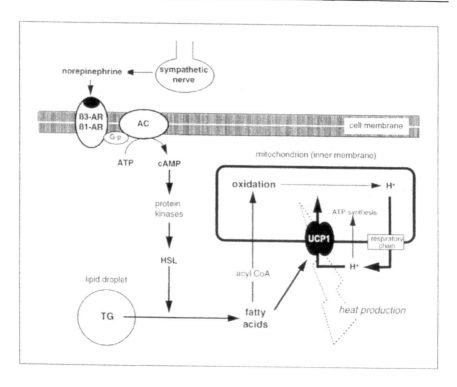

Fig. 4.5 Acute activation of brown adipocyte thermogenesis. Upon stimulation of the sympathetic nervous system norepinephrine is released, acting on β-adrenergic receptors (β-AR), predominantly the β₃-AR. G-protein mediated activation of adenylate cyclase (AC) leads through cAMP and protein kinase cascade to activation of the hormone sensitive lipase (HSL), hydrolyzing stored triglycerides (TG) to liberate fatty acids. Fatty acids serve as both substrate for thermogenesis and activators of the uncoupling protein-1 (UCP1).

Fatty acids are the main fuel for BAT thermogenesis with other substrates like lactate, ketones, and amino acids playing only a minor role.[102] Although glucose is probably not a major substrate for thermogenesis, glucose uptake of brown fat is more than 50-fold increased during cold exposure, an effect which can be mimicked by chronic norepinephrine infusion.[103] The increased glucose uptake during acute thermogenesis is thought to prevent the depletion of citric acid cycle intermediates assuring a high catabolic capacity.[104] It is also important for anaerobic, glycolytic generation of ATP during uncoupled respiration.[28]

Fatty acids originating from endogenous lipid stores are oxidized during initial BAT thermogenesis. However, during prolonged cold exposure endogenous stores would be rapidly exhausted. Fatty acids can therefore also be derived from circulating chylomicrons and lipoproteins through the action of lipoprotein lipase (LPL). LPL activity and also gene expression of brown fat is stimulated by cold via norepinephrine and β-adrenergic receptor activation which is in marked contrast to white fat where LPL activity is decreased by cold exposure.[25,27,105]

Intact fatty acid oxidation seems to be essential for the function of brown fat. Mice carrying mutations which result in a disruption of this pathway were shown to be highly sensitive to cold and to display an attenuated response to cold exposure regarding the expression of thermogenic relevant genes such as UCP1 and type II deiodinase.[106]

Adaptive Responses

Cold is the most important environmental stimulus for adaptive responses of brown adipose tissue. However, also photoperiod and diet have been shown to influence the thermogenic capacity of BAT. Nicotine was also shown to increase BAT thermogenesis and UCP1 expression in BAT and even within WAT depots.[107,108] Cold exposure however provokes the most drastic improvement in BAT thermogenesis, and is also the best investigated (Table 4.3.). As most adaptive responses are mediated by the sympathetic nervous systems, photoperiod and diet-induced improvements of BAT thermogenesis are also very similar to the effect of cold. The role of brown fat as an energy dissipating organ in diet-induced thermogenesis will be discussed in the next chapter (Chapter 5) in the context of BAT in energy metabolism and obesity. Photoperiodic influences on BAT function are important for certain rodent species but probably not significant in humans. However, as they are very rarely mentioned in reviews on BAT function, I will briefly outline the general significance and pathway of photoperiodic effects.

For seasonal acclimatization of small mammals from Northern latitudes, photoperiod, i.e., changes in day length or better night length are important environmental cues for triggering numerous physiological adaptive processes including NST capacity. For free-living animals it is important to be prepared for the winter before cold actually hits them. The shortening of day length precedes the actual onset of winter and is also a very predictable environmental cue. Exposure to short photoperiods (i.e., long nights), even at thermoneutrality, is able to increase the capacity for NST and also BAT thermogenesis in many rodent species (reviewed in 109). However, there seem to be large inter-species variations in the ability of short photoperiods to increase BAT thermogenic capacity. In laboratory rats the influence of photoperiod is negligible,[110] in Siberian hamsters (*Phodopus sungorus*) it amounts to about one third,[38] and in root voles (*Microtus oeconomus*) to as much as 60% of the cold-induced response.[111] Information on day length is provided by melatonin, the hormone of the pineal gland which is only synthesized during dark.[112] Indeed, injection or implantation of melatonin was found to mimic the effects of short photoperiods in rodent species which are photoperiod sensitive.[113,114] It is not yet clear if the photoperiod i.e., melatonin-induced adaptive changes of BAT function, are mediated predominantly by the sympathetic nervous system and norepinephrine. In mice and rats BAT thermogenesis is primarily controlled by sympathetic innervation; however there is evidence that this is not true for the Syrian hamster.[115] As the innervation of brown fat will be discussed in more detail by T. Bartness in this volume, the reader is referred to Chapter 6 for more information on this topic.

Improvement of BAT thermogenic capacity as induced by cold exposure involves numerous changes on the organ level, e.g., cellularity, innervation, vascularization but also on the level of the brown adipocyte such as increased mitochondrial density, increased UCP1 expression etc. (reviewed in ref. 6). Table 4.3. summarizes the most important changes especially in gene expressions accompanying increased BAT thermogenic function induced by cold acclimation.[116-148] Obviously, most of these changes are related to the increased thermogenic capacity of BAT in cold. This implies not only increase of the respiratory capacity and UCP1 content of brown adipocytes, but also a recruitment of new adipocytes and improvement of innervation and vascularization of the tissue. Lipid metabolism also markedly changes during cold adaptation as both lipolysis and de novo lipogenesis are increased (reviewed in ref. 149). As pointed out by Bukowiecki, this is only an apparent paradox because it is important for the physiological function of brown fat that lipid stores are replenished simultaneously to their oxidation during thermogenesis.[149]

The results reported here have mostly been obtained using rats and mice as animal models. But it should also be noted that not all species follow the same strategy of improving their BAT thermogenesis. Rats and mice for example show a cold-induced hyperplasia, i.e., proliferation of preadipocytes and recruitment of new brown adipocytes. This is not true for the Siberian hamster where no increase in DNA content could be observed after cold exposure. The BAT mass of cold adapted Siberian hamsters actually decreases, due to lipid depletion, although the respiratory capacity of BAT and its mitochondrial protein is largely increased.[38] This means that one should be careful to generalize findings obtained using just one particular animal model.

So far I have referred to "cold" in a very general way. In many studies cold exposure means putting an animal at 5 °C for a certain period of time. This is quite drastic for small rodents like mice. Many studies also do not take into account that the thermoneutral temperature for small mammals is higher than the normal 23 °C of most animal facilities. The thermoneutral temperature for mice is between 28 °C and 32 °C and for rats around 25-28 °C. Below thermoneutrality animals display increased metabolic rates and thus increased BAT activity (see ref. 6 for review). A mouse kept at 22 °C is thus already cold adapted to a certain extent. If cold acclimated animals were to be compared with animals kept at thermoneutrality, many effects of cold would become even more pronounced. Even intermittent cold exposure is quite effective in eliciting adaptive responses. Siberian hamsters subjected to 2 times 2 hours of cold (5 °C) daily over a period of 2 weeks improved their NST capacity comparable to animals of chronic cold exposure and also displayed a significant increase in BAT thermogenesis and UCP1 expression.[150]

As listed in Table 4.3., there are several transcription factors, like members of the C/EBP and PPAR family, whose expression change during cold exposure. Their significance for adipocyte differentiation and UCP1 expression will be discussed in the following chapter.

Adrenergic Control of BAT Thermogenic Activity

As illustrated above, acute thermogenesis is under control of the sympathetic nervous system and its neurotransmitter norepinephrine (NE). Sympathetic stimulation is also the most important cue for triggering the adaptive response of BAT to cold and also diet. The predominant β-adrenergic receptor in mature brown adipocytes is the β_3-AR, an atypical β-AR which was first cloned from a human genomic library in 1989 and which is mainly expressed in white and brown adipose tissue. The most abundant expression of β_3-AR is found in BAT, although it can also be detected in white fat and might also be present in other tissues, mainly the gastrointestinal tract (see ref. 101,151,152 for review). The presence of a functional β_3-AR in human tissues is still under debate. Using sensitive techniques like RT-PCR, the mRNA could be detected in many fat depots, white and brown and also other tissues; however this does not mean that a functional receptor is present (reviewed in ref. 151,152). Binding studies using a selective agonist indicate the presence of the β_3-AR at least in infant brown adipose tissue.[153] Only very recently a monoclonal antibody against the human β_3-AR has been raised, which demonstrated the expression of the β_3-AR in a variety of human tissues, including muscle and adipose tissue.[154]

Binding properties of the β_3-AR are quite distinct from the β_1 and β_2-AR. It has a relatively low affinity to norepinephrine and is insensitive to many classical β-AR antagonists. Furthermore, β-AR antagonists (like CGP 12177) can function as partial agonists of the β_3-AR. By now numerous novel selective β_3-AR agonists are available, notably BRL 37344 and CL 316,243 which show high affinity for rodent β_3-AR, and CGP 12177 which is so far the most effective compound in human adipose tissue (see ref. 151,152 for more information on pharmacokinetics of the β_3-AR). For over 15 years pharmaceutical industries have been trying to develop β_3-AR agonists for treatment of obesity and diabetes. However, although results were very promising in animal studies (rodents and even dogs), so far human clinical trials have not been very successful (for review see ref. 158).

The β_3-AR is a late differentiation marker of brown adipocytes; preadipocytes express mainly the β_1-AR whereas the β_3-AR mRNA was found to be greatly increased during adipocyte differentiation.[98,155] Cold leads to increased NE release which in turn promotes proliferation of brown preadipocytes probably through its action on β_1-AR, as demonstrated for mouse preadipocytes.[156] Over-expression of β_1-AR in transgenic mice induced the abundant appearance of brown adipocytes in subcutaneous WAT, confirming the role of β_1-AR in brown preadipocyte proliferation.[157] In mature brown adipocytes NE acts predominantly on β_3-AR inducing the expression of UCP1 gene. Both β_1 and β_3-AR are coupled to G_s protein and adenylate cyclase resulting in an increase in cAMP. But it is still controversially discussed if the β_3-AR is coupled to more than one second messenger. Bronnikov and coworkers suggest that the switch from β_1 to β_3-AR during differentiation is sufficient to explain the distinct β-AR subtype effect on proliferation and differentiation.[159] Chaudry and Granneman on the other hand suggest that the β-AR subtypes regulate different physiological responses in part by differential signal transduction to subcellular compartments.[160] Indeed it is puzzling what the physiological importance of the β_3-AR might be, if its only physiological ligand is norepinephrine and if the intracellular signal transduction is identical to the β_1-AR. Targeted disruption of the β_3-AR has been performed by two groups in order to study the physiological significance of the β_3-AR.[161,162] Both studies found no dramatic effects on phenotype except a slight increase in body fat which was accentuated by high fat feeding.[162] Cold induced UCP1 accumulation and BAT growth seemed to be normal in these mice lacking the β_3-AR. One apparent difference between the two studies is that while Susulic et al[161] found a compensatory increase in the expression of β_1-AR especially in BAT, Revelli et al[162] report rather a decrease in β_1-AR gene expression. The reason for this difference is not quite clear but could be related to mouse strain differences.

Although the β_3-AR is important for BAT thermogenesis, it is obviously not the only adrenergic receptor involved in BAT thermogenesis and its adaptation. Treatment of rats in vivo with CL316,243, a very selective β_3-AR agonist, did not reproduce all the effects of norepinephrine. Although mitochondria and UCP1 expression in BAT were markedly increased, no hyperplasia occurred as is the case with norepinephrine treatment.[163] This confirms the findings in cell cultures reported above implicating the β_1-AR in brown fat hyperplasia.[156] Nevertheless, CL316,243 treatment is able to reverse or prevent obesity in several rat and mice models (reviewed in 28); however this is probably not exclusively due to its effect on brown adipose tissue. In an elegant study using transgenic and gene knockout mice, Grujic et al could demonstrate that expression of β_3-AR in BAT only is not sufficient to restore all effects of β_3-agonist treatment on energy metabolism in β_3-knockout mice. Rather the expression of β_3-AR in BAT and WAT was necessary.[164]

It has been speculated for many years that norepinephrine which displays a rather low affinity to the β_3-AR might not be the only physiological ligand. Recently it was shown that octopamine, a naturally occurring amine was able to stimulate lipolysis in fat cells from several species which are known to express a functional β_3-AR.[165] In brown adipocytes it could also increase oxygen consumption similar to β_3-AR agonist treatment. The pharmacological profile also indicated that octopamine can be considered the most selective for β_3-AR among the biogenic amines.[165] However, it is still completely unknown if this has any physiological significance. It can also not be excluded that other as yet unknown β-adrenoceptors are present in BAT. This was suggested by pharmacological experiments using β_3-AR knockout mice.[166] The same group also found pharmacological evidence for the presence of a putative fourth β-AR in human adipose tissue.[167]

Brown adipocytes also possess α-adrenergic receptors (α-ARs) (for review see ref. 168). Indeed it was shown that the first response of brown adipocytes to adrenergic stimulation is an α-AR mediated rapid depolarization due to an increase of chloride permeability.[169] Apparently both $\alpha 1$-AR (coupled to G_q-protein leading to Ca^{2+} entry) and $\alpha 2$-AR (coupled to G_i-protein leading to cAMP decrease) are present on brown adipocytes. Although stimulation with an α1-AR selective agonist (phenylephrine) was found to increase respiration of brown adipocytes in several studies,[170,171] no physiologically significant function of the α1-AR has been unequivocally demonstrated so far.

Endocrine Control of BAT Thermogenic Activity

Endocrine mechanisms involved in adaptive changes of BAT function as elucidated mainly by in vivo experiments have been reviewed in detail by Himms-Hagen in 1986.[6] Since then research has focused more on cellular and molecular mechanisms and especially on the regulation of the UCP1 gene expression which will be discussed in the next chapter (Chapter 5). Here I will give only a short overview on the most important endocrine mechanisms controlling brown fat physiological function.

Thyroid Hormones

The thermogenic effect of thyroid hormones has been recognized for a long time; they are known to increase obligatory thermogenesis through numerous metabolic pathways (see ref. 172 for review). Early studies on regulation of brown fat function therefore also investigated BAT thermogenesis in hypothyroid animals (reviewed in ref. 173). These studies were somewhat controversial, because although thyroid hormones were found to be required for cold-induced thermogenesis, BAT did not seem to be a target organ and high doses of T_4 or T_3 could even decrease BAT thermogenesis. The role of thyroid hormones was thus considered a merely permissive one (see ref. 6,173 for an overview). In the mid-1980s however, it was discovered that BAT contains a type II T_4 5'-deiodinase (5'-D-II) activity which was largely increased by cold exposure or exogenous catecholamines.[174] This enzyme is responsible for deiodination of thyroxin (T_4) to the active metabolite 3,5,3'-triiodothyronine (T_3) and belongs to the deiodinase family of selenoproteins. So far three different types of the 5'-deiodinase are known with distinct tissue specific expression, the 5'-D-II being expressed mostly in brain, pituitary and BAT.[175] Cold was found to increase 5'-D-II activity in BAT of rats and hamsters accompanied by a parallel increase in serum T_3 concentration.[176] BAT is apparently not only a source of local but also of plasma, i.e., systemic T_3.[177,178] Moreover, experiments in different animal species confirmed that the local production of T_3 in brown fat is essential for maximum thermogenic function of BAT including expression of UCP1.[179-181] For example, inhibition of the 5'-D-II activity in vivo by injection of iopanoic acid also abolished part of the cold-induced increase in UCP1 mRNA levels.[181] In unilateral denervated athyreotic rats only stimulation by norepinephrine and T_3 together led to maximum stimulation of UCP1 gene expression.[182] Together these results point to a synergistic action of norepinephrine and T_3 in adaptive BAT function. A recent report also provides evidence that T_3 alone without interference of the sympathetic nervous system is able to significantly stimulate UCP1 gene expression in vivo.[183] This confirms findings in primary cultures of brown adipocytes. In brown adipocytes from both Siberian hamster and fetal rats, T_3 significantly increased UCP1 gene expression and protein amount even in the absence of adrenergic stimulation.[184,185]

Pancreatic Hormones (Insulin and Glucagon)

Insulin has been implicated in control of brown fat thermogenesis. However, the action of insulin is rather complex and difficult to elucidate because it might have direct effects on the tissue itself as well as indirect effects through hypothalamic influences on the sympathetic nervous system. Diabetic rats display reduced BAT thermogenic response to cold or norepinephrine due to progressive atrophy of BAT as shown in several studies. Prolonged but not acute insulin treatment was able to restore the thermogenic answer of BAT which indicates that the role of insulin in BAT thermogenesis is rather a permissive one through its general anabolic action on protein synthesis.[186,189] It was also shown that insulin is most probably not required for acute thermogenic action of catecholamines.[186] The permissive nature of insulin on BAT thermogenic properties was also confirmed by in vitro studies on cultured brown adipocytes.[187,188] Like adipose tissue in general, BAT is a highly insulin sensitive tissue. Glucose uptake of isolated brown adipocytes is stimulated by insulin; however, in the absence of insulin, norepinephrine is also able to stimulate glucose uptake, and it potentiates synergistically the effect of submaximal doses of insulin on glucose transport.[190] In this respect BAT differs from WAT: in both adipose tissues insulin increases glucose transport by increasing transcription and amount of the glucose transporter GLUT4. But in BAT only GLUT4 expression is also modulated by catecholamines (see ref. 191 for review).

A stimulatory effect of glucagon on BAT thermogenesis has repeatedly been demonstrated both in vivo and in vitro; it was found to increase respiration of isolated brown fat cells, as well as brown adipose tissue blood flow and also overall oxygen consumption (reviewed in ref. 6). More recent studies also confirmed that exogenous glucagon in physiological concentrations was able to stimulate brown fat thermogenesis in vivo.[192,193] Similar to insulin, glucagon can act directly on brown adipocytes or indirectly through hypothalamic influences on sympathetic nerve activity. Microinjection of glucagon into the lateral hypothalamus was shown to increase the sympathetic nerve activity to brown fat, and the thermogenic response to glucagon could be abolished by a β-adrenergic blocker.[194,195] However, isolated brown adipocytes at least from rats possess glucagon responsive receptors, which means that a direct effect of glucagon on BAT thermogenesis can not be definitely excluded.[195]

Conclusions

Brown adipose tissue function is subject to physiological regulation involving nervous and humoral signals, and adaptation of BAT thermogenesis can be triggered by environmental cues such as temperature (cold) and photoperiod (short days). For small mammals including the human neonate BAT thermogenesis is a crucial mechanism for survival in the cold. The possible role of BAT in fever and thermoregulatory feeding awaits experimental confirmation.

The biochemical mechanism of heat production of BAT mitochondria via UCP1 has been largely unraveled since the identification of this protein some 25 years ago, even if there remain some controversies about the exact molecular mechanisms of proton translocation by UCP1. The complete elucidation of UCP1 structure/function relationship will probably rely on crystallization of this protein which has not been achieved so far.

References

1. Jansky L. Nonshivering thermogenesis and its thermoregulatory significance. Biol Rev 1970; 48:86-132.
2. Hsieh ACL, Carlson LD, Gray G. Role of the sympathetic nervous system in the control of chemical regulation of heat production. Am J Physiol 1957; 190:247-251.
3. Zeissberger E, Brück K. Quantitative Beziehung zwischen Noradenalin-Effekt und Ausmaß der zitterfreien Thermogenese beim Meerschweinchen. Pflügers Arch 1967; 296:263-275.
4. Heldmaier G. Zitterfreie Wärmebildung und Körpergröße bei Säugetieren. Z Vergl Physiologie 1971; 73:222-248.
5. Klaus S, Heldmaier G, Ricquier D. Seasonal acclimation of bank voles and wood mice: non-shivering thermogenesis and thermogenic properties of brown adipose tissue mitochondria. J Comp Physiol 1988; 158:157-164.
6. Himms-Hagen J. Brown adipose tissue and cold-acclimation. In: Trayhurn P, Nicholls DG, eds. Brown Adipose Tissue. London: E. Arnold Ltd. 1986: 214-268.

7. Foster DO, Frydman ML. Tissue distribution of cold-induced thermogenesis in conscious warm- or cold-acclimated rats recalculated from changes in tissue and blood flow: The dominant role of brown adipose tissue in the replacement of shivering by nonshivering thermogenesis. Can J Physiol Pharmacol 1979; 56:741-746.

8. Heldmaier G, Buchberger A. Sources of heat during nonshivering thermogenesis in Djungarian hamsters: A dominant role of brown adipose tissue during cold adaptation. J Comp Physiol B 1985; 156:237-245.

9. Himms-Hagen J. Role of thermogenesis in brown adipose tissue in the regulation of energy balance. In: Hollenberg CH, Roncari DAK, eds. The Adipose and Obesity: Cellular and Molecular Mechanisms. New York: Raven Press 1983:259-270.

10. Cannon B, Houstek J, Nedergaard J. Brown adipose tissue. More than an effector of thermogenesis? Ann N Y Acad Sci 1998; 856:171-187.

11. Himms-Hagen J. Role of brown adipose tissue thermogenesis in control of thermoregulatory feeding in rats: a new hypothesis that links thermostatic and glucostatic hypothesis for control of food intake. Proc Soc Exp Biol Med 1995; 208:159-169.

12. Himms-Hagen J. Does thermoregulatory feeding occur in newborn infants? A novel view of the role of brown adipose tissue thermogenesis in control of food intake. Obesity Res 1995; 3:361-369.

13. Székely M, Szelényi Z, Sümegi I. Brown adipose tissue as a source of heat during pyrogen-induced fever. Acta Physiol Acad Sci Hung 1973; 43:83-88

14. Rasmussen AT. The 'so-called' hibernating gland. J Morphol 1923; 38:147-205.

15. Smith RE. Thermogenic activity of the hibernating gland in the cold acclimated rat. Physiologist 1961; 4:113-118.

16. Lindberg O, ed. Brown Adipose Tissue. New York: American Elsevier, 1970.

17. Néchad M. Structure and development of brown adipose tissue. In: Trayhurn P, Nicholls DG, eds. Brown Adipose Tissue. London: E. Arnold Ltd. 1986:1-30.

18. Ailhaud G, Grimaldi P, Negrel R. Cellular and molecular aspects of adipose tissue development. Ann Rev Nutr 1992; 12:207-233.

19. Flatmark T, Pedersen JI. Brown adipose tissue mitochondria. Biochim Biophys Acta 1975; 416:53-103.

20. Nicholls D, Locke RM. Thermogenic mechanisms in brown fat. Physiol Rev 1984; 64(1):1-103.

21. Ricquier D, Bouillaud F. The brown adipose tissue mitochondrial uncoupling protein. In: Trayhurn P, Nicholls DG, eds. Brown Adipose Tissue. London: E. Arnold Ltd. 1986:86-104.

22. Rial E, Gonzalez-Barroso MM, Fleury C et al. The structure and function of the brown fat uncoupling protein UCP1: current status. Biofactors 1998; 8(3-4):209-219.

23. Klain GJ, Hannon JP. Differential response of rat brown and white adipose tissue to environmental or nutritional stress. Comp Biochem Physiol [B] 1977;58(3):227-230

24. Bertin R, Andriamihajy M, Portet R. Glycerokinase acitivity in brown and white adipose tissues of cold-adapted obese Zucker rats. Biochem 1981; 66:569-572.

25. Radomski MW, Orme T. Response of lipoprotein lipase in various tissues to cold exposure. Am J Physiol 1971; 220:1852-1856.

26. Carneheim CJ, Nedergaard J, Cannon B. β-adrenergic stimulation of lipoprotein lipase in rat brown adipose tissue during acclimation to cold. Am J Physiol 246:E327-E333.

27. Klingenspor M, Klaus S, Wiesinger H et al. Short photoperiod and cold activate brown fat lipoprotein lipase in the Djungarian hamster. Am J Physiol 1989; 257(5 Pt 2):R1123-1127.

28. Himms-Hagen J, Ricquier D. Brown adipose tissue. In: Bray GA, Bouchard C, James WPT, eds. Handbook of Obesity. New York: Marcel Dekker Inc., 1998:415-441.

29. Guerra C, Koza RA, Yamashita H et al. Emergence of brown adipocytes in white fat in mice is under genetic control. J Clin Invest 1998; 102:412-420.

30. Digby JE, Montague CT, Sewter CP et al. Thiazolidinedione exposure increases the expression of uncoupling protein 1 in cultured human preadipocytes. Diabetes 1998; 47: 138-141.

31. Schneider T, Klaus S, Klingenspor M. Identification of β-adrenergically controlled mRNAs in H1B1B brown adipocytes. Int J Obes 1999; 23 Suppl 5:S23.

32. Boeuf S, Klingenspor M, Klaus S. Identification of differentially espressed genes in white and brown adipocytes. (abstract). Kidney Blood Press Res 2000; 23:67.

33. Trayhurn P. Brown adipose tissue: from thermal physiology to bioenergetics. J Biosci 1993; 18:161-173.

34. Hope PJ, Pyle D, Daniels CB et al. Identification of brown fat and mechanisms for energy balance in the marsupial *Sminthopsis crassicaudata*. Am J Physiol 1997; 273:R161-R167.

35. Nicol SC, Pavlides D, Andersen NA. Nonshivering thermogenesis in marsupials: absence of thermogenic response to beta 3-adrenergic agonists. Comp Biochem Physiol A Physiol 1997; 117(3):399-405.

36. Rose RW. Nonshivering thermogenesis in a marsupial (the Tasmanian Bettong, *Bettongia gaimardi*) is not attributable to brown adipose tissue. Physiol Biochem Zool 1999; 72(6):699-704.

37. Fons R, Sicart R. Contribution à la connaisance du métabolisme énergétique chez deux Crocidurinae: Suncus etruscus (Savi, 1822) et Crocidura russula (Hermann, 1780) (Insectivora, Soricidae). Mammalia 1976; 40:299-311.

38. Rafael J, Vsiansky P, Heldmaier G. Seasonal adaptation of brown adipose tissue in the Djungarian Hamster. J Comp Physiol [B] 1985; 155(4):521-528.

39. Girardier L. Brown fat: An energy dissipating tissue. In: Girardier L, Stock M, eds. Mammalian Thermogenesis. London: Chapman and Hall, 1983:50-98.

40. Trayhurn P. Species distribution of brown adipose tissue: Characterization of adipose tissues from uncoupling protein and ist mRNA. In: Carey C, Florant BA, Wunder BA, Horwitz B, eds. Life In the Cold: Ecological Physiological and Molecular Mechanisms. Boulder: Westview Press, 1993:361-368.

41. Klaus S, Casteilla L, Bouillaud F et al. The uncoupling protein UCP: A membraneous mitochondrial ion carrier exclusively expressed in brown adipose tissue. Int J Biochem 1991; 23:791-801.

42. Trayhurn P, Temple NJ, Van Aerde J. Evidence from immunoblotting studies on uncoupling protein that brown adipose tissue is not present in the domestic pig. Can J Physiol Pharmacol 1989; 67:1480-1485.

43. Casteilla L, Champigny O, Bouillaud F et al. Sequential changes in the expression of mitochondrial protein mRNA during the development of brown adipose tissue in bovine and ovine species. Biochem J 1989; 257:665-671.

44. Soppela P, Nieminen M, Saarela S et al. Brown fat-specific mitochondrial uncoupling protein in adipose tissue of newborn reindeer. Am J Physiol 1991; 260:R1229-R1234.

45. Finn D, Lomax MA, Trayhurn P. An immunohistochemical and in situ hybridization study of the postnatal development of uncoupling protein-1 and uncoupling protein-1 mRNA in lamb perirenal adipose tissue. Cell Tissue Res 1998; 294(3):461-6.

46. Lean MEJ, James WPT. Brown adipose tissue in man. In: Trayhurn P, Nicholls DG, eds. Brown Adipose Tissue. London: E. Arnold Ltd., 1986:339-365.

47. Hatai S. On the presence in human embryos of an interscapular gland corresponding to the so-called hibernating gland of lower mammals. Anat Anzeiger 1902; 21:369-373.

48. Sidlo J, Zaviacic M, Trutzova H. Brown adipose tissue. I. Morphology (review). Cs Patol 1996; 32:41-44.

49. Heaton JM. The distribution of brown adipose tissue in the human. J Anat 1972; 112:35-39.

50. Zancanaro C, Carnielli VP, Moretti C et al. An ultrastructural study of brown adipose tissue in pre-term human new-borns. Tissue Cell 1995; 27(3):339-348.

51. Houstek J, Vizek K, Pavelka S et al. Type II iodothyronine 5'-deiodinase and uncoupling protein in brown adipose tissue of human newborns. J Clin Endocrinol Metab 1993; 77(2):382-387.

52. Lean ME, James WP, Jennings G et al. Brown adipose tissue uncoupling protein content in human infants, children and adults. Clin Sci 1986; 71:291-297.

53. Cunningham S, Leslie P, Hopwood D et al. The characterization and energetic potential of brown adipose tissue in man. Clin Sci 1985; 69:343-348.

54. Bouillaud F, Villarroya F, Hentz E et al. Detection of brown adipose tissue uncoupling protein mRNA in adult patients by a human genomic probe. Clin Sci 1988; 75:21-27.

55. Lean ME, James WP, Jennings G et al. Brown adipose tissue in patients with phaeochromocytoma. Int J Obes 1986; 10:219-227.

56. Gisselsson D, Hoglund M, Mertens F et al. Hibernomas are characterized by homozygous deletions in the multiple endocrine neoplasia type 1 region. Metaphase fluorescence in situ hybridization reveals complex rearrangements not detected by conventional cytogenetics. Am J Pathol 1999; 155(1):61-66.

57. Zancanaro C, Pelosi G, Accordini C et al. Immunohistochemical identification of the uncoupling protein in human hibernoma. Biol Cell 1994; 80(1):75-78.

58. Garruti G, Ricquier D. Analysis of uncoupling protein and its mRNA in adipose tissue deposits of adult humans. Int J Obes Relat Metab Disord 1992; 16:383-390.

59. Kortelainen ML, Pelletier G, Ricquier D et al. Immunohistochemical detection of human brown adipose tissue uncoupling protein in an autopsy series. J Histochem Cytochem 1993; 41(5):759-764.

60. Champigny O, Ricquier D. Evidence from in vitro differentiating cells that adrenoceptor agonists can increase uncoupling protein mRNA level in adipocytes of adult humans: an RT-PCR study. J Lipid Res 1996; 37(9):1907-1914

61. Reid G, Tervit H. Sudden infant death syndrome (SIDS): Disordered brown fat metabolism and thermogenesis. Med Hypotheses 1994; 42(4):245-249.

62. Douglas RJ. Could a lowered level of uncoupling protein in brown adipose tissue mitochondria play a role in SIDS aetiology? Med Hypotheses 1992; 37(2):100-102.

63. Pearson RD, Greenaway AC. Sudden infant death syndrome and hibernation: Is there a link? Med Hypotheses 1990; 31(2):131-134.

64. Durlach J, Durlach V, Rayssiguier Y et al. Is sudden infant death syndrome a magnesium-dependent disease of the transition from chemical to physical thermoregulation? Magnes Res 1991; 4(3-4):137-152

65. Nicholls DG, Cunningham S, Rial E. The bioenergetic mechanisms of brown adipose tissue. In: Trayhurn P, Nicholls DG, eds. Brown Adipose Tissue. London: E. Arnold Ltd., 1986:52-85.

66. Klingenberg M, Huang S-G. Structure and function of the uncoupling protein from brown adipose tissue. Biochim Biophys Acta 1999; 1415:271-296.

67. Rial E, Gonzalez-Barroso M, Fleury C et al. Retinoids activate proton transport by the uncoupling proteins UCP1 and UCP2. EMBO J 1999; 18(21):5827-5833.

68. Ricquier D, Kader JC. Mitochondrial protein alteration in active brown fat. A sodium dodecyl sulfate-polyacrylamide gel electrophoretic study. Biochem Biophys Res Commun 1976; 73:577-583.

69. Fleury C, Neverova M, Collins S et al. Uncoupling protein-2: A novel gene linked to obesity and hyperinsulinemia. Nature Genet 1997; 15:269-273.

70. Boss O, Samec S, Paolino-Giacobino A et al. Uncoupling protein-3: A new member of the mitochondrial carrier family with tissue specific expression. FEBS Lett. 1997; 408:39-42.

71. Laloi M, Klein M, Riesmeier JW et al. A plant cold-induced uncoupling protein. Nature 1997; 389(6647):135-136.

72. Jarmuszkiewicz W, Sluse-Goffart CM, Hryniewiecka L et al. Identification and characterization of a protozoan uncoupling protein in acanthamoeba castellanii. J Biol Chem 1999; 274(33): 23198-23202

73. Mao W, Yu XX, Zhong A et al. UCP4, a novel brain-specific mitochondrial protein that reduces membrane potential in mammalian cells. FEBS Lett 1999; 443(3):326-330

74. Kozak LP, Britton JH, Kozak UC et al. The mitochondrial uncoupling protein gene. Correlation of exon structures to transmembrane domains. J Biol Chem 1988; 263:1274-1277.

75. Bouillaud F, Weissenbach J, Ricquier D. Complete cDNA derived amino acid sequence of rat brown adipose tissue uncoupling protein. J Biol Chem 1986: 261:1487-1490.

76. Ridley RG, Patel HV, Gerber GE et al. Complete nucleotide sequence and derived amino acid sequence of cDNA encoding the mitochondrial uncoupling protein of rat brown adipose tissue: lack of mitochondrial targeting pre-sequence. Nuc Acid Res 1986; 14:4025-4035.

77. Aquila H, Link TA, Klingenberg M. The uncoupling protein from brown fat mitochondria is related to the mitochondrial ADP/ATP carrier. Analysis of sequence homologies and folding of the protein in the membrane. EMBO J 1985; 4:2369-2376.

78. Balogh AG, Ridley RG, Patel HV et al. Rabbit brown adipose tissue uncoupling protein mRNA: Use of only one of the two polyadenylation signals in its processing. Biochem Biophys Res Comm 1989; 161:156-161.

79. Cassard AM, Bouillaud F, Mattei MG et al. Human uncoupling protein gene: Structure, comparison with rat gene, and assignment to the long arm of the chromosome 4. J Cell Biochem 1990; 43:255-264.

80. Casteilla L, Bouillaud F, Forest C et al. Nucleotide sequence of a cDNA encoding bovine brown fat uncoupling protein. Homology with ADP binding site of ADP/ATP carrier. Nucl Acids Res 1989; 17:2131.

81. Klaus S, Raimbault S, Bouillaud F et al. Sequence of the brown adipose tissue specific uncoupling protein UCP from the Etruscan shrew (*Suncus etruscus*). In: Geiser F, Hulbert AJ, Nicol SC eds. Adaptations To the Cold. Tenth international Hibernation Symposium. Armidale: University of New England Press, 1996:293-298.

82. Ricquier D, Bouillaud F. The mitochondrial uncoupling protein: structural and genetic studies. Prog Nucleic Acid Res Mol Biol 1997; 56:83-108.

83. Aquila H, Link TA, Klingenberg M. Solute carriers involved in energy transfer of mitochondria form a homologous protein family. FEBS Lett 1987; 212(1):1-9.

84. Walker JE, Runswick MJ. The mitochondrial transport protein superfamily. J Bioenerg Biomembr 1993; 25(5):435-446.

85. Kuan J, Saier MH Jr. The mitochondrial carrier family of transport proteins: structural, functional, and evolutionary relationships. Crit Rev Biochem Mol Biol 1993; 28(3):209-233.

86. Miroux B, Frossard V, Raimbault S et al. The topology of the brown adipose tissue mitochondrial uncoupling protein determined with antibodies against its antigenic sites revealed by a library of fusion proteins. EMBO J 1993; 12:3739-3745.

87. Klaus S, Casteilla L, Bouillaud F et al. Expression of brown fat uncoupling protein in Xenopus oocytes and import into mitochondrial membranes. Biochem Biophys Res Comm 1990; 167:784-789.

88. Casteilla L, Blondel O, Klaus S et al. Permanent expression of functional mitochondrial uncoupling protein in Chinese Hamster ovary cells. Proc Natl Acad Sci USA 1990; 87:5124-5128.

89. Murdza-Inglis DL, Patel HV, Freeman KB et al. Functional reconstitution of rat uncoupling protein following its high level expression in yeast. J Biol Chem 1991; 266(18):11871-11875.

90. Bathgate B, Freebairn EM, Greenland AJ et al. Functional expression of the rat brown adipose tissue uncoupling protein in *Saccharomyces cerevisiae*. Mol Microbiol 1992; 6(3):363-370.

91. Arechaga I, Raimbault S, Prieto S, et al. Cysteine residues are not essential for uncoupling protein function. Biochem J 1993; 296(Pt 3):693-700.

92. Jezek P, Engstova H, Zackova M et al. Fatty acid cycling mechanism and mitochondrial uncoupling proteins. Biochim Biophys Acta 1998; 1365(1-2):319-327.

93. Gonzalez-Barroso MM, Fleury C, Bouillaud F et al. The uncoupling protein UCP1 does not increase the proton conductance of the inner mitochondrial membrane by functioning as a fatty acid anion transporter. J Biol Chem 1998; 273(25):15528-15532.

94. Jezek P, Garlid KD. Mammalian mitochondrial uncoupling proteins. Int J Biochem Cell Biol 1998; 30(11):1163-1168.

95 Winkler E, Klingenberg M. Effect of fatty acids on H+ transport activity of the reconstituted ucoupling protein. J Biol Chem 1994; 2508-2515.

96. Gonzalez-Barroso MM, Fleury C, Jimenez MA et al. Structural and functional study of a conserved region in the uncoupling protein UCP1: The three matrix loops are involved in the control of transport. J Mol Biol 1999; 292(1):137-149.

97. Boss O, Muzzin P, Giacobino JP. The uncoupling proteins, a review. Eur J Endocrinol 1998; 139(1):1-9

98. Nedergaard J, Herron D, Jacobsson A et al. Norepinephrine as a morphogen? Its unique interaction with brown adipose tissue. Int J Dev Biol 1995; 39:827-837.

99. Desautels M, Heal S. Differentiation-dependent inhibition of proteolysis by norepinephrine in brow adipocytes. Am J Physiol 1999; 277:E215-E222.

100. Nedergaard J, Cannon B. Brown Adipose tissue: Development and function. In: Polin RA, Fox WW, eds. Fetal and Neonatal Physiology. 2nd ed. Philadelphia: WB Saunders Company, 1998:478-489.

101. Lowell BB, Flier JS. Brown adipose tissue, β3-adrenergic receptors, and obesity. Ann Rev Med 1997; 48:307-316.

102. Trayhurn P. Fuel selection in brown adipose tissue. Proc Nutr Soc 1995; 54:39-47.

103. Liu X, Perusse F, Bukowiecki LJ. Chronic norepinephrine infusion stimulates glucose uptake in white and brown adipose tissues. Am J Physiol 1994; 266(3Pt2):R914-920.

104. Cannon B, Nedergaard J. The physiological role of pyruvate carboxylation in hamster brown adipose tissue. Eur J Biochem 1979; 94:419-426.

105. Carneheim CJ, Nedergaard J, Cannon B. β-adrenergic stimulation of lipoprotein lipase in rat brown adipose tissue during acclimation to cold. Am J Physiol 1984; 246:E327-E333.
106. Guerra C, Koza RA, Walsh K et al. Abnormal nonshivering termogenesis in mice with inherited defects of fatty acid oxidation. J Clin Invest 1998; 102:1724-1731.
107. Sidlo J, Zaviacic M, Trutzova H. Brown adipose tissue. III. Effect of ethanol, nicotine and caffeine exposure. Soud Lek 1996; 41(2):20-22.
108. Yoshida T, Sakane N, Umekawa T et al. Nicotine induces uncoupling protein 1 in white adipose tissue of obese mice. Int J Obes Relat Metab Disord 1999; 23(6):570-575.
109. Heldmaier G, Böckler H, Buchberger et al. Seasonal acclimation and thermogenesis. In: Gilles R, ed, Circulation, Respiration and Metabolism. Berlin, Heidelberg: Springer-Verlag, 1985:490-501.
110. Kott KS, Horwitz BA. Photoperiod and pinealectomy do not affect cold-induced deposition of brown adipose tissue in the Long-Evans rat. Cryobiology 1983; 20(1):100-105.
111. Wang D, Sun R, Wang Z et al. Effects of temperature and photoperiod on thermogenesis in plateau pikas (*Ochotona curzoniae*) and root voles (*Microtus oeconomus*). J Comp Physiol [B] 1999; 169(1):77-83.
112. Goldman BD. The pyhsiology of melatonin in mammals. In: Reiter R, ed, Pineal Research Reviews. New York: Allen R Liss 1983; 1:145-182.
113. Heldmaier G, Steinlechner S, Rafael J et al. Photoperiodic control and effects of melatonin on nonshivering thermogenesis and brown adipose tissue. Science 1981; 212(4497):917-919.
114. Bartness TJ, Wade GN. Photoperiodic control of body weight and energy metabolism in Syrian hamsters (*Mesocricetus auratus*): Role of pineal gland, melatonin, gonads, and diet. Endocrinology 1984; 114(2):492-498.
115. Himms-Hagen J. Neural control of brown adipose tissue thermogenesis, hypertrophy, and atrophy. Frontiers Neuroendocrinol 1991; 12(1):38-93.
116. Klingenspor M, Ivemeyer M, Wiesinger H et al. Biogenesis of thermogenic mitochondria in brown adipose tissue of Djungarian hamsters during cold adaptation. Biochem J 1996; 316 (Pt 2):607-613
117. Holm C, Fredrikson G, Cannon B et al. Hormone-sensitive lipase in brown adipose tissue: identification and effect of cold exposure. Biosci Rep 1987; 7(11):897-904.
118. Madar Z, Harel A. Does the glycogen synthase (EC 2.4.1.21) of brown adipose tissue play a regulatory role in glucose homeostasis? Br J Nutr 1991; 66(1):95-104.
119. Rehnmark S, Antonson P, Xanthopoulos KG et al. Differential adrenergic regulation of C/EBP alpha and C/EBP beta in brown adipose tissue. FEBS Lett 1993; 318(3):235-241.
120. Nehme CL, Fayos BE, Bartles JR. Distribution of the integral plasma membrane glycoprotein CE9 (MRC OX-47) among rat tissues and its induction by diverse stimuli of metabolic activation. Biochem J 1995; 310 (Pt 2):693-698.
121. Bourova L, Novotny J, Svoboda P. Resolution and identification of Gq/G11alpha and Gialpha/Goalpha proteins in brown adipose tissue: effect of cold acclimation. J Mol Endocrinol 1999; 23(2):223-229.
122. Ricquier D, Bouillaud F, Toumelin P et al. Expression of uncoupling protein mRNA in thermogenic or weakly thermogenic brown adipose tissue. Evidence for a rapid beta-adrenoreceptor-mediated and transcriptionally regulated step during activation of thermogenesis. J Biol Chem 1986; 261(30):13905-13910.
123. Boss O, Samec S, Dulloo A et al. Tissue-dependent upregulation of rat uncoupling protein-2 expression in response to fasting or cold. FEBS Lett 1997; 412(1):111-114.
124. Carmona MC, Valmaseda A, Brun S et al. Differential regulation of uncoupling protein-2 and uncoupling protein-3 gene expression in brown adipose tissue during development and cold exposure. Biochem Biophys Res Commun 1998; 243(1):224-228.
125. Larkin S, Mull E, Miao W et al. Regulation of the third member of the uncoupling protein family, UCP3, by cold and thyroid hormone. Biochem Biophys Res Commun 1997; 240:222-227.
126. Denjean F, Lachuer J, Geloen A et al. Differential regulation of uncoupling protein-1, -2 and -3 gene expression by sympathetic innervation in brown adipose tissue of thermoneutral or cold-exposed rats. FEBS Lett1999; 444(2-3):181-185.
127. Klingenspor M, Ebbinghaus C, Hulshorst G et al. Multiple regulatory steps are involved in the control of lipoprotein lipase activity in brown adipose tissue. J Lipid Res 1996; 37(8):1685-1695.

128. Freake HC, Oppenheimer JH. Stimulation of S14 mRNA and lipogenesis in brown fat by hypothyroidism, cold exposure, and cafeteria feeding: evidence supporting a general role for S14 in lipogenesis and lipogenesis in the maintenance of thermogenesis. Proc Natl Acad Sci USA 1987; 84(9):3070-3077.

129. Olichon-Berthe C, Van Obberghen E, Le Marchand-Brustel Y. Effect of cold acclimation on the expression of glucose transporter Glut 4. Mol Cell Endocrinol 1992; 89(1-2):11-18.

130. Duchamp C, Burton KA, Geloen A et al. Transient upregulation of IGF-I gene expression in brown adipose tissue of cold-exposed rats. Am J Physiol 1997; 272(3Pt1):E453-460.

131. Shimizu Y, Nikami H, Tsukazaki K et al. Increased expression of glucose transporter GLUT-4 in brown adipose tissue of fasted rats after cold exposure. Am J Physiol 1993; 264(6Pt1):E890-E895.

132. Matz JM, LaVoi KP, Epstein PN et al. Thermoregulatory and heat-shock protein response deficits in cold-exposed diabetic mice. Am J Physiol 1996; 270(3Pt2):R525-R532.

133. Beattie JH, Black DJ, Wood AM et al. Cold-induced expression of the metallothionein-1 gene in brown adipose tissue of rats. Am J Physiol 1996; 270(5Pt2):R971-R977.

134. Daikoku T, Shinohara Y, Shima A et al. Dramatic enhancement of the specific expression of the heart-type fatty acid binding protein in rat brown adipose tissue by cold exposure. FEBS Lett 1997; 410(2-3):383-386.

135. Asano A, Kimura K, Saito M. Cold-induced mRNA expression of angiogenic factors in rat brown adipose tissue. J Vet Med Sci 1999; 61(4):403-409.

136. Tonello C, Giordano A, Cozzi V et al. Role of sympathetic activity in controlling the expression of vascular endothelial growth factor in brown fat cells of lean and genetically obese rats. FEBS Lett1999; 442(2-3):167-172.

137. Salvatore D, Bartha T, Larsen R. The Guanosine Monophosphate Reductase Gene is conserved in rats and ist expression increases rapidly in brown adipose tissue during cold exposure. J Biol Chem 1998; 273:31092-31096.

138. Koza RA, Kozak UC, Brown LJ et al. Sequence and tissue-dependent RNA expression of mouse FAD-linked glycerol-3-phosphate dehydrogenase. Arch Biochem Biophys1996; 336(1):97-104.

139. Tvrdik P, Asadi A, Kozak LP et al. Cig30, a mouse member of a novel membrane protein gene family, is involved in the recruitment of brown adipose tissue. J Biol Chem. 1997; 272(50):31738-31746.

140. Guardiola-Diaz HM, Rehnmark S, Usuda N et al. Rat peroxisome proliferator-activated receptors and brown adipose tissue function during cold acclimatization. J Biol Chem 1999; 274(33):23368-23377.

141. Puigserver P, Wu Z, Park CW et al. A cold-inducible coactivator of nuclear receptors linked to adaptive thermogenesis. Cell 1998; 92:829-839.

142. Schneider T, Klaus S, Klingenspor M. Nur77 mRNA is strongly induced in brown adipose tissue of cold exposed mice. Keystone Symposium: Molecular control of adipogenesis and obesity. 2000 (abstract).

143. Burysek L, Tvrdik P, Houstek J. Expression of interleukin-1 alpha and interleukin-1 receptor type I genes in murine brown adipose tissue. FEBS Lett 1993; 334(2):229-232.

144. Moinat M, Deng C, Muzzin P, Assimacopoulos-Jeannet F, Seydoux J, Dulloo AG, Giacobino JP. Modulation of obese gene expression in rat brown and white adipose tissues. FEBS Lett 1995; 373(2):131-134.

145. Evans BA, Agar L, Summers RJ. The role of the sympathetic nervous system in the regulation of leptin synthesis in C57BL/6 mice. FEBS Lett 1999; 444(2-3):149-154.

146. Bengtsson T, Redegren K, Strosberg AD et al. Down-regulation of beta3 adrenoreceptor gene expression in brown fat cells is transient and recovery is dependent upon a short-lived protein factor. J Biol Chem 1996; 271(52):33366-33375.

147. Scarpace PJ, Matheny M, Temer N. Differential down-regulation of beta3-adrenergic receptor mRNA and signal transduction by cold exposure in brown adipose tissue of young and senescent rats. Pflugers Arch 1999; 437(3):479-483.

148. Nisoli E, Tonello C, Benarese M et al. Expression of nerve growth factor in brown adipose tissue: implications for thermogenesis and obesity. Endocrinology 1996; 137(2):495-503.

149. Bukowiecki L. Lipid metabolism in brown adipose tissue. In: Trayhurn P, Nicholls DG, eds. Brown Adipose Tissue. London: E. Arnold Ltd. 1986:105-121.

150. Wiesinger H, Klaus S, Heldmaier G et al. Increased nonshivering thermogenesis, brown fat cytochrome c oxidase activity, GDP-binding and uncoupling protein mRNA levels after short daily cold exposure of *Phodopus sungorus*. Can J Physiol Pharmacol. 1990; 68:195-200.

151. Giacobino JP. Beta3-adrenoreceptor: An update. Eur J Endocrinol 1995; 132:377-385.

152. Strosberg D. Structure and function of the beta3-adrenergic receptor. Pharmacol Toxicol 1997; 37:421-450.

153. Deng C, Paoloni-Giacobino A, Kuehne F et al. Respective degree of expression of beta 1-, beta 2- and beta 3-adrenoceptors in human brown and white adipose tissues. Br J Pharmacol 1996; 118(4):929-934.

154. Chamberlain PD, Jennings KH, Paul F et al. The tissue distribution of the human beta3-adrenoceptor studied using a monoclonal antibody: Direct evidence of the beta3-adrenoceptor in human adipose tissue, atrium and skeletal muscle. Int J Obes Relat Metab Disord 1999; 23(10):1057-1065.

155. Klaus S, Muzzin P, Revelli JP et al. Control of β3-adrenergic receptor gene expression in brown adipocytes in culture. Mol Cell Endocrinol 1995; 109:189-195.

156. Bronnikov G, Houstek J, Nedergaard J. β-adrenergic, cAMP-mediated stimulation of proliferation of brown fat cells in primary culture. Mediation via beta1 but not via beta3 receptors. J Biol Chem 1992; 267:2006-2013.

157. Soloveva V, Graves RA, Rasenick MM et al. Transgenic mice over-expressing the beta 1-adrenergic receptor in adipose tissue are resistant to obesity. Mol Endocrinol 1997; 11(1):27-38.

158. Weyer C, Gautier JF, Danforth E. Development of beta3-adrenoceptor agonists for the treament of obesity and diabetes—an update. Diab Metabol (Paris) 1999; 25:11-21.

159. Bronnikov G, Bengtsson T, Kramarova L et al. Beta1 to beta3 switch in control of cyclic adenosine monophosphate during brown adipocyte development explains distinct beta-adrenoceptor subtype mediation of proliferation and differentiation. Endocrinology 1999; 140(9):4185-4197.

160. Chaudhry A, Granneman JG. Differential regulation of functional responses by beta-adrenergic receptor subtypes in brown adipocytes. Am J Physiol 1999; 277(1 Pt 2):R147-153.

161. Susulic VS, Frederich RC, Lawitts J et al. Targeted disruption of the beta 3-adrenergic receptor gene. J Biol Chem 1995; 270(49):29483-29492

162. Revelli JP, Preitner F, Samec S et al. Targeted gene disruption reveals a leptin-independent role for the mouse beta 3-adrenoceptor in the regulation of body composition. J Clin Invest 1997; 100(5):1098-1106.

163. Himms-Hagen J, Cui J, Danforth E Jr, et al. Effect of CL-316,243, a thermogenic beta 3-agonist, on energy balance and brown and white adipose tissues in rats. Am J Physiol 1994; 266:R1371-R1382.

164. Grujic D, Susulic VS, Harper M et al. Beta3-adrenergic receptors on white and brown adipocytes mediate beta3-selective agonist-induced effects on energy expenditure, insulin secretion, and food intake. A study using transgenic and gene knockout mice. J Biol Chem 1997; 272(28):17686-17693.

165. Carpene C, Galitzky J, Fontana E et al. Selective activation of beta3-adrenoceptors by octopamine: Comparative studies in mammalian fat cells. Naunyn Schmiedebergs Arch Pharmacol 1999; 359(4):310-321.

166. Preitner F, Muzzin P, Revelli JP et al. Metabolic response to various beta-adrenoceptor agonists in beta3-adrenoceptor knockout mice: Evidence for a new beta-adrenergic receptor in brown adipose tissue. Br J Pharmacol 1998; 124(8):1684-1688.

167. Galitzky J, Langin D, Montastruc JL et al. On the presence of a putative fourth beta-adrenoceptor in human adipose tissue. Trends Pharmacol Sci 1998; 19(5):164-166.

168. Cannon B, Jacobsson A, Rehnmark S et al. Signal transduction in brown adipose tissue recruitment: Noradrenaline and beyond. Int J Obes 1996; 20(suppl 3):S36-S42.

169. Pappone PA, Lee SC. Alpha-adrenergic stimulation activates a calcium-sensitive chloride current in brown fat cells. J Gen Physiol 1995; 106(2):231-258.

170. Schimmel RJ, Elliott ME, Dehmel VC. Interactions between adenosine and alpha 1-adrenergic agonists in regulation of respiration in hamster brown adipocytes. Mol Pharmacol 1987; 32(1):26-33.

171. Borst SE, Oliver RJ, Sego RL et al. Alpha-adrenergic receptor-mediated thermogenesis in brown adipose tissue of rat. Gen Pharmacol 1994; 25(8):1703-1710.

172. Silva JE. Thyroid hormone control of thermogenesis and energy balance. Thyroid 1995; 5(6):481-492.
173. Siva JE, Rabelo R. Regulation of the uncoupling protein gene expression. Eur J Endocrinol 1997; 136:251-264.
174. Silva JE, Larsen PR. Adrenergic activation of triiodothyronine production in brown adipose tissue. Nature 1983; 305(5936):712-713.
175. Kohrle J. Local activation and inactivation of thyroid hormones: The deiodinase family. Mol Cell Endocrinol 1999; 151(1-2):103-119.
176. Kopecky J, Sigurdson L, Park IR et al. Thyroxine 5'-deiodinase in hamster and rat brown adipose tissue: effect of cold and diet. Am J Physiol 1986; 251(1Pt1):E1-7.
177. Silva JE, Larsen PR. Potential of brown adipose tissue type II thyroxine 5'-deiodinase as a local and systemic source of triiodothyronine in rats. J Clin Invest 1985; 76(6):2296-2305.
178. Fernandez JA, Mampel T, Villarroya F et al. Direct assessment of brown adipose tissue as a site of systemic triiodothyronine production in the rat. Biochem J 1987; 243(1):281-284.
179. Bianco AC, Silva JE. Intracellular conversion of thyroxine to triiodothyronine is required for the optimal thermogenic function of brown adipose tissue. J Clin Invest 1987; 79(1):295-300.
180. Silva JE. Full expression of uncoupling protein gene requires the concurrence of norepinephrine and triiodothyronine. Mol Endocrinol 1988; 2(8):706-713.
181. Reiter RJ, Klaus S, Ebbinghaus C et al. Inhibition of 5'-deiodination of thyroxine suppresses the cold-induced increase in brown adipose tissue messenger ribonucleic acid for mitochondrial uncoupling protein without influencing lipoprotein lipase activity. Endocrinology 1990; 126(5):2550-2554.
182. Bianco AC, Sheng XY, Silva JE. Triiodothyronine amplifies norepinephrine stimulation of uncoupling protein gene transcription by a mechanism not requiring protein synthesis. J Biol Chem 1988; 263(34):18168-18175.
183. Branco M, Ribeiro M, Negrao N et al. 3,5,3'-Triiodothyronine actively stimulates UCP in brown fat under minimal sympathetic activity. Am J Physiol 1999; 276(1Pt1):E179-E187.
184. Klaus S, Ely M, Encke D et al. Functional assessment of white and brown adipocyte development and energy metabolism in cell culture: Dissociation of terminal differentiation and thermogenesis in brown adipocytes. J Cell Science 1995; 108:3171-3180.
185. Guerra C, Roncero C, Porras A et al. Triiodothyronine induces the transcription of the uncoupling protein gene and stabilizes its mRNA in fetal rat brown adipocyte primary cultures. J Biol Chem 1996; 271(4):2076-2081.
186. Shibata H, Perusse F, Bukowiecki LJ. The role of insulin in nonshivering thermogenesis. Can J Physiol Pharmacol 1987; 65:452-458.
187. Kopecky J, Baudysova M, Zanotti F et al. Synthesis of mitochondrial uncoupling protein in brown adipocytes differentiated in cell culture. J Biol. Chem 1990; 265:22204-22209.
188. Klaus S, Cassard-Doulcier A-M, Ricquier D. Development of *Phodopus sungorus* brown adipocytes in primary cell culture: Effect of atypical beta-adrenergic agonist, insulin and triiodothyronin on differentiation, mitochondrial development and expression of the uncoupling protein UCP. J Cell Biol 1991; 115:1783-1790.
189. Geloen A, Trayhurn P. Regulation of the level of uncoupling protein in brown adipose tissue by insulin. Am J Physiol 1990; 258(2 Pt 2):R418-424.
190. Marette A, Bukowiecki LJ. Stimulation of glucose transport by insulin and norepinephrine in isolated rat brown adipocytes. Am J Physiol 1989; 257(4 Pt 1):C714-C721.
191. Assimacopoulos-Jeannet F, Cusin I, Greco-Perotto RM et al. Glucose transporters: structure, function, and regulation. Biochimie 1991; 73(1):67-70.
192. Billington CJ, Bartness TJ, Briggs J et al. Glucagon stimulation of brown adipose tissue growth and thermogenesis. Am J Physiol 1987; 252(1 Pt 2):R160-R165.
193. Billington CJ, Briggs JE, Link JG et al. Glucagon in physiological concentrations stimulates brown fat thermogenesis in vivo. Am J Physiol 1991; 261(2 Pt 2):R501-R507.
194. Shimizu H, Egawa M, Yoshimatsu H et al. Glucagon injected in the lateral hypothalamus stimulates sympathetic activity and suppresses monoamine metabolism. Brain Res 1993; 630(1-2):95-100.
195. Dicker A, Zhao J, Cannon B et al. Apparent thermogenic effect of injected glucagon is not due to a direct effect on brown fat cells. Am J Physiol 1998; 275(5 Pt 2):R1674-R1682.

CHAPTER 5

Brown Adipocyte Differentiation and Function in Energy Metabolism

Susanne Klaus

B rown adipose tissue (BAT) is a specialized thermogenic adipose tissue with great importance for thermoregulation and cold defense of small mammals. As outlined in the previous chapter, the thermogenic function of BAT is intrinsically associated with the uncoupling of the mitochondrial respiratory chain by the uncoupling protein 1 (UCP1) which is unique for brown adipocytes. In the '70's and '80's research was primarily focused on the biochemical mechanisms of BAT thermogenesis and its physiological regulation. The finding that BAT thermogenesis can also be activated after excess dietary energy intake has enhanced research efforts especially from the obesity field. The last decade has witnessed enormous progress in molecular biology techniques including transgenic animals. The availability of brown preadipocyte cell culture systems and increasing numbers of transgenic animal models with altered BAT function have contributed much to our current knowledge on molecular mechanisms of brown fat function and its role in energy homeostasis.

This chapter will focus on molecular mechanisms of brown adipocyte differentiation and possible functions of BAT in energy metabolism, reviewing the latest developments with respect to brown adipocyte cell culture systems and transgenic animals models displaying altered brown fat function.

Brown Preadipocyte Cell Culture Systems

Adipocyte differentiation has been studied extensively since the '70's when the first preadipocyte cell lines were established by Green and coworkers.[1,2] The original 3T3 L1 and 3T3 F442A cell lines derived from segregated mouse embryo are still widely used, and numerous other white preadipocyte cell lines have been established since then (see ref. 3-6 and Chapter 3 of this volume for reviews). However, these were all white preadipocyte cell lines which could not be induced to show any characteristics of brown adipocytes, notably expression of UCP1. First attempts to obtain brown preadipocyte cell lines by conventional methods of immortalization were not successful because the resulting cell lines lost the ability of UCP expression and could thus not be considered as brown preadipocytes.[7] Only in the last decade have brown preadipocyte cell lines been established.[8-18] First in vitro studies on brown adipocyte development and UCP1 gene regulation relied therefore on primary cultures of brown preadipocytes using stromal vascular fraction isolated from BAT depots of young mice and hamsters (see ref. 19 for review). Since then several primary culture systems including various

---

rodent, but also rabbit, lamb and human brown preadipocytes have been established as indicated in Table 5.1.[20-29]

Most of the brown preadipocyte cell lines established so far have their origin in BAT tumors, also called hibernomas, obtained from transgenic mice expressing simian virus 40 (SV40) transforming genes,[8,13,17] or directly by transformation of brown preadipocytes with SV40.[12,16] Exceptions are the RBM-Ad cells which were derived from bone marrow,[14] and HB2 cells derived from BAT of p53 knockout mice.[18] Protein 53 (p53) is a tumor suppressor gene critical in control of cell cycle and apoptosis and various cell types of p53 knockout mice were shown to maintain a high proliferative potential in culture. As p53 is a target of the SV40 T-antigens, immortalization of brown adipocytes lacking p53 is not too surprising.[18] As shown in Table 5.1, in most of the brown preadipocyte cell culture systems UCP1 gene expression could be induced or stimulated via β-adrenergic receptor activation, confirming the in vivo findings which pointed to sympathetic activation as the main stimulus for improving BAT thermogenic function. Using brown adipocyte culture systems, a critical role of retinoids in UCP1 gene expression could also be identified.[25] Interestingly, thiazolidinediones (TZD) are also potent stimulators of UCP1 gene expression as will be discussed later.

The adipocyte lineage is of mesodermal origin derived from embryonic stem cells as illustrated in Fig. 5.1. Increasing commitment of totipotent embryonic stem cells gives rise to mesodermal multipotent stem cells from which adipocytes, chondrocytes and myocytes origin (discussed in detail by G. Ailhaud in Chapter 3 of this volume). This was elegantly demonstrated by studies using the multipotent murine stem cell line C3H10T1/2. Treatment of these cells with 5-azacytidine, an inhibitor of DNA methylation, resulted in different cell lines committed to either the adipocyte, myocyte or chondrocyte lineage.[30] C3H10T1/2 cells thus represent a noncommitted, i.e., multipotent mesenchymal stem cell line. In a recent study it was shown that treatment of these cells with insulin and thiazolidinediones directed them towards differentiation into brown adipocytes: they accumulated lipid and expressed UCP1 and increased mitochondrial mass.[15] This provides direct evidence that brown adipocytes originate from the same mesodermal stem cells as white adipocytes.

Bone marrow also contains adipose precursor cells and from those a cell line (RBM-Ad) could be selected which also differentiates into adipocytes resembling brown adipocytes which are able to express UCP1.[14] This suggests a close relationship between white and brown preadipocytes. From a comparative study on primary cultured stromal vascular fraction of white and brown adipose tissue, we know however, that preadipocytes are already committed to either the white or brown adipocyte lineage.[29] Together these results suggest that at some time point after commitment of multipotent stem cell towards the adipose lineage a further commitment occurs to either white or brown adipocyte lineage. However, at what point this occurs and what the underlying molecular mechanisms are is still completely unknown. Currently in our laboratory we are identifying genes differentially expressed in white or brown preadipocytes from primary cell cultures.[31] Hopefully this will help to further elucidate the molecular and cellular mechanisms underlying brown adipocyte differentiation and UCP1 gene expression.

Regulation of Brown Adipocyte Differentiation

As already mentioned in the previous chapter, UCP1 expression, i.e., the presence of brown adipocytes in classical white depots, can be obtained by cold exposure[32-34] or treatment with β3-adrenergic agonists.[35,36] This has led to the proposition of terms like "convertible adipose tissue," "masked brown adipocytes" and "dormant brown adipocyte" (reviewed in ref. 37) to describe this phenomenon. Furthermore it has been shown that thiazolidinediones (TZD), a class of insulin-sensitizing compounds, are also able to stimulate UCP1 gene expression in

Table 5.1. Brown preadipocyte culture systems

I. Primary cultures of brown preadipocytes

Species	Specifics	UCP1 expression stimulated by	Ref.
mouse	3-4 week old	β3-adrenergic stimulation	23
Siberian hamster	4-6 week old	β3-adrenergic stimulation, cAMP, T₃	29
rat	3-4 week old (Wistar)	β3-adrenergic stimulation, retinoic acid, carotenoids	24, 25, 26
rat	newborn	β3-adrenergic stimulation, T₃	22
rabbit	newborn / young	β3-adrenergic stimulation, retinoic acid, TZD	27
lamb	newborn: perirenal	dexamethasone	28
human	perirenal	β3-adrenergic stimulation	21
human	perirenal, omental, subcutaneous	TZD	20

continued on next page

several preadipocyte culture systems, including human preadipocytes from subcutaneous depots.[10,15,18,21,27] Experiments using preadipocytes isolated from rabbit fat depots however, confirmed previous findings in rodents that preadipocytes are already determined as brown or white, and that TZD triggers UCP1 expression only in brown preadipocytes.[27] As described above, there seems to be a close relation of white and brown preadipocytes but to our current knowledge they still represent two differently determined precursor cells even if a certain number of brown preadipocytes seem to be present in almost all white fat depots. This makes all the more intriguing the question of the molecular differences and mechanisms in white and brown preadipocytes which finally lead to the differentiation into two cell types of very different function.

Interestingly, BAT possesses a high tumor potential as pointed out before by Zennaro and coworkers.[17] Several transgenic animal approaches using SV40 T antigen expression driven by different promoters resulted in early hibernoma development.[8,13,17,38,39] As described above, several of these hibernomas could be used to develop brown preadipocyte cell lines (Table 5.1).

So far white and brown preadipocytes cultured in vitro can not be distinguished either by morphological, metabolic, or molecular criteria. However even in cell culture, differentiated brown adipocytes can be recognized by the presence of UCP1 and a higher respiratory activity, i.e., higher mitochondrial content.[19] Differences between white and brown adipocytes in vivo which are much more prominent than in cultured cells are examined in more detail in Chapter 2 in this volume.

II. brown preadipocyte cell lines

Cell lines	Origin	UCP1 expression stimulated by	Ref.
HIB 1B	BAT tumor of transgenic mice	β-adrenergic stimulation, cAMP, retinoic acid, TZD	8, 9, 10, 11
brown adipocyte cell lines	SV40 transfected fetal rat brown adipocytes	IGF-1	12
brown adipocyte cell lines	BAT tumor of transgenic mice	β-adrenergic stimulation, cAMP	13
RBM-Ad	rat bone marrow	β3-adrenergic stimulation	14
C3H10T1/2	Pluripotent mesodermal stem-cell line	TZD	15
PAZ6	SV40 transformed human brown preadipocytes	norepinephrine	16
T37i	BAT tumor of transgenic mice	isoproterenol, retinoic acid	17
HB2	BAT of p53-knockout mice	β-adrenergic stimulation, retinoic acid, TZD (permissive)	18

IGF-1, insulin like growth factor 1; TZD, thiazolidinedione; T_3, triiodthyronine

The molecular mechanism of adipocyte differentiation has been studied by many groups during the last years and much insight has been gained as evidenced by the recent appearance of numerous reviews on this topic.[6,40-44] The vast majority of these studies however, concern white adipocyte development and differentiation which is described in detail by G. Aihaud in Chapter 3 of this volume. Key transcription factors include members of the C/EBP (CAAT enhancer binding protein) and PPAR (peroxisome proliferator activated protein) families. Functional differentiation of brown adipose tissue includes two aspects: adipose conversion of preadipocytes and development of thermogenic function. In vitro, these two aspects could be dissociated, implying that they are not necessarily coupled. It seems that mechanisms leading to acquisition of adipogenic properties are quite similar in white and brown adipocytes (see ref. 19 for review).

UCP1 Gene Transcription

Regarding the development of thermogenic function, best investigated is the regulation of UCP1 gene expression since after all, UCP1 activity is the basis of BAT thermogenesis. As shown in Figure 5.1, thermogenic function of BAT, including UCP1 gene transcription is stimulated by norepinephrine via cAMP pathway, thyroid hormones, retinoic acid and PPAR ligands. After UCP1 genes from mouse, rat and humans had been cloned, several groups started to search for regulatory elements in the UCP1 gene responsible for this activation (for review see ref. 45,46). Figure 5.2 shows the organization of the rat UCP1 gene. Analyzing the rat gene, the group of D. Ricquier identified a 211 bp enhancer region at −2.4 kb in the upstream region of the transcription start site which was also found to be present in the mouse gene.[47,48] Using transgenic mice, the same group could later demonstrate that the 211-bp enhancer alone is sufficient to direct and restrict expression to brown fat and that it also mediates β-adrenergic induced activation of transcription.[49] The distal promoter contains binding sites for CCAAT/enhancer binding proteins (C/EBP). C/EBPs, especially C/EBPα and C/EBPβ, were found to be important for UCP1 gene expression and also lipid accumulation.[50] Cold exposure or treatment with norepinephrine decreased levels of C/EBPα while increasing C/EBPβ (see also the previous chapter, Table 4.3).[51] Studies using knockout mouse models showed that both C/EBPs and also C/EBPδ play a role in brown adipocyte differentiation and UCP gene expression but are not essential for the determination of brown adipocytes (see ref. 52 for review). So far experimental evidence shows that the most important regulatory cis-elements in the UCP1 gene are located within the 211bp enhancer element and include binding sites for NF1, CREB, ets1, Sp1, Jun, RXR, RAR, TR, and PPAR (Fig. 5.2).

Figure 5.3. illustrates transcription factors known to be involved in regulation of thermogenic function of brown adipocytes, i.e., UCP1 gene expression as well as mitochondrial biogenesis and activity. Activation of the UCP1 gene by cAMP has been clearly demonstrated; however, the exact molecular mechanism is still somewhat elusive. Although several putative cAMP response elements (CREs) have been identified, the transactivation of the UCP1 gene by CREB could so far not be demonstrated. Silva and coworkers observed a synergism between cAMP and T_3 in UCP1 gene activation, mediated by a 39 bp sequence in the proximal part of the 211 bp enhancer.[46] Yubero and coworkers propose that CREs located in the proximal promoter (minimal promoter, see Fig. 5.2.) are responsible for the differentiation dependent activation of UCP1 gene by norepinephrine.[53] In undifferentiated brown preadipocytes they observed a high expression of c-Jun which they found to be a negative regulator of basal and cAMP mediated transcription of the UCP1 gene.

The strong activation of UCP1 gene transcription by retinoic acid (RA) (all-trans-RA or 9-cis-RA) has been discovered rather recently and was also confirmed in vivo.[25,54-56] It was further shown that dietary supplementation with vitamin A (of which retinoic acid is the main biological active metabolite) also led to an increase in UCP1 gene expression in rats.[57] Several groups could demonstrate that RA activates UCP1 gene transcription through a retinoid responsive region within the 211 bp enhancer (see ref. 58 for review). The exact architecture of this region is not yet clear and might involve three different sites. Two subfamilies of nuclear receptors, retinoic acid receptors (RAR) and retinoid-x receptors (RXR) can mediate the biological effects of RA and both seem to be involved in UCP1 gene activation, possibly by RAR/RXR heterodimer formation.[46,58] There are several subtypes of RAR and RXR receptors which show distinct expression patterns during brown adipocyte differentiation. Retinoid-induced induction of UCP1 gene relies apparently on a complex combination of the RAR and RXR subtypes.[59]

In recent years, the role of PPARγ as a central regulator of fat cell differentiation has been clearly established.[60] PPARγ is a member of the PPAR subfamily of nuclear hormone receptors which has two other members: PPARα and PPARδ (also called PPARβ, Nuc-1 or FAAR). PPARγ is predominantly expressed in adipose tissue and heterodimerizes with RXR (like the other PPARs). PPARγ displays a high affinity for thiazolidinedione (TZD), a new class of antidiabetic, insulin sensitizing drugs. TZD was shown to increase brown fat mass in vivo in

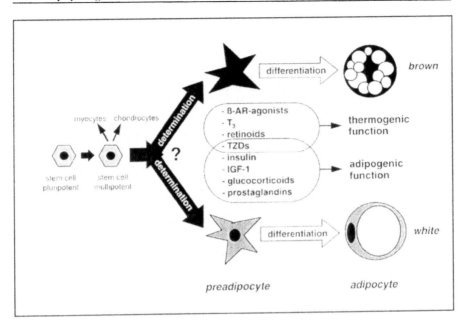

Fig. 5.1 Differentiation of brown and white adipocytes. At what time point determination into brown or white lineage occurs is still unknown. Factors inducing adipogenic conversion are similar for brown and white preadipocytes, whereas inducers of thermogenic function are only functional on brown preadipocytes. TZD, thiazolidinedione.

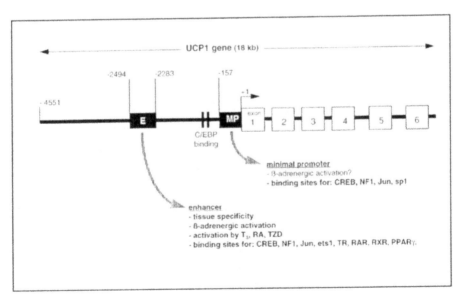

Fig. 5.2 Organization of the rat UCP1 gene. Indicated are the 5 exons (white boxes) as well as regulatory regions within the promoter (black boxes) and their regulatory function. Adapted from ref. 45, 46.

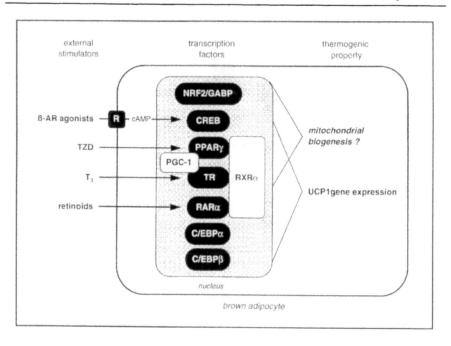

Fig. 5.3 Transcription factors involved in development and activation of brown adipocyte thermogenic properties. Indicated is the possible heterodimerization of PPARγ, TR or RARα with RXRα. See text for further explanations. NRF2/GABP, nuclear respiratory factor-2/GA-binding protein, PPARγ, peroxisome proliferator activated receptor gamma, TZD thiazolidinediones, PGC-1, PPARγ Coactivator-1, C/EBP, CCAAT enhancer binding protein, RAR, retinoic acid receptor, RXR, retinoid-x receptor; TR, thyroid receptor, CREB, cyclic AMP response element binding protein.

rats and to induce brown preadipocyte differentiation and UCP1 expression in different cell models including human preadipocytes (see Table 5.1).[10,11,15,18,20,27,61,62] A putative PPAR response element within the 211 bp enhancer of UCP1 gene has also been identified, and PPARγ was shown to activate the enhancer element of UCP1 gene in HIB-1B cells.[11,46,61] The different PPARs are differentially regulated during cold acclimatization, which also suggests an involvement of PPARs in cold-induced differentiation and activation of brown adipose tissue (see also Table 4.3 in the previous chapter).[63]

Mitochondrial Biogenesis

Activation of UCP1 gene expression is undoubtedly a key component of adaptive thermogenesis in BAT. However, without a concomitant increase of mitochondrial respiratory capacity, this would not considerably increase maximal thermogenesis. Of all mammalian tissues brown fat is probably the one undergoing the most pronounced changes in mitochondrial content in response to various physiological situations and environmental changes.[64] In contrast to UCP1 gene regulation, still very little is known about the cellular and molecular mechanisms involved in mitochondrial biogenesis in BAT which requires a coordinated regulation of mitochondrial and nuclear genome encoded genes.[64] There is a discrepancy between findings in vivo and in vitro. Although in vivo β-adrenergic agonists induce an increase in mitochondrial activity, this could not be reproduced in vitro using cell cultures. So far, only T$_3$ has been shown to significantly increase respiratory capacity and thus mitochondrial content in primary

Table 5.2. Transgenic animals with changes in BAT function or UCP1 expression

Type of transgene	Effect on BAT	Phenotype/effect on energy homeostasis	Ref.
UCP-DTA mouse: diphteria toxin A chain expression directed into brown adipocytes	decreased UCP1 content, increased fat accumulation, reduced thermogenesis	obesity, type II diabetes, hyperphagia/no hyperphagia, decreased resting and total metabolic rate, decreased body tempurature	75, 78, 79
aP2-UCP mouse: expression of UCP1 driven by aP2 promoter (UCP1 expression in WAT)	Reduction of endogenous UCP1 expression, reduced thermogenesis, (but increased thermogenesis in WAT)	reduction of genetic and dietary obesity	76, 82, 83,
UCP-1 knock out	increased fat accumulation, increased UCP2 expression, reduced thermogenesis	increased cold sensitivity, no obesity, no hyperphagia	77, 84
disruption of RIIß regulatory subunit of protein kinase A	five fold increase in UCP1 content	obesity resistance, increased resting and total metabolic rate, increased body temperature	80, 81
Inactivation of dopamine ß-hydroxylase (disruption of epinephrine and norepinephrine synthesis)	reduced UCP1 mRNA, hypertrophy (increased fat accumulation), reduced thermogenesis	increased cold sensitivity, no obesity, hyperphagia, increased basal metabolic rate	85
Overexpression of β1-AR in adipose tissue	Appearance of brown adipocytes in subcutaneous WAT	partial resistance to diet-induced obesity	86
A-ZIP/F-1 Dominant negative protein expression preventing DNA binding of B/ZIP transcription factors	90% reduction, reduced UCP1 mRNA, unilocular appearance (WAT abolished)	lipodystrophy, type II diabetes, reduced BAT thermogenesis	87
Over-expression of nSREBB-1c in adipose tissue (aP2-promoter)	hypertrophy, unilocular appearance abolished UCP1 expression (WAT greatly reduced)	lipodystrophy, type II diabetes	88
Heterozygous PPARγ deficiency (PPARγ +/-)	decreased weight, decreased cell size, increased UCP1 expression	partial resistance to diet-induced obesity and insulin resistance, increased body temperature	89

cultures of brown preadipocytes.[29] It is thus possible that an unknown yet crucial factor is missing under cell culture conditions.

Villarroya and coworkers investigated the transcriptional regulation of the FoF1 ATP synthase (ATPsynβ) as a model gene for elucidating the mechanisms of mitochondrial biogenesis associated with brown adipocyte differentiation. They identified nuclear respiratory factor-2/GA-binding protein (NRF2/GABP) as the transcription factor which plays the major role in regulation of gene expression of ATPsynβ. They also suggest NRF2/GABP to be an important element for enhanced mitochondrial biogenesis during differentiation of brown adipocytes.[64] It still remains to be established if it also plays a role in adaptive thermogenesis of BAT, e.g., during cold exposure.

Another mechanism possibly involved in adaptive thermogenesis of BAT and also muscle was reported by B. Spiegelman and coworkers.[65] They cloned a novel transcriptional coactivator of nuclear receptors which they termed PPAR Gamma Coactivator (PGC)-1. PGC-1 expression is greatly increased upon cold exposure in brown fat (see also Table 4.3 in the previous chapter) and muscle, and it was found to be a powerful transcriptional coactivator for PPARγ and the thyroid receptor.[65] The same group showed later that in muscle cells PGC-1 stimulates NRF1 and NRF2 gene expression and furthermore binds to and coactivates NRF1 thereby stimulating mitochondrial biogenesis.[66] Very recently they also proposed a mechanism by which PGC-1 transcriptional activity is induced, that is by docking of PGC-1 to PPARγ which in turn stimulates a conformational change of PGC-1 which then allows binding of other coactivators.[67] However, the exact role of PGC-1 in brown fat mitochondrial biogenesis and possible involvement of other factors still needs to be further investigated.

Brown Fat in Energy Regulation and Obesity (Transgenic Animals)

It is now well established that—at least in rodents—BAT plays an important role in total energy homeostasis and body weight regulation. First evidence were the experiments of Rothwell and Stock in the late '70s describing a diet induced thermogenesis (DIT) in BAT.[68,69] They found that feeding of a highly palatable "cafeteria diet" to rats resulted in an increased energy intake but not an appropriate increase in body fat. The animals rather increased their energy expenditure by increasing brown fat activity.[68,69] Since then it was shown that certain dietary components are able to increase UCP1 expression in BAT, such as polyunsaturated fatty acids,[70] vitamin A and carotenoids,[57,26] caffeine and ethanol.[71]

Further experimental evidence strongly points to a role of BAT in energy homeostasis. Many animal models of obesity display a diminished BAT function.[72] Abolishment of BAT function by denervation and excision of interscapular BAT results in an increase in total body fat.[73] Diet-induced obesity in rats could be reversed by treatment with the ß3-AR agonist CL316,243 which markedly increased resting metabolic rate and UCP1 expression in brown adipocytes.[74] It has therefore been hypothesized for a long time that BAT plays a causal role in the development of obesity in rodents although this could not be proved directly (see ref. 37 for review). But the recent years have seen a tremendous progression in transgenic animal technology and today several transgenic animal models have confirmed the importance of brown fat in regulation of body weight and energy homeostasis—at least in mice (Table 5.2.).[75-89]

Transgenic animals which were purposely designed to change BAT function or UCP1 expression include UCP-DTA mice with largely abolished brown fat,[75] aP2-UCP mice with directed expression of UCP1 into white fat,[76] and UCP1 knock out mice.[77] Abolishment of BAT function in UCP-DTA was achieved by using the promoter of the UCP1 gene to drive BAT-specific expression of a gene for diphtheria toxin A chain. UCP-DTA mice were found to have 70% decreased UCP1 and BAT content; they are less cold resistant and develop obesity and insulin resistance.[75,78] Physiological characterization of these mice revealed that they have decreased basal and total metabolic rates in comparison to wild-type mice and that they also maintain a lower body temperature.[79] This indicates that BAT contributes significantly to setting of body temperature level in rodents, and that a down-shift of body temperature of less than 1 °C can significantly affect the overall energy budget.

A causal link between the level of body temperature and BAT functionality is also suggested from other transgenic mice. McKnight and coworkers created genetically lean mice by targeted disruption of the RIIß subunit of protein kinase A. These mice are resistant to dietary induced obesity due to chronic activation of BAT thermogenesis resulting in elevated body temperature and thus an increased metabolic rate.[80,81] A similar phenotype was found in heterozygous PPARγ deficient (PPARγ +/-) transgenic mice. These animals are partially resistant to obesity and insulin resistance induced by a high fat diet. Interestingly, they display increased UCP expression in BAT, possibly indicating increased BAT thermogenesis; and they also show an elevated body-temperature.[89] Forced expression of UCP1 in white adipose tissue using aP2 promoter also resulted in a certain resistance to genetic as well as dietary-induced obesity. However, these animals show a partial involution of BAT, i.e., reduced BAT thermogenesis.[76,82,83]

Transgenic mice lacking UCP1 were found to display increased cold-sensitivity as expected, but contrary to the BAT-ablated UCP-DTA mice they did not develop obesity.[77] Interestingly, the BAT of these mice showed a large cold-inducible over-expression of UCP2 and an apparent hyper-recruitment of BAT after cold exposure.[84]

Another transgenic mouse model lacking norepinephrine and adrenaline (epinephrine) was reported to have hypoactive BAT thermogenesis. These mice were hyperphagic but did not become obese due to an elevated basal metabolic rate.[85] It could be speculated that brown fat has an effect on overall metabolic rate independent of the sympathetic nervous system / UCP1 axis, maybe due to unknown secreted factors acting on other tissues.

From transgenic animal studies as listed in Table 5.2 it can be concluded that brown fat is essential for cold resistance and body temperature regulation in mice. However, although over-expression of UCP1 or hyperactive BAT thermogenesis can obviously prevent development of obesity, a decreased BAT activity does not necessarily lead to obesity. Neither UCP1 knock out mice nor norepinephrine deficient mice developed obesity. Only the BAT ablated UCP-DTA mice become obese and it can not be completely ruled out that this might be due to a residual "leaky" expression of the diphtheria toxin A chain in brain or hypothalamic regions affecting central appetite or energy regulation.

Recently two different groups presented transgenic animals displaying a phenotype resembling that of congenital generalized lipodystrophy, i.e., almost completely abolished white adipose tissue, insulin resistance and type II diabetes.[87,88] Interestingly, in both animal models BAT could still be detected although virtually no white fat was present any more. The transgenic aP2-SREBP mice even displayed a BAT hypertrophy.[88] In both animal models UCP1 expression in BAT was markedly reduced or even completely absent, and BAT histologically resembled WAT. This underlines that although BAT and WAT share a certain number of molecular pathways regulating adipocytes differentiation, there must still exist some different mechanisms, because the effects of the transgenes were not the same on white and brown adipose tissue in these transgenic mice.

Conclusion and Perspectives

The establishment of brown preadipocyte cell cultures and cell lines in the last decade has greatly increased our knowledge of brown adipocyte differentiation and molecular regulation of thermogenic properties like UCP1 gene expression. However, the molecular regulation of other thermogenic properties like mitochondrial biogenesis is still poorly understood. Also the key question is not yet answered: "what are the crucial mechanisms directing a precursor cell towards brown as opposed to white adipocyte differentiation?"

Transgenic animals models with modified BAT function have confirmed that brown adipose tissue is important for energy regulation in small mammals i.e., rodents. However, its role in the adult human is by no means clear. It has been hypothesized that nonshivering, i.e., BAT thermogenesis, is only necessary for mammals below 10 kg body weight,[90] a theoretical limit which of course adult humans exceed by far. According to our current knowledge it is not very likely that BAT contributes significantly to overall energy expenditure of adults in nonpathological situations (see also Chapter 4). But the obvious presence of "dormant" brown adipocytes or preadipocytes in various human fat depots is intriguing, and possibly these cells could be a target for anti-obesity drugs. If proliferation of these cells and differentiation of thermogenic properties could be stimulated specifically, they could elicit a small but still significant increase in overall energy expenditure with beneficial effect on body weight development. Hopefully, the future years will therefore bring elucidation of the molecular mechanisms leading to brown adipocyte development and differentiation.

References

1. Green H, Kehinde O. Spontaneous heritable changes leading to increased adipose conversion in 3T3 cells. Cell 1976; 7:105-113.
2. Green H, Kehinde O. Sublines of mouse 3T3 cells that accumulate lipid. Cell 1974; 1:113-116.
3. Ailhaud G, Grimaldi P, Negrel R. Cellular and molecular aspects of adipose tissue development. Ann Rev Nutr 1992; 12:207-233.
4. Hwang C-S, Loftus TM, Mandrup S et al. Adipocyte differentiation and leptin expression. Annu Rev Cell Dev Biol 1997; 13:231-259.
5. Gregoire FM, Smas CM, Sul H. Understanding adipocyte differentiation. Physiol Rev 1998; 78(3):783-809.
6. Gimble JM, Robinson CE, Clarke SL et al. Nuclear hormone receptors and adipogenesis. Crit Rev Euk Gene Expr 1998; 8(2):141-168.
7. Forest C, Doglio A, Ricquier D et al. A preadipocyte clonal line from mouse brown adipose tissue. Exp Cell Res 1987; 168:218-232.
8. Ross S, Choy L, Graves RA et al. Hibernoma formation in transgenic mice and isolation of a brown adipocyte cell line expressing the uncoupling protein gene. Proc Natl Sci USA 1992; 89:7561-7565.
9. Klaus S, Choy L, Champigny O et al. Characterization of the novel brown adipocyte cell line HIB 1B. Adrenergic pathways involved in regulation of uncoupling protein gene expression. J Cell Science 1994; 107:313-319.
10. Rabelo R, Camirand A, Silva JE. 3',5'-cyclic adenosine monophosphate-response sequences of the uncoupling protein gene are sequentially recruited during darglitazone-induced brown adipocyte differentiation. Endocrinology 1997 Dec;138(12):5325-32
11. Tai TAC, Jennermann C, Brown KK et al. Activation of the nuclear receptor peroxisome proliferator-activated receptor gamma promotes brown adipocyte differentiation. J Biol Chem 1996; 271(47):29909-29914.
12. Benito M, Porras A, Santos E. Establishment of permanent brown adipocyte cell lines achieved by transfection with SV40 large T antigen and ras genes. Exp Cell Res 1993; 209(2):248-254
13. Kozak UC, Kozak LP. Norepinephrine-dependent selection of brown adipocyte cell lines. Endocrinol. 1994; 134:906-913.
14. Marko O, Cascieri MA, Ayad N et al. Isolation of a preadipocyte cell line from rat bone marrow and differentiation to adipocytes. Endocrinology 1995; 136(10):4582-4588.
15. Paulik MA, Lenhard JM. Thiazolidinediones inhibit alkaline phosphatase activity while increasing expression of uncoupling protein, deiodinase, and increasing mitochondrial mass in C3H11T1/2 cells. Cell Tissue Res 1997; 290(1):79-87.
16. Zilberfarb V, Pietri-Rouxel F, Jockers R et al. Human immortalized brown adipocytes express functional beta3-adrenoceptor coupled to lipolysis. J Cell Sci 1997; 110(Pt7):801-807.

17. Zennaro M-C, LeMenuet D, Viengchareun S et al. Hibernoma development in transgenic mice identifies brown adipose tissue as a novel target of aldosterone action. J Clin Invest 1998; 101:1254-1260.
18. Irie Y, Asano A, Canas X et al. Immortal brown adipocytes from p53-knockout mice: differentiation and expression of uncoupling proteins. Biochem Biophys Res Commun 1999; 255(2):221-225.
19. Klaus S. Functional differentiation of white and brown adipocytes. BioEssays 1997; 19:215-223.
20. Digby JE, Montague CT, Sewter CP et al. Thiazolidinedione exposure increases the expression of uncoupling protein 1 in cultured human preadipocytes. Diabetes 1998; 47:138-141.
21. Champigny O, Ricquier D. Evidence from in vitro differentiating cells that adrenoceptor agonists can increase uncoupling protein mRNA level in adipocytes of adult humans: An RT-PCR study. J Lipid Res 1996; 37(9):1907-1914
22. Guerra C, Roncero C, Porras A et al. Triiodothyronine induces the transcription of the uncoupling protein gene and stabilizes its mRNA in fetal rat brown adipocyte primary cultures. J Biol Chem 1996; 271(4):2076-2081.
23. Rehnmark S, Kopecky J, Jacobsson A et al. Brown adipocytes differentiated in vitro can express the gene for the uncoupling protein thermogenin: effects of hypothyroidism and norepinephrine. Exp Cell Res 1989; 182:75-83.
24. Champigny O, Holloway BR, Ricquier D. Regulation of UCP gene expression in brown adipocytes differentiated in primary culture: effects of a new β-adrenoceptor agonist. Mol. Cell. Endocrinol. 1992; 86:73-82.
25. Larose M, Cassard-Doulcier AM, Fleury C et al. Essential cis-acting elements in rat uncoupling protein gene are in an enhancer containing a complex retinoic acid response domain. J Biol Chem 1996; 271(49):31533-31542.
26. Serra F, Bonet ML, Puigserver P et al. Stimulation of uncoupling protein 1 expression in brown adipocytes by naturally occurring carotenoids. Int J Obes Relat Metab Disord 1999; 23(6):650-655.
27. Cambon B, Reyne Y, Nougues J. In vitro induction of UCP1 mRNA in preadipocytes from rabbit considered as a model of large mammals brown adipose tissue development: importance of PPARgamma agonists for cells isolated in the postnatal period. Mol Cell Endocrinol 1998; 146(1-2):49-58.
28. Casteilla L, Nougués J, Reyne Y et al. Differentiation of ovine brown adipocyte precursor cells in a chemically defined serum-free medium. Importance of glucocorticoids and age of animals. Eur J Biochem 1991; 198:195-199.
29. Klaus S, Ely M, Encke D et al. Functional assessment of white and brown adipocyte development and energy metabolism in cell culture: Dissociation of terminal differentiation and thermogenesis in brown adipocytes. J Cell Science 1995; 108:3171-3180.
30. Taylor SM, Jones PA. Multiple new phenotypes induced in 10T1/2 and 3T3 cells treated with 5-azacytidine. Cell 1979; 17:772-779.
31. Boeuf S, Klingenspor M, Klaus S. Identification of differentially espressed genes in white and brown adipocytes. (abstract) Kidney Blood Press Res 2000; 23:67.
32. Loncar D. Convertible adipose tissue in mice. Cell Tiss Res 1991; 266:149-161.
33. Cousin B, Cinti S, Morroni M et al. Occurence of brown adipocytes in rat white adipose tissue: molecular and morphological characterization. J Cell Science 1992; 103:931-942.
34. Cousin B, Bascands-Viguerie N, Kassis N et al. Cellular changes during cold acclimatation in adipose tissues. J Cell Physiol 1996; 167(2):285-289.
35. Champigny O, Ricquier D, Blondel O et al. Beta 3-adrenergic receptor stimulation restores message and expression of brown-fat mitochondrial uncoupling protein in adult dogs. Proc Natl Acad Sci U S A 1991; 88(23):10774-10777.
36. Himms-Hagen J, Cui J, Danforth E Jr, et al. Effect of CL-316,243, a thermogenic beta 3-agonist, on energy balance and brown and white adipose tissues in rats. Am J Physiol 1994; 266:R1371-R1382.
37. Himms-Hagen J, Ricquier D. Brown adipose tissue. In: Bray GA, Bouchard C, James WPT, eds. Handbook of obesity. New York: Marcel Dekker Inc., 1998:415-441.
38. Fox N, Crooke R, Hwang LH et al. Metastatic hibernomas in transgenic mice expressing an alpha-amylase-SV40 T antigen hybrid gene. Science 1989; 244(4903):460-463.

39. Perez-Stable C, Altman NH, Brown J et al. Prostate, adrenocortical, and brown adipose tumors in fetal globin/T antigen transgenic mice. Lab Invest 1996; 74(2):363-373.
40. Loftus TM, Lane MD. Modulating the transcriptional control of adipogenesis. Curr Opin Genet Dev 1997; 7(5):603-608.
41. Gregoire FM, Smas CM, Sul HS. Understanding adipocyte differentiation. Physiol Rev 1998; 78(3):783-809.
42. Fajas L, Fruchart JC, Auwerx J. Transcriptional control of adipogenesis. Curr Opin Cell Biol 1998; 10(2):165-173.
43. Cowherd RM, Lyle RE, McGehee RE Jr. Molecular regulation of adipocyte differentiation. Semin Cell Dev Biol 1999; 10(1):3-10.
44. Ailhaud G. Cell surface receptors, nuclear receptors and ligands that regulate adipose tissue development. Clin Chim Acta 1999; 286(1-2):181-190.
45. Ricquier D, Bouillaud F. The mitochondrial uncoupling protein: structural and genetic studies. Prog Nucleic Acid Res Mol Biol 1997; 56:83-108.
46. Siva JE, Rabelo R. Regulation of the uncoupling protein gene expression. Eur J Endocrinol 1997; 136:251-264.
47. Cassard-Doulcier AM, Gelly C, Fox N et al. Tissue-specific and beta-adrenergic regulation of the mitochondrial uncoupling protein gene: Control by cis-acting elements in the 5' flanking region. Mol Endocrinol 1993; 7:496-506.
48. Kozak UC, Kopecky J, Teisinger J et al. An upstream enhancer regulating brown-fat-specific expression of the mitochondrial uncoupling protein gene. Mol Cell Biol 1994; 14(1):59-67.
49. Cassard-Doulcier AM, Gelly C, Bouillaud F et al. A 211-bp enhancer of the rat uncoupling protein-1 (UCP-1) gene controls specific and regulated expression in brown adipose tissue. Biochem J 1998; 333 (Pt 2):243-246.
50. Yubero P, Manchado C, Cassard-Doulcier AM et al. CCAAT/enhancer binding protein alpha and beta are transcriptional activators of the brown fat uncoupling protein gene promoter. Biochem Biophys Res Commun 1994; 198: 653-659.
51. Rehnmark S, Antonson P, Xanthopoulos KG et al. Differential adrenergic regulation of C/EBP alpha and C/EBP beta in brown adipose tissue. FEBS Lett 1993; 318(3):235-241.
52. Darlington GJ, Ross SE, MacDougald OA. The role of C/EBP genes in adipocyte differentiation. J Biol Chem 1998; 273(46):30057-30060.
53. Yubero P, Barbera MJ, Alvarez R et al. Dominant negative regulation by c-Jun of transcription of the uncoupling protein-1 gene through a proximal cAMP-regulatory element: A mechanism for repressing basal and norepinephrine-induced expression of the gene before brown adipocyte differentiation. Mol Endocrinol 1998; 12(7):1023-1037.
54. Alvarez R, de Andrés J, Yubero P et al. A novel regulatory pathway of brown fat thermogenesis. J Biol Chem 1995; 10:5666-5673.
55. Cassard-Doulcier AM, Larose M, Matamala JC et al. In vitro Interactions between nuclear proteins and uncoupling protein gene promoter reveal several putative transactivating factors including Ets1, retinoid X receptor, thyroid hormone receptor, and a CACCC box-binding protein. J Biol Chem 1994; 270:24335-24342.
56. Puigserver P, Vazquez F, Bonet ML et al. In vitro and in vivo induction of brown adipocyte uncoupling protein (thermogenin) by retinoic acid. Biochem J 1996; 317:827-833.
57. Kumar MV, Sunvold GD, Scarpace PJ. Dietary vitamin A supplementation in rats. Suppression of leptin and induction of ucp1 mRNA. J Lipid Res 1999; 40(5):824-829.
58. Villarroya F, Giralt M, Iglesias R. Retinoids and adipose tissues: Metabolism, cell differentiation and gene expression. Int J Obes Relat Metab Disord 1999; 23(1):1-6.
59. Alvarez R, Checa M, Brun S et al. Both retinoic-acid-receptor- and retinoid-X-receptor-dependent signaling pathways mediate the induction of the brown-adipose-tissue-uncoupling-protein-1 gene by retinoids. Biochem J 2000; 345(1):91-97.
60. Spiegelman BM. PPAR-gamma: adipogenic regulator and thiazolidinedione receptor. Diabetes 1998; 47(4):507-514.
61. Sears IB, MacGinnitie MA, Kovacs LG et al. Differentiation-dependent expression of the brown adipocyte uncoupling protein gene: regulation by peroxisome proliferator-activated receptor gamma. Mol Cell Biol 1996; 16(7): 3410-3419.

62. Kelly LJ, Vicario PP, Thompson GM et al. Peroxisome proliferator-activated receptors gamma and alpha mediate in vivo regulation of uncoupling protein (UCP-1, UCP-2, UCP-3) gene expression. Endocrinology 1998; 139(12):4920-4927.

63. Guardiola-Diaz HM, Rehnmark S, Usuda N et al. Rat peroxisome proliferator-activated receptors and brown adipose tissue function during cold acclimatization. J Biol Chem 1999; 274(33):23368-23377.

64. Villena JA, Vinas O, Mampel T et al. Regulation of mitochondrial biogenesis in brown adipose tissue: Nuclear respiratory factor-2/GA-binding protein is responsible for the transcriptional regulation of the gene for the mitochondrial ATP synthase beta subunit. Biochem J 1998; 331(Pt1):121-127.

65. Puigserver P, Wu Z, Park CW et al. A cold-inducible coactivator of nuclear receptors linked to adaptive thermogenesis. Cell 1998; 92:829-839.

66. Wu Z, Puigserver P, Andersson U et al. Mechanisms controlling mitochondrial biogenesis and respiration through the thermogenic coactivator PGC-1. Cell 1999; 98(1):115-124.

67. Puigserver P, Adelmant G, Wu Z et al. Activation of PPARgamma coactivator-1 through transcription factor docking. Science 1999; 286(5443):1368-1371.

68. Rothwell N, Stock M. A role for brown adipose tissue in diet-induced thermogenesis. Nature 1979; 281:31-33.

69. Rothwell N, Stock M. Brown adipose tissue and diet-induced thermogenesis. In: Trayhurn P, Nicholls DG, eds. Brown Adipose Tissue. London: E. Arnold Ltd., 1986:299-338.

70. Sadurskis A, Dicker A, Cannon B et al. Polyunsaturated fatty acids recruit brown adipose tissue: increased UCP content and NST capacity. Am J Physiol 1995; 269(2 Pt 1):E351-E360.

71. Sidlo J, Zaviacic M, Trutzova H. Brown adipose tissue. III. Effect of ethanol, nicotine and caffeine exposure. Soud Lek 1996; 41(2):20-22.

72. Himms-Hagen J. Brown adipose tissue: Interdisciplinary studies. FASEB J 1990; 4:2890-2898.

73. Dullo AG, Miller DS. Energy balance following sympathetic denervation of brown adipose tissue. Can J Physiol Pharmacol 1984; 62:235-240.

74. Ghorbani M, Claus TH, Himms-Hagen J. Hypertrophy of brown adipocytes in brown and white adipose tissues and reversal of diet-induced obesity in rats treated with a beta3-adrenoceptor agonist. Biochem Pharmacol 1997; 54(1):121-131.

75. Lowell BB, Susulic SV, Hamann A et al. Development of obesity in transgenic mice after genetic ablation of brown adipose tissue. Nature 1993; 366:740-742.

76. Kopecky J, Clarke G, Enerback S et al. Expression of the mitochondrial uncoupling protein gene from the aP2 gene promoter prevents genetic obesity. J Clin Invest 1995; 96(6):2914-2923.

77. Enerback S, Jacobsson A, Simpson EM et al. Mice lacking mitochondrial uncoupling protein are cold-sensitive but not obese. Nature 1997; 387(6628):90-94.

78. Hamann A, Flier JS, Lowell BB. Decreased brown fat markedly enhances susceptibility to diet-induced obesity, diabetes, and hyperlipedemia. Endocrinology 1996; 137:21-29.

79. Klaus S, Münzberg H, Trüloff C et al. Physiology of transgenic mice with brown fat ablation: obesity is due to lowered body temperature. Am J Physiol 1998; 274:R287-R293.

80. Cummings DE, Brandon EP, Planas JV et al. Genetically lean mice result from targeted disruption of the RIIβ subunit of protein kinase A. Nature 1996; 382:622-626.

81. McKnight GS, Cummings DE, Amieux PS et al. Cyclic AMP, PKA, and the physiological regulation of adiposity. Recent Prog Horm Res 1998; 53:139-159.

82. Kopecky J, Hodny Z, Rossmeisl M et al. Reduction of dietary obesity in aP2-UCP transgenic mice: Physiology and adipose tissue distribution. Am J Physiol 1996; 270(5 Pt 1):E768-E775.

83. Kopecky J, Rossmeisl M, Hodny Z et al. Reduction of dietary obesity in aP2-UCP transgenic mice: mechanism and adipose tissue morphology. Am J Physiol 1996; 270(5 Pt 1):E776-E786

84. Cannon B, Matthias A, Golozoubova V et al. Gene expression in UCP-1 ablated mice: the phenomenon of hyper-recruitment. Int J Obes 1999; 23(suppl 5):S40.

85. Thomas SA, Palmiter RD. Thermoregulatory and metabolic phenotypes of mice lacking noradrenaline and adrenaline. Nature 1997; 387:94-97.

86. Soloveva V, Graves RA, Rasenick MM et al. Transgenic mice over-expressing the beta 1-adrenergic receptor in adipose tissue are resistant to obesity. Mol Endocrinol 1997; 11(1):27-38.

87. Moitra J, Mason MM, Olive M et al. Life without white fat: A transgenic mouse. Genes Dev 1998; 12:3165-3181.
88. Shimomura I, Hammer RE, Richardson JA et al. Insulin resistance and diabetes mellitus in transgenic mice expressing nuclear SREBP-1c in adipose tissue: Model for congenital generalized lipodystrophy. Genes Dev 1998; 12(20):3182-3194.
89. Kubota N, Terauchi Y, Miki H et al. PPAR gamma mediates high-fat diet-induced adipocyte hypertrophy and insulin resistance. Mol Cell 1999; 4(4):597-609.
90. Heldmaier G. Zitterfreie Wärmebildung und Körpergrösse bei Säugetieren. Z Vergl Physiologie 1971; 73:222-248.

CHAPTER 6

Central Nervous System Innervation of Brown Adipose Tissue

Timothy J. Bartness, C. Kay Song, Gregory E. Demas

Environmental factors such as ambient temperature and food availability can affect the survival and reproductive success of animals. This is especially important for small rodents with their increased metabolic rates compared with larger animals, and because of their increased heat loss due to high surface-to-volume ratios. A variety of physiological adaptations have evolved that buffer small mammals, and other animals including humans, against the inhibitory influences of cold exposure and food scarcity on reproductive functioning. Adipose tissues, in particular, appear to play a critical role in neutralizing the suppression of reproductive responses by cold temperatures or starvation. Specifically, brown adipose tissue (BAT) is the primary site of nonshivering thermogenesis, whereas white adipose tissue (WAT) is the largest source of internally stored energy in most mammals. Control of these and other tissues occurs via humoral or neural influences, or both. Thus, the growth and function of these adipose tissues is vital to the regulation of energy balance and therefore ultimately critical for reproductive success.

The innervation of WAT was reviewed by us recently.[1] The purpose of this chapter, however, is to review one of the mechanisms regulating BAT biology—the innervation of BAT by sensory and sympathetic nervous system (SNS) nerves. Although the functions of BAT and WAT are significantly different, there are many similarities in the neuroanatomy of their motor (SNS) and sensory innervation, especially in terms of the SNS outflow from the brain to BAT (see below) and to WAT.[1] These anatomical similarities notwithstanding, there are conditions where one adipose tissue, but not the other, is activated by the SNS. For example, in response to starvation or severe food restriction the SNS drive on BAT is decreased, while the SNS drive on WAT is increased.[2] Therefore, some of the issues related to the differential control of WAT and BAT also will be addressed at the end of the chapter. In addition, it also is important to note what topics relevant to the innervation of BAT will not be discussed here, or that will be discussed in greater detail elsewhere in this volume. For example, the neuroanatomy of the innervation of BAT will be discussed here, but also will be discussed by S. Cinti in this volume (Chapter 2). Moreover, there will be little discussion of adrenergic receptor and function in this review, but this topic has been reviewed by M. Lafontan and colleagues (see ref. 3,4). Although we recognize that manipulations of the parasympathetic nervous system can indirectly affect BAT (e.g., vagotomy-induced changes in insulin secretion),[5] we are unaware of any convincing data supporting innervation of BAT by this branch of the autonomic nervous system; therefore it will not be discussed further. Finally, this review builds on two reviews of the innervation of BAT that preceded it.[6,7] We especially recommend the review by J. Himms-Hagen for a comprehensive treatise on the neural control of BAT and review of the literature before 1991.[6]

Adipose Tissues, edited by Susanne Klaus. ©2001 Eurekah.com.

Innervation of BAT: Neuroanatomy at the Level of the Fat Pad

BAT receives dense innervation by the SNS (for review see ref. 6,7). The BAT pad that receives almost exclusive attention is interscapular BAT (IBAT) because of its large size (~20-25% of all BAT of mice and rats), accessibility, and clear SNS innervation (see below). Other BAT pads are located around major blood vessels such as the aorta, renal arteries, and in the axillary regions and, moreover, there appears to be brown adipocytes located within WAT pads and that either are dormant or convertible from white adipocytes.[8-11] Unless otherwise stated, however, references to BAT will be synonymous with IBAT. Each IBAT pad is innervated by five intercostal nerve bundles in rats, emerging from the intercostal muscles lying ventral to the pad.[12,13] The nerve fibers composing these bundles are heterogeneous in both their diameter [14] and in their distribution of innervation within the pad.[12,13] The anterior four of the five intercostal nerves appear to supply most, if not all, of the SNS innervation of IBAT, as measured by norepinephrine (NE) content; [13] the function of the fifth nerve is unclear.

The innervation of IBAT by the SNS has long been established, aided by the advent of the histofluorescence technique for visualizing catecholamines.[15] It was reported originally that only the blood vessels in BAT, but not adipocytes, receive noradrenergic innervation.[16] The same laboratory, however, later reported NE varicosities in the parenchymal space and among blood vessels,[17] findings confirmed subsequently by others,[18-22] and verified by electron microscopy (EM).[18,19,23-25] It should be noted that there is rich innervation of individual adipocytes in BAT, rather than just the vasculature as originally believed (see e.g., ref. 16). This individual innervation of brown adipocytes by the SNS is most readily apparent using an improved glyoxylic acid technique for visualization of catecholamines in BAT.[26] Treatment of rats with 6-hydroxy-dopamine (6OHDA) peripherally, a neurotoxin that destroys catecholaminergic neurons, eliminates the visualization of catecholamine innervation from both parenchymal and vascular components of the BAT, and markedly depletes BAT of catecholamine content.[18,19,26,27] There is one report that claims elimination of only parenchymal noradrenergic fibers, however.[25]

Additional putative neurotransmitters besides NE have been detected immunocytochemically in BAT pads or among BAT adipocytes residing in periovarian WAT pads (see e.g., ref. 9,10 and Chapter 2 in this volume). For example, neuropeptide Y (NPY) is frequently colocalized with NE in sympathetic neurons[28] and this also is seen in the innervation of BAT vasculature[29] and BAT adipocytes.[10,25,30-32] Other neuropeptides found in BAT associated with the vasculature are calcitonin gene-related peptide (CGRP), and substance P (SP), which will be discussed below with regard to sensory innervation. Most recently, vasoactive intestinal peptide (VIP) has been detected in the BAT vasculature of rats and is especially apparent after cold exposure.[10]

Sensory Innervation of BAT

BAT contains immunoreactive fibers for CGRP and SP.[10,30-34] More specifically, CGRP, but not SP, has been reported in BAT parenchyma.[10,31] Because both SP and CGRP are classically associated with sensory nerves, it is assumed that they perform a similar function in BAT. Some credence is lent to this notion because treatment of rats with capsaicin, the principal piquant of chili peppers that destroys unmyelinated sensory nerves,[35,36] eliminates CGRP immunoreactivity in BAT.[33] In contrast, treatment with 6OHDA spares CGRP immunoreactivity in BAT but depletes NE in the catecholaminergic nerve fibers.[27] The function of the sensory innervation of BAT will be discussed below in the context of BAT denervation experiments.

Peripheral Denervation of BAT

Effects on Growth and Normal Function

One strategy used to test the role of the BAT SNS innervation in a variety of conditions is to denervate the IBAT. The SNS denervation of BAT has been done globally by systemic injections of the catecholaminergic neurotoxins 6OHDA or guanethidine. There are several drawbacks to these denervation methods. For example, all peripheral organs are denervated; moreover, the SNS denervations are not permanent for 6OHDA.[37,38] Unilateral surgical denervation of BAT can be readily done using each animal as a within animal control because the two lobes of the IBAT pad are innervated unilaterally by the SNS in rats,[12] mice,[32] Syrian hamsters,[39,40] and Siberian hamsters.[41,7,42] Although surgical denervation is permanent, a major drawback of this approach is the inability to selectively sever only the SNS innervation and leave the accompanying sensory innervation intact. In addition, it is not possible to surgically denervate the parenchymal SNS innervation selectively and spare the SNS innervation of the vasculature, although there are claims that 6OHDA eliminates sympathetics to both components of IBAT, whereas surgical denervation spares the vasculature.[18,19] Nonetheless our understanding of the regulation of BAT physiological function by its innervation clearly have been aided by the use of the unilateral surgical denervation model.

A variety of functions have been ascribed to the SNS innervation of BAT through the use of the unilateral surgical denervation model (for review see ref. 6). In most animals, decreases in pad mass and protein content, as well as decreases in the binding of the purine nucleotide guanosine 5'-diphosphate to brown adipose tissue mitochondria (GDP binding i.e., the capacity of proton conductance pathway[43]) and UCP concentration (i.e., thermogenic capacity)[43] along with the decrease in NE content occur in the denervated pad, but not the contralateral innervated pad.[44-46] These atrophic responses are exaggerated between the denervated and neurally intact pads under conditions that stimulate BAT growth and thermogenic activity, such as cold exposure/acclimation (see directly below).

Mediation of the Responses to Cold Exposure/Acclimation

Perhaps the most studied of the factors that affect BAT and its thermogenic response is cold exposure or cold acclimation (for review see ref. 47). Cold causes impressive increases in BAT thermogenesis revealed as an ensemble of coordinated responses. Specifically, blood flow is increased to nearly 200% over basal levels.[48-50] The number of mitochondria, the density of their cristae, along with the gene expression of the family of UCPs found in BAT (i.e., UCP1-3) all are increased in the cold.[51] In addition, other components or reflections of these increases such as total cytochrome oxidase activity (COA, roughly equivalent to mitochondrial mass), GDP-binding and thyroxine 5'-deiodinase activity (an enzyme responsible for in situ conversion of thyroxine to triiodothyronine, a hormone involved in thermogenesis) also are markedly increased by the cold.[52] All these changes are amplified further because of the increase in growth of the tissue in response to cold exposure, as reflected in increased mass and protein content (see e.g., ref. 45,53), as well as increased hyperplasia (increased brown fat cell number). The latter is exhibited as increases in the mitotic activity of BAT precursor cells and endothelial cells from capillaries[54-57] and includes increases in vascular endothelial growth factor gene expression that supports angiogenesis.[58] The above ensemble of BAT responses to the cold, as well as increased glucose uptake, glucose utilization and glucose transporter number all are greatly diminished or blocked in denervated IBAT compared with innervated IBAT with a notable exception.[59-61] Specifically, the cold-induced increase in UCP3 is not affected by denervation, as are the cold-induced increases in UCP1 and UCP2.[51,59-61] Collectively, these studies highlight the importance of the innervation of IBAT for the thermogenic responses to the cold and the importance of the SNS innervation of the tissue.

Many or most of the BAT cold exposure/acclimation responses can be mimicked by treatment with the classically defined postganglionic neurotransmitter of the SNS, NE. Indeed, NE treatment engenders thermogenic responses that prepare animals for the cold exposure seemingly exactly as does cold acclimation itself.[62] Moreover, NE produces similar changes in BAT to those produced by the cold including the increased proliferation of BAT precursor cells (preadipocytes, interstitial and capillary endothelial cells.)[57] It is not surprising, therefore, that infusions or injections of NE restore, or nearly restore the effects of IBAT denervation in cold exposed animals including proliferation/mitotic activity.[57]

Mediation of Hormonal Effects

The role of circulating hormones in the physiological control of BAT and their modulation by the SNS is fraught with the difficulties associated with in vivo studies and with the caveats associated with in vitro experiments. Simple administration of these hormones in vivo results in pharmacology, rather than physiology. Because of the widespread effects of many of these circulating factors (e.g., insulin), their removal without replacement is an interpretational nightmare, not the least of which may involve the means of their removal (e.g., side effects of the toxin streptozotocin) and the route of administration and pattern of their replacement. Nevertheless, some findings seem clear and notable (see directly below).

Experimentally-induced diabetes mellitus (i.e., streptozotocin-induced diabetes) results in decreases in BAT mass, COA and DNA content,[63] as well as UCP content.[64] It is not surprising that stimulators of thermogenesis such as access to a high energy and varied diet ('cafeteria' diet-feeding), or NE treatment, do not increase oxygen consumption in these diabetic animals possessing atrophied BAT.[65] The inability of these stimuli to increase oxygen consumption is not reversible with chronic insulin replacement, although acute insulin treatment does restore the thermogenic effects of a single dose of NE.[65] Insulin treatment partially restores the diabetes-related decreases in IBAT mass, and protein and DNA content even when IBAT is denervated,[63] whereas the restoration of UCP levels in IBAT by insulin replacement is inhibited with denervation.[64] Thus, streptozotocin-induced diabetes appears to have long-lasting, sometimes irreversible deleterious effects on BAT thermogenic functions that are not always restored with insulin replacement, but insulin does appear to work through or in conjunction with the innervation of BAT for the maintenance and restoration of UCP levels.

Glucagon is a thermogenically active hormone that increases whole animal oxygen consumption when administered exogenously.[66,67] Glucagon administration resulting in physiological concentrations of the hormone increases GDP-binding in BAT,[68] as do pharmacological doses that increase BAT mass, protein, DNA and COA contents, as well as lipoprotein lipase (LPL) activity,[69] the latter responsible for the uptake of free fatty acids from the vasculature to cells in many tissues.[70] The effects of glucagon are mostly maintained in rats with IBAT axotomy, although their magnitude is somewhat diminished.[69]

Finally, leptin originally thought to only be synthesized and secreted from WAT (for review see ref. 71) but now found in a variety of tissues including BAT, appears to affect BAT thermogenesis via the SNS.[72-74] Although the role of the SNS in leptin gene expression and secretion by BAT and WAT is discussed at length in this volume (see Chapter 9), it seems appropriate to discuss the effect of IBAT axotomy on leptin expression. Leptin administration increases UCP1 mRNA in rat and mouse IBAT,[75,76] an effect blocked by denervation.[76] Leptin also increases LPL gene in rat IBAT, but does so independently of its innervation.[76] The leptin-induced increase in UCP1 via the SNS innervation of BAT was confirmed recently in globally chemically sympathectomized rats.[77] Together, these findings suggest that the leptin induction of BAT UCP1 gene expression is dependent on SNS innervation, whereas leptin induction of BAT LPL gene expression is independent of this innervation.[78] It is tempting to speculate that the effects of leptin on other tissues may be via the SNS. For example, the inhibition of insulin secretion by leptin also is blocked in globally sympathectomized rats.[77]

Mediation of the Effects of Food Intake/Fasting

Overfeeding induced by access to a cafeteria diet increases IBAT mass, protein content and GDP-binding in neurally intact pads, effects blocked by surgical denervation.[45] Overeating induced by supplemental sucrose solution stimulates BAT lipogenesis, an effect also blocked by axotomy.[77] In contrast, fasting causes an atrophy of BAT reflected in decreased mass, protein and UCP contents, but these effects are not blocked by IBAT or global denervation,[44,80] suggesting that fasting-induced BAT atrophy is not mediated via its innervation. This is not surprising because fasting decreases NE turnover (NETO, an indicator of SNS drive) in BAT;[81] therefore denervation of an already inhibited neural drive would not be expected to block these responses that are sustained by innervation (see above). Although it has been stated repeatedly in the literature that food deprivation or fasting decreases SNS activity, caution should be exerted in classifying the state of SNS activity from examination of only one or two peripheral target tissues. Although Cannon envisioned the SNS working as a single unit in 'fight or flight' situations,[82] at least among the adipose tissues, differential SNS drives can occur. For example, fasting decreases the SNS drive (NETO) on BAT, but increases its drive on WAT.[2] It is now apparent that integrated circuits are activated under a variety of conditions resulting in many combinations of differential drives in several SNS target tissues.[83]

Sensory Denervation of BAT

Because it is impossible to surgically denervate only the sensory innervation of BAT, global sensory denervation (capsaicin desensitization) has been used to begin to test the functional significance of this innervation (see ref. 6 for review). Unfortunately, peripheral application of capsaicin not only destroys small unmyelinated sensory nerves that are capsaicin-sensitive, but also appears toxic for several areas of the brain.[84] Perhaps these neurons are destroyed centrally because they receive one pole of the bipolar dorsal root ganglion neurons, or because they may be inexplicably vulnerable to capsaicin toxicity.[6] In addition to markedly decreasing or abolishing the immunoreactivity of the sensory peptides SP and CGRP in BAT after capsaicin treatment,[11,33] other morphological and some functional changes occur. Specifically, capsaicin desensitization induces atrophy of rat IBAT manifested in decreased DNA (decreased cell number), protein content, mitochondrial number (cytochrome oxidase activity), as well as thermogenic capacity (total UCP).[33,85-87] These effects occur without apparent changes in SNS innervation or capacity of the proton conductance pathway (concentration of UCP per mitochondrion/ mitochondrial protein content).[33] Functionally, sensory denervated IBAT does not show normal growth in mass or thermogenic capacity with cafeteria diet feeding, nor with cold exposure.[33,85] It has been hypothesized recently that the sensory innervation of BAT dampens the SNS-stimulated thermogenic activity of the tissue through the release of CGRP.[88]

SNS Outflow from Brain to BAT

Initial attempts at identifying the neuroanatomical basis for the CNS connections from brain to BAT involved injections of the monosynaptic retrograde tract-tracer horseradish peroxidase into laboratory rat IBAT pads and labeling postganglionic neurons.[7] In addition to defining the relative distribution of postganglionic sympathetic innervation of BAT, their results suggested that the two IBAT lobes were bilaterally, rather than unilaterally innervated,[89,90] the latter supporting their previous findings.[42] Others have found support only for the unilateral innervation of IBAT (see above). The rostral brain origins of the innervation of the preganglionic sympathetic neurons in the spinal cord that, in turn project to these postganglionic neurons, remained to be defined neuroanatomically until recently.[91] That is, although the results of dozens of physiological studies suggest that chains of neurons, originating in rostral portions of the neuroaxis, form descending polysynaptic pathways that terminate in BAT (see below and ref. 6), no technique was available until recently to define such monosynaptic connections within the same animal. This was accomplished using a viral transneuronal tract tracer, the Bartha's K strain of the pseudorabies virus (PRV) that has been used to trace the SNS outflow from brain to peripheral sympathetic targets (see e.g., ref. 91-95). Specifically, some viruses, like the PRV, are taken up into neurons following binding to viral attachment protein molecules located on the surface of neuronal membranes. These protein surface molecules act as 'viral receptors'. Neurons synapsing on infected cells become exposed to relatively high concentrations of the virus particles that have been exocytosed. The virus particles are then taken up at synaptic contact sites and this process continues causing an infection along the neuronal chain from the periphery to higher CNS sites.[96,97] These infected neurons are visualized using standard immunocytochemistry. The major advantage of using viruses as transneuronal markers, over some of the more traditionally used tracers, is their ability to replicate within the neuron performing as a self-amplifying cell marker. Furthermore, because the transfer of the virus only is by a transsynaptic mechanism, rather than by lateral spread to adjacent but unrelated neurons or by a nonsynaptic mechanism,[96-98] the transsynaptic transfer of the Bartha's K strain of the PRV after injection into BAT yields a hierarchical chain of functionally connected neurons from brain to BAT. Therefore, we will review the functional studies of CNS-BAT connections including electrophysiological recording, chemical and electrical stimulation, and lesion studies in light of the only neuroanatomically defined SNS outflow from brain to BAT.[91]

We adapted the PRV methodology for the determination of the CNS sympathetic outflow to BAT in Siberian hamsters and laboratory rats.[91] PRV-infected neurons were found in the spinal cord in the sympathetic preganglionic cells in well-defined clusters located in the intermediolateral (IML) cell group and the central autonomic nucleus of the spinal cord ipsilateral to the unilateral IBAT injection site. In the brainstem, most of the classically-defined autonomic areas had PRV-infected neurons and these were bilateral and uniform including several portions of the reticular area (lateral, intermediate, parvocellular, medullary and gigantocellular reticular nuclei), the caudal raphe area (raphe pallidus and raphe obscurus nuclei), C1 adrenaline cells and rostroventrolateral medullary regions, nucleus of the solitary tract, and lateral paragigantocellular nucleus, raphe magnus. In the midbrain, the infection also was bilateral and uniform with the central gray being the most heavily infected area, but also including the dorsal raphe. In the forebrain, infections were bilateral, but a greater density of labeling was seen ipsilateral to the injection site. PRV immunoreactivity was seen in the paraventricular hypothalamic nucleus (PVN), medial preoptic area (MPOA), lateral hypothalamus (LH), suprachiasmatic nucleus (SCN), dorsal hypothalamic area, zona incerta, arcuate nucleus, retrochiasmatic area, lateral septal region, bed nucleus of the stria terminalis (BNST), and very few (1.6% of all infected neurons in the forebrain) in the ventromedial hypothalamic nucleus (VMH) and, in the case of the latter, only on the peripheral edges of the nucleus. Figure 6.1.shows a schematic diagram of the origins of SNS innervation of IBAT.

The expression of the immediate-early gene, c-fos, has been used as a marker of neuronal activation through the detection of its protein product using immunocytochemistry (for review see ref. 99). Using this technique, several CNS areas known to be activated with cold exposure/acclimation also show PRV-infected neurons after IBAT injections including the PVN, nucleus of the solitary tract, lateral septum,[100] preoptic-anterior hypothalamic area,[101] zona incerta, central gray, [102] and raphe pallidus and raphe obscurus.[103] Unfortunately, the c-fos technique cannot be combined with the PRV methodology because PRV induces c-fos expression.[104]

Electrophysiological Studies of BAT

Many recordings of the activity of the nerves, presumably SNS nerves, innervating IBAT have been done in conjunction with environmental, humoral, and CNS manipulations (for review see ref. 6). Of these studies, the majority have focused on manipulations of the VMH and recordings from the sympathetic inputs to IBAT (see e.g., ref. 105-109). Because the VMH does not appear to be connected to IBAT via the SNS (see below), these and related studies will not be discussed here. Of the remaining studies recording from the sympathetic nerves innervating IBAT, the studies consistent with the now known connections of the brain to BAT via the SNS,[91] are those involving the raphe pallidus of the brainstem.[110-112] The results of these studies will be discussed below because they include chemical stimulation of the brain in conjunction with the electrophysiological recording from the SNS innervation of BAT.

Manipulations of the VMH/PVN and the Responses of BAT

The VMH has been implicated as a pivotal center for the SNS outflow from the brain to the periphery including BAT and WAT (see below). Because of this emphasis and because this view *is not* based on neuroanatomical evidence, rather on the results of stimulation or lesion studies directed towards the VMH, we will devote considerable discussion to this point. Recall that we found few or no infected cells in the VMH after injections of PRV into BAT; therefore there appears to be little neuroanatomical support for the notion of a VMH-SNS-IBAT circuitry. Others have found a similar lack of neuroanatomical support for putative VMH-SNS-pancreas[113] and VMH-SNS-adrenal medulla connections[92,93] using the PRV transsynaptic tract tracer methodology. Thus, we conclude that the VMH is not part of the SNS outflow to the periphery generally,[1] or to BAT or WAT specifically.[91,114] One likely reason for the discrepancy between the results of the stimulation/lesion studies, and those of the PRV circuit tracing experiment,[91] is that, although these manipulations were targeted for VMH, neighboring neural elements known to be a part of the SNS outflow from the brain to BAT also were affected. These neural elements undoubtedly included the DMN and PVN of the hypothalamus, arcuate nucleus, periventricular nucleus, retrochiasmatic area or their efferents because they are part of the CNS-SNS-BAT circuitry.[91] This misplaced focus on the VMH is reminiscent of the 'myth of the VMH' [115] that resulted originally from the misguided emphasis on these nuclei in relation to the control of body fat and food intake (see e.g., ref. 116,117). This focus on the VMH has been subsequently redirected to other brain regions, especially the PVN and its descending projections that course just lateral to the VMH and that make direct and indirect connections with the spinal preganglionics (see e.g., ref. 118,119). Thus, BAT thermogenesis may have been altered unintentionally via lesions or stimulation affecting the PVN or its caudal projections. It cannot be the case, however, that the VMH has a role in the SNS outflow by indirectly affecting other brain sites, either 'up' or 'downstream' from itself

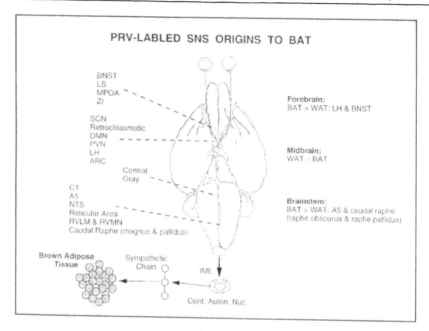

Fig. 6.1.Schematic diagram of origins of the SNS innervation of interscapular brown adipose tissue (BAT). Summary of major structures showing infections after injection of the transneuronal tract tracer, the pseudorabies virus (PRV), into interscapular BAT (IBAT) (left side). Significant differences in major structures infected with the PRV after injection into IBAT versus white adipose tissue (WAT; inguinal or epididymal WAT) (right side). BNST, bed nucleus of the stria terminalis; LS, lateral septal area; MPOA, medial preoptic area; ZI, zona incerta; SCN, suprachiasmatic nucleus; retrochiasmatic, retrochiasmatic area; DMN, dorsomedial hypothalamic nucleus; PVN, hypothalamic paraventricular nucleus; LH, lateral hypothalamic nucleus; ARC, arcuate nucleus; Central Gray, midbrain central gray; C1, adrenergic area; A5, noradrenergic area; reticular Area; includes: lateral, intermediate, parvocellular, medullary and gigantocellular reticular nuclei [ventral and alpha parts]; NTS, nucleus of the solitary tract; RVLM, rostral ventrolateral medullary region; RVMM, rostral ventromedial medullary area; caudal raphe, includes raphe magnus and raphe pallidus; IML, intermediolateral horn of the spinal cord, cent. auton. nuc., central autonomic nucleus of the spinal cord.

because these connections should have been labeled by this transneuronal viral technique.[91] Furthermore, just because surgical denervation of IBAT blocks the ability of VMH stimulation to increase IBAT temperature, glucose uptake, or blood flow does not necessarily support the VMH-SNS-IBAT connections.[120-126] Instead, these data could just as easily be interpreted as an axotomy-induced blockade of the ancillary stimulation of PVN descending pathways coursing next to, or through, the VMH. Another possibility, albeit somewhat complex, is that VMH manipulations might alter the secretion of a humoral factor (e.g., glucocorticoid secretion from the adrenal cortex) that, in turn affects CNS sites that have been identified as bone fide components of the SNS outflow from brain to BAT (e.g., PVN). Still another alternative way of accounting for the lack of a VMH-SNS-IBAT connection as revealed by the PRV technique is that VMH cells are not susceptible to infection by this virus, although there is no reason to suspect this. One final, and more tenable, conclusion is that the few scattered infected neurons seen largely in the peripheral aspects of the VMH after PRV injections into IBAT are sufficient to support the control of BAT thermogenesis. For example, in a different system, only a few gonadotropin-releasing hormone (GnRH) neurons are required to elicit LH, as is evidenced by hypogonadal mice, genetically devoid of GnRH neurons, that are rendered reproductively functional after receiving fetal grafts containing GnRH neurons from normal mice.[127] By analogy then, perhaps only a few VMH neurons might be required to participate in the SNS outflow from the VMH to adipose tissues, as incredulous as this might seem.

If the focus on the VMH should be redirected towards the PVN, as suggested above, then manipulations of the PVN should be especially likely to affect BAT thermogenic responses. Unfortunately this is not necessarily the case. Clearly, lesions of the PVN or its descending projections block diet-induced thermogenesis thought to be mediated by BAT,[128,129] whereas electrical[130] or glutamate[131] stimulation of the PVN increases IBAT temperature. It is claimed, however, that electrical stimulation of only the VMH and not the PVN increases IBAT temperature,[132] and accelerates NETO in this tissue.[133] Moreover, cold-induced thermogenesis is not blocked by severing the caudal projections of the PVN.[128] Although glutamate injections into the PVN elicit an increase in IBAT SNS activity, similar injections into the VMH elicit an even greater increase.[109] Furthermore kainic acid lesions of the VMH decrease the activity of the SNS nerves to IBAT more so than do PVN lesions acutely, and only the VMH lesions affect this response chronically.[107] Therefore, the results of some of these functional studies appear to conflict with the defined SNS outflow from brain to BAT requiring a reassessment of the specificity of these lesion or stimulation techniques or adoption of one or more of the alternative possibilities for the a role of the VMH in BAT thermogenesis.

Manipulations of the SCN and the Responses of BAT

Connections of the suprachiasmatic nucleus (SCN), the 'biological clock', to BAT via the SNS have been suggested by the increase in IBAT thermogenesis after electrical stimulation of the retinohypothalamic tract that innervates the SCN,[134] or by glutamate injections directly into the SCN.[135] Indeed, the SCN now appears to be a principal component of the general SNS outflow from brain to peripheral tissues including BAT. That is, once again using the viral transneuronal tract tracer PRV, we first reported that the SCN is a component of the general SNS outflow from the brain to WAT,[1,114] and more recently to BAT.[91] Moreover, it was the dorsomedial aspects of the SCN that were most highly infected after virus injections into BAT or WAT from Siberian hamsters and laboratory rats.[91,114] In addition, it was the arginine vasopressin containing neurons in this region that were frequently labeled.[136-138] The participation of the SCN in the CNS origins of the sympathetic outflow to peripheral targets was confirmed for a wide range of target tissues recently by others.[139-143] With regard to BAT, significant numbers of infected neurons occur in the SCN after PRV injections into IBAT,[91] and SCN neurons project to PVN neurons that, in turn project to the IML.[144] Together, these results suggest that the relation between BAT thermogenesis and the nycthemeral cycles of body temperature, as well as the sleep-wake cycle,[6] may have SCN-BAT connection. Indeed, the circadian rhythm of heart rate, locomotor activity, and body temperature all are blocked by SCN lesions,[145] and SCN lesions block circadian rhythms of NETO in the adrenal and pineal glands, heart, kidney, and liver.[146] Furthermore the SCN-SNS-BAT circuit may be involved in the circadian timing of torpor bouts in Siberian hamsters because SCN lesions block the expression of the torpor-associated rhythmic daily decreases in body temperature in this species.[147]

Manipulations of Other CNS Sites and the Responses of BAT

There is extensive labeling of neurons in the medial preoptic area (MPOA) after PRV injections into IBAT.[91] It is tempting to speculate that this MPOA-SNS-IBAT connection underlies the increases of IBAT thermogenesis and the firing rate of the sympathetic nerves that innervate IBAT after electrical,[148] or chemical[105] stimulation of the MPOA.

Prepontine knife cuts disinhibit part of the SNS regulation of IBAT and body temperature. Specifically, knife cuts at the prepontine, but not post mammillary body levels, markedly increase (3-4 °C) IBAT temperature (followed by colonic temperature) in urethane-anesthetized rats,[149] an effect mimicked by injections of procaine into this area.[150] This prepontine knife cut-induced hyperthermia is associated with a 12-fold increase in blood flow to IBAT[149] that is blocked by propanolol (β adrenergic receptor blocker)[149] and suppressed by chemical sympathectomy (6OHDA),[151] but oddly not suppressed by IBAT axotomy.[151] Neurons in the rostral raphe pallidus appear to be the neurons critical for this medullary brainstem control of IBAT thermogenesis.[111,112] That is, application of the GABA$_A$ receptor blocker, bicuculline, into the rostral raphe pallidus triggers a large and rapid increase in sympathetic nerve activity to IBAT,[111,112] an area also showing c-fos immunoreactivity suggesting neuronal activation after cold exposure,[103] and labeled with infected neurons after IBAT injections of PRV.[91] Moreover, stimulation of the rostral raphe pallidus appears to activate a specific brainstem-SNS-IBAT pathway. Thus, the sympathetic nerves innervating IBAT show increased firing rates without affecting SNS activity in the splanchnic nerve.[112]

Another brain site likely to be involved with the SNS outflow to BAT is the arcuate nucleus, a brain locus that also shows infected neurons after PRV injections into IBAT.[91] This area, among others, has been strongly implicated in the control of energy balance (for review see ref. 152). Indeed, among the many neurotransmitters found in the arcuate, NPY-containing neurons originating in the arcuate and projecting to the PVN[153,154] appear pivotal in both the CNS control of food intake and of thermogenesis.[155,156] Neurons in the arcuate nucleus also have been shown to directly innervate SNS thoracic preganglionic neurons in the IML, making them likely neuroanatomically candidates for modulating IBAT thermogenesis.[157] Destruction of the arcuate nucleus neonatally by monosodium glutamate (MSG), including the NPY-producing neurons, results in obesity associated with decreased food intake suggesting impaired energy expenditure.[158] Additional indirect support for an arcuate-SNS-BAT connection is that cold exposed MSG-treated mice have inhibited UCP gene expression in IBAT that is normalized with NE injections.[159] These data, and those showing decreases in NETO in BAT after MSG treatment,[160] suggest impaired SNS innervation or NE release in BAT. Thus, a dysfunctional SNS innervation and/or function in BAT is implicated in the development of obesity in this model,[161] perhaps due to destruction of arcuate neurons that have spinally projecting axons innervating the preganglionic sympathetic neurons in the cord that, in turn, ultimately synaps on post-ganglionic neurons innervating BAT adipocytes.

Centrally-Administered Leptin and the Responses of BAT

Finally, with regard to the CNS control of BAT thermogenic and growth associated responses, we would be remiss in not briefly discussing possible brain sites stimulated by leptin that in turn stimulate BAT via the SNS (see also Chapter 9 of this volume for additional leptin-SNS-BAT/WAT discussions). Injection of leptin into the VMH, but not the LH, increases glucose uptake in IBAT as well as several other tissues (but not WAT), an effect blocked by surgical denervation of IBAT.[162] In addition, leptin activates neurons (induces c-fos expression) in the retrochiasmatic area and arcuate nucleus that synthesize cocaine- and amphetamine-regulated transcript (CART) and proopiomelanocortin (POMC) and that innervate thoracic spinal cord preganglionic neurons in the IML.[157,114] Thus, the intriguing possibility exists that the leptin-induced increase in energy expenditure may be via the activation of this pathway.[157] In support of this notion, both the retrochiasmatic area and arcuate nucleus contain infected neurons after PRV injections into the IBAT pad.[91] Finally, leptin also activates neurons in the PVN and to a lesser extent the zona incerta (ZI) and arcuate nucleus.[163]

These three areas possess infected neurons after injections of PRV into IBAT[91] also suggesting these virally laden neurons may be part of the SNS outflow that stimulates BAT thermogenesis.

Some Species Differences in BAT SNS Mediation of BAT Function

The majority of the information regarding SNS innervation of BAT has been obtained from studies in laboratory rats. In rats, NE release by the SNS is the critical for BAT thermogenic activity and growth (for review see ref. 6). The role of the SNS innervation of BAT, and NE specifically, is less clear in other species. For example, in Siberian hamsters, NE treatment stimulates BAT LPL activity to a similar extent as cold exposure.[164] Cold-induced activity of BAT LPL activity, however, is not affected by sympathetic denervation.[164] In addition, IBAT denervation in Siberian hamsters reduces, but does not prevent, BAT 5'-deiodinase activity,[41] or cytochrome oxidase activity.[165] NE appears to play a minimal role in regulating BAT function in Syrian hamsters as well. For example, BAT in Syrian and Siberian hamsters also possesses both SNS and sensory innervation and NE release by sympathetic nerves regulates thermogenesis both in vitro and in vivo similar to rats.[166,40] Unlike rats, however, the increases in BAT mass in response to cold, diet, or short photoperiods do not appear to be mediated by NE. For example, chronic administration of NE does not increase BAT mass in Syrian hamsters as it does in rats.[40,167,168] Furthermore, the normal covariance of increases in sympathetic drive (NETO) with increases in GDP-binding or mitochondrial mass (cytochrome oxidase activity) in rat BAT mitochondria is dissociated in Syrian hamsters. Specifically, there are increases in Syrian hamster BAT GDP-binding, but not NETO in short photoperiods.[169] There also are increases in mitochondrial mass, but not NETO when Syrian hamsters are fed a high-fat diet.[170] Furthermore, reductions in BAT mass after fasting[171] or during pregnancy and lactation[172] also are not correlated with SNS activity in Syrian hamsters. Surgical denervation of Syrian hamster BAT can trigger atrophy of the pad as it does in rats, but unlike rats, atrophy cannot be prevented by administration of NE.[39,40] In addition, chemical sympathectomy using 6-OHDA does not affect BAT mass in Syrian hamsters as it does in rats.[39] Collectively these results suggest that the innervation of BAT is the primary mechanism for triggering increases in BAT thermogenesis in both rats and hamsters; in hamsters, however, trophic effects on IBAT can occur in the apparent absence of changes in the sympathetic drive on the tissue.

Directions for Future Research

Although we are attaining a better understanding of the innervation of BAT and WAT,[91] there still are substantial gaps in our knowledge. Some of these gaps are applicable to more general issues of SNS functioning, such as the differential control of SNS drives on target tissues. For example, BAT SNS drive is markedly decreased with prolonged fasting, whereas WAT SNS drive is increased.[2] How does this occur? Where is the differential control exerted— from brain sites relatively rostral in the neuroaxis, or is the control exerted at the level of the postganglionic SNS neurons as appears to be the case for differential drives on WAT pads?[173]

The exact connectivity among the CNS sites that are part of the SNS outflow from brain to BAT or WAT, as revealed by the PRV methodology,[91,114] is not known. That is, for many of the forebrain structures, there are known monosynaptic connections from these sites to the SNS spinal preganglionics in the IML, as well as for brainstem structures. Thus, it almost certainly is not the situation that all the SNS outflow to BAT or WAT starts in the forebrain structures that synapse on midbrain structures that, in turn, synapse on brainstem structures that ultimately project to the spinal preganglionics. It also seems likely that, despite striking similarities in the patterns of infected structures across the neuroaxis after PRV injections in adipose or other SNS innervated tissues,[91,92,95,114,174,175] there are important differences in the hierarchical chains of neurons involved in the thermogenic responses of BAT and the lipolytic responses of WAT, for example. Indeed, there are some marked differences (see below and Fig. 6.1 [right side]) in the SNS innervation of BAT and WAT by the CNS. Furthermore, we may be naïve in thinking that the same exact neurocircuitry is involved with the many conditions that stimulate thermogenesis by BAT (e.g., cold exposure/acclimation, overeating, infections) or that stimulate lipolysis by WAT (e.g., fasting, cold exposure/acclimation, pregnancy). Moreover, the functions of the sensory innervation of both types of adipose tissues (e.g., trophic influences) may share common neurotransmitters (e.g., CGRP, substance P), but perhaps no more. We also do not have a full appreciation for the role of the SNS innervation of BAT and WAT in terms of maintenance of mass or growth (e.g., NE stimulates fat cell proliferation in BAT,[56] and inhibits fat cell proliferation in WAT).[176] Finally, as mentioned repeatedly above, although the vast majority of the data supporting VMH-SNS-BAT or -WAT innervation most likely can be ascribed to the PVN, rather than the VMH, all the data cannot be explained by the misdirected emphasis on the VMH and the nonselectivity of the lesion and stimulation techniques.

Unfortunately, to our knowledge, there is no 'viscerotopic' organization to the SNS innervation of BAT or WAT specifically, or any SNS target tissue generally. There are, however, some clear differences between the patterns of infections seen across the structures after injections of the PRV into BAT or WAT. These include, but are not restricted to, a larger percentage of infected cells in the A5 and caudal raphe region (raphe obscurus & raphe pallidus nuclei) in the brainstem, and a larger percentage of infected cells in the lateral hypothalamus and BNST after PRV injections into BAT versus WAT (Fig. 6.1 [right side]).[91,114] This comparison does not address the more difficult issue of differences in the pattern of infections within the structures after BAT or WAT injections of the PRV. As more specific techniques and better experimental ingenuity find one another, our understanding of the innervation of adipose tissues should blossom.

Acknowledgments

This research is supported, in part, from NIH research grant RO1 DK35254 and NIMH Research Scientist Development Award KO2 MH00841 to TJB.

References

1. Bartness TJ, Bamshad M. Innervation of mammalian white adipose tissue: Implications for the regulation of total body fat. Am J Physiol 1998; 275:R1399-R1411.
2. Migliorini RH, Garofalo MAR, Kettelhut IC. Increased sympathetic activity in rat white adipose tissue during prolonged fasting. Am J Physiol 1997; 272:R656-R661.
3. Lafontan M. Differential recruitment and differential regulation by physiological amines of fat cell β-1, β-2 and β-3 adrenergic receptors expressed in native fat cells and in transfected cell lines. Cell Signal 1994; 6:363-392.
4. Lafontan M, Berlan M. Fat cell α_2-adrenoceptors: The regulation of fat cell function and lipolysis. Endocrine Rev 1995; 16(6):716-738.
5. Andrews PLR, Rothwell NJ, Stock MJ. Influence of subdiaphragmatic vagotomy and brown fat sympathectomy on thermogenesis in rats. Am J Physiol 1985; 249:E239-E243.
6. Himms-Hagen J. Neural control of brown adipose tissue thermogenesis, hypertrophy, and atrophy. Front Neuroendocrinol 1991; 12(1):38-93.
7. Girardier L, Seydoux J. Neural control of brown adipose tissue. In: Trayhurn P, Nicholls DG, eds. Brown Adipose Tissue. London: Edward Arnold, 1986:122-151.
8. Cousin B, Bascands-Viguerie N, Kassis N et al. Cellular changes during cold acclimation in adipose tissues. J Cell Physiol 1996; 167:285-289.

9. Cousin B, Cinti S, Morroni M et al. Occurrence of brown adipocytes in rat white adipose tissue: molecular and morphological characterization. J Cell Sci 1992; 103:931-942.

10. Giordano A, Morroni M, Santone G et al. Tyrosine hydroxylase, neuropeptide Y, substance P, calcitonin gene-related peptide and vasoactive intestinal peptide in nerves of rat periovarian adipose tissue: An immunohistochemical and ultrastructural investigation. J Neurocytol 1996; 25:125-136.

11. Giordano A, Morroni M, Carle F et al. Sensory nerves affect the recruitment and differentiation of rat periovarian brown adipocytes during cold acclimation. J Cell Sci 1998; 111:2587-2594.

12. Foster DO, Depocas F, Zaror-Behrens G. Unilaterality of the sympathetic innervation of each pad of interscapular brown adipose tissue. Can J Physiol 1982; 60:107-113.

13. Foster DO, Depocas F, Zuker M. Heterogeneity of the sympathetic innervation of rat interscapular brown adipose tissue via intercostal nerves. Can J Physiol Pharmacol 1982; 60:747-754.

14. Flaim KE, Horowitz JM, Horwitz BA. Functional and anatomical characteristics of the nerve—Brown adipose tissue interaction in the rat. Pflugers Arch 1976; 365:9-14.

15. Falck B, Hillarp NA, Thieme G et al. Fluorescence of catecholamines and related compounds condensed with formaldehyde. J Histochem Cytochem 1962; 10:348-354.

16. Wirsen C. Adrenergic innervation of adipose tissue examined by fluorescence microscopy. Nature 1964; 202:913.

17. Wirsen C. Studies in lipid mobilization with special reference to morphological and histochemical aspects. Acta Physiol Scand 1965; 65(Suppl. 252):1-46.

18. Thureson-Klein A, Mill-Hyde B, Barnard T et al. Ultrastructural effects of chemical sympathectomy on brown adipose tissue. J Neurocytol 1976; 5:677-690.

19. Thureson-Klein A, Lagercrantz H, Barnard T. Chemical sympathectomy of interscapular brown adipose tissue. Acta Physiol Scand 1976; 98:8-18.

20. Derry DM, Schonbaum E, Steiner G. Two sympathetic nerve supplies to brown adipose tissue of the rat. Can J Physiol Pharmacol 1969; 47:57-63.

21. Daniel H, Derry DM. Criteria for differentiation of brown and white fat in the rat. Can J Physiol Pharmacol 1969; 47:941-945.

22. Derry DM, Daniel H. Sympathetic nerve development in the brown adipose tissue of the rat. Can J Physiol Pharmacol 1970; 48:160-168.

23. Bargmann W, Hehn GV, Linder E. Über die Zellen des braunen Fettgewebes und ihre Innervation. Z Zellforsch Mikroskop Anat 1968; 85:601-613.

24. Lever JD, Jung RT, Nnodim JO et al. Demonstration of a catecholaminergic innervation in human perirenal brown adipose tissue at various ages in the adult. Anat Rec 1986; 215:251-255.

25. Nnodim JO, Lever JD. Neural and vascular provisions of rat interscapular brown adipose tissue. Am J Anat 1988; 182:283-293.

26. Cottle MK, Cottle WH, Perusse F et al. An improved glyoxylic acid technique for the histochemical localization of catecholamine in brown adipose tissue. Histochem J 1985; 17:1279-1288.

27. Mukherjee S, Lever JD, Norman D et al. A comparison of the effects of 6-hydroxydopamine and reserpine on noradrenergic and peptidergic nerves in rat brown adipose tissue. J Anat 1989; 167:189-193.

28. Lundberg JM, Terenius L, Hokfelt T et al. Neuropeptide Y (NPY)-like immunoreactivity in peripheral noradrenergic neurons and effects of NPY on sympathetic function. Acta Physiol Scand 1982; 116:477-480.

29. Cannon B, Nedergaard J, Lundberg JM et al. Neuropeptide tyrosine (NPY) is costored with noradrenaline in vascular but not in parenchymal sympathetic nerves of brown adipose tissue. Exp Cell Res 1986; 164:546-550.

30. Norman D, Mukherjee S, Symons D et al. Neuropeptides in interscapular and perirenal brown adipose tissue in the rat: A plurality of innervation. J Neurocytol 1988; 17:305-311.

31. Lever JD, Mukherjee S, Norman D et al. Neuropeptide and noradrenaline distributions in rat interscapular brown fat and in its intact and obstructed nerves of supply. J Auton Nerv Syst 1988; 25:15-25.

32. Wantanabe J, Mishiro K, Amatsu T et al. Absence of paravascular nerve projection and cross-innervation in interscapular brown adipose tissue of mice. J Auton Nerv Syst 1994; 49:269-276.

33. Himms-Hagen J, Cui J, Sigurdson SL. Sympathetic and sensory nerves in control of growth of brown adipose tissue: Effects of denervation and of capsaicin. Neurochem Int 1990; 17:271-279.

34. Melnyk A, Himms-Hagen J. Leanness in capsaicin-desensitized rats one year after treatment. Int J Ob 1994; 18(Suppl. 2):130.

35. Jansco G, Kiraly E, Jansco-Gabor A. Direct evidence for an axonal site of action of capsaicin. Nauyn-Schmiedeberg's Arch Pharmacol 1980; 31:91-94.

36. Jansco G, Kiraly E, Joo F et al. Selective degeneration by capsaicin of a subpopulation of primary sensory neurons in the adult rat. Neurosci Lett 1985; 59:209-214.

37. Seydoux J, Mory G, Girardier L. Short-lived denervation of brown adipose tissue of the rat induced by chemical sympathetic denervation. J Physiol 1981; 77:1017-1022.

38. Depocas F, Foster DO, Zaror-Behrens G et al. Recovery of function in sympathetic nerves of interscapular brown adipose tissue of rats treated with 6-hydroxydopamine. Can J Physiol Pharmacol 1984; 62:1327-1332.

39. Desautels M, Dulos RA. Role of neural input in photoperiod-induced changes in hamster brown adipose tissue. Can J Physiol Pharmacol 1990; 68:677-681.

40. Hamilton JM, Bartness TJ, Wade GN. Effects of norepinephrine and denervation on brown adipose tissue in Syrian hamsters. Am J Physiol 1989; 257:R396-R404.

41. Meywirth A, Redlin U, Steinlechner S et al. Role of the sympathetic innervation in the cold-induced activation of 5'-deiodinase in brown adipose tissue of the Djungarian hamster. Can J Physiol Pharmacol 1990; 69:1896-1900.

42. Seydoux J, Constantinidis H, Tsacopoulos H et al. In vitro study of the control of the metabolic activity of brown adipose tissue by the sympathetic nervous system. J Physiol (Paris) 1977; 73:985-996.

43. Trayhurn P, Milner RE. A commentary on the interpretation of in vitro biochemical measures of brown adipose tissue thermogenesis. Can J Physiol Pharmacol 1989; 67:811-819.

44. Desautels M, Dulos RA, Mozaffari B. Selective loss of uncoupling protein from mitochondria of surgically denervated brown adipose tissue of cold-acclimated mice. Biochem Cell Biol 1986; 64:1125-1134.

45. Rothwell NJ, Stock MJ. Effects of denervating brown adipose tissue on the responses to cold, hyperphagia and noradrenaline treatment in the rat. J Physiol 1984; 355:457-463.

46. Gong TWL, Horwitz BA, Stern JS. The effects of 2-deoxy-D-glucose and sympathetic denervation on brown fat GDP binding in Sprague-Dawley rats. Life Sci 1990; 46:1037-1044.

47. Himms-Hagen J. Brown adipose tissue and cold-acclimation. In: Trayhurn P, Nicholls DG, editors. Brown Adipose Tissue. London: Arnold, 1986: 214-268.

48. Foster DO, Frydman ML. Nonshivering thermogenesis in the rat II. Measurements of blood flow with microspheres point to brown adipose tissue as the dominant site of the calorigenesis induced by noradrenaline. Can J Physiol Pharmacol 1978; 56:110-122.

49. Foster DO. Quantitative role of brown adipose tissue in thermogenesis. In: Trayhurn P, Nicholls DG, eds. Brown Adipose Tissue. London: Edward Arnold, 1986:31-51.

50. Foster DO, Frydman ML. Tissue distribution of cold-induced thermogenesis in conscious warm- or cold-acclimated rats reevaluated from changes in tissue blood flow: The dominant role of brown adipose tissue in the replacement of shivering by nonshivering thermogenesis. Can J Physiol Pharmacol 1979; 57:257-270.

51. Denjean F, Lachuer J, Geloen A et al. Differential regulation of uncoupling protein-1, -2, and -3 gene expression by sympathetic innervation in brown adipose tissue of thermoneutral or cold-exposed rats. FEBS Lett 1999; 444:181-185.

52. Silva JE, Larsen PR. Adrenergic activation of triiodothyronine production in brown adipose tissue. Nature 1983; 305:712-713.

53. Park IRA, Himms-Hagen J. Neural influences on trophic changes in brown adipose tissue during cold acclimation. Am J Physiol 1988; 255:R874-R881.

54. Bukowiecki LJ, Collet AJ, Follea N et al. Brown adipose tissue hyperplasia: A fundamental mechanism of adaptation to cold and hyperphagia. Am J Physiol 1982; 242:E353-E359.

55. Bukowiecki LJ, Geloen A, Collet AJ. Proliferation and differentiation of brown adipocytes from interstitial cells during cold acclimation. Am J Physiol 1986; 250:C880-C887.

56. Geloen A, Collet AJ, Guay G et al. Beta-adrenergic stimulation of brown adipocyte proliferation. Am J Physiol 1988; 254:C175-C182.

57. Geloen. A, Collet AJ, Bukowiecki LJ. Role of sympathetic innervation in brown adipocyte proliferation. Am J Physiol 1992; 263:R1176-R1181.

58. Asano A, Morimatsu M, Nikami H et al. Adrenergic activation of vascular endothelial growth factor mRNA expression in rat brown adipose tissue: Implication in cold-induced thermogenesis. Biochm J 1997; 328:179-183.

59. Takahashi A, Shimazu T, Maruyama Y. Importance of sympathetic nerves for the stimulatory effect of cold exposure on glucose utilization in brown adipose tissue. Jpn J Physiol 1992; 42:653-664.

60. Tsukazaki K, Nikami H, Shimizu Y et al. Chronic administration of beta-adrenergic agonists can mimic the stimulative effect of cold exposure on protein synthesis in rat brown adipose tissue. J Biochem (Tokyo) 1995; 117:96-100.

61. Shimizu Y, Nikami H, Saito M. Sympathetic activation of glucose utilization in brown adipose tissue in rats. J Biochem 1991; 110:688-692.

62. Bocker H, Steinlechner S, Heldmaier G. Complete cold substitution of noradrenaline-induced thermogenesis in the Djungarian hamster, Phodopus sungorus. Experientia 1982; 38:262-262.

63. Bartness TJ, Billington CJ, Levine AS et al. Insulin and metabolic efficiency in rats. I: Effect of sucrose feeding and BAT axotomy. Am J Physiol 1986; 251:R1109-R1117.

64. Geloen A, Trayhurn P. Regulation of the level of uncoupling protein in brown adipose tissue by insulin requires the mediation of the sympathetic nervous system. FEBS Lett 1990; 267:265-267.

65. Rothwell NJ, Stock MJ. A role for insulin in the diet-induced thermogenesis of cafeteria-fed rats. Metabolism 1981; 30:673-678.

66. Doi K, Kuroshima A. Modified metabolic responsiveness to glucagon in cold-acclimated and heat-acclimated rats. Life Sci 1982; 30:785-791.

67. Howland RJ. Acute cold exposure increases the glucagon sensitivity of thermogenic metabolism in the rat. Experientia 1986; 42:162-163.

68. Billington CJ, Briggs JE, Link JG et al. Glucagon in physiological concentrations stimulates brown fat thermogenesis in vivo. Am J Physiol 1991; 261:R501-R507.

69. Billington CJ, Bartness TJ, Briggs J et al. Glucagon stimulation of brown adipose tissue growth and thermogenesis. Am J Physiol 1987; 252:R160-R165.

70. Newsholme EA, Leech AR. Biochemistry for the Medical Sciences. Chichester: John Wiley, 1983.

71. Friedman JM. Leptin, leptin receptors, and the control of body weight. Nutr Rev 1998; 56(2 Pt 2):S38-S46.

72. Tsuruo Y, Sato I, Iida M et al. Immunohistochemical detection of the *ob* gene product (leptin) in rat white and brown adipocytes. Horm Metab Res 1996; 28(12):753-755.

73. Cinti S, Frederich RC, Zingaretti MC et al. Immunohistochemical localization of leptin and uncoupling protein in white and brown adipose tissue. Endocrinology 1997; 138:797-804.

74. Dessolin S, Schalling M, Champigny O et al. Leptin gene is expressed in rat brown adipose tissue at birth. FASEB J 1997; 11:382-387.

75. Trayhurn P, Duncan JS, Rayner DV. Acute cold-induced suppression of *ob* (obese) gene expression in white adipose tissue of mice: mediation by the sympathetic nervous system. Biochm J 1995; 311:729-733.

76. Scarpace PJ, Matheny M, Pollock BH et al. Leptin increases uncoupling protein expression and energy expenditure. Am J Physiol 1997; 273:E226-E230.

77. Mizuno A, Murakami T, Otani S et al. Leptin affects pancreatic endocrine functions through the sympathetic nervous system. Endocrinology 1998; 139:3863-3870.

78. Scarpace PJ, Matheny M. Leptin induction of UCP1 gene expression is dependent on sympathetic innervation. Am J Physiol 1998; 275:E259-E264.

79. Granneman JG, Campbell RG. Effects of sucrose feeding and denervation on lipogenesis in brown adipose tissue. Metabolism 1984; 33(3):257-261.

80. Desautels M, Dulos RA. Importance of neural input and thyroid hormones in the control of brown fat atrophy in mice. Can J Physiol Pharmacol 1990; 68:1100-1107.

81. Young JB, Landsberg L. Suppression of the sympathetic nervous system during fasting. Science 1977; 196:1473-1475.

82. Cannon WB. The James-Lange theory of emotions: A critical examination and an alternative. Am J Psych 1927; 29:444-454.

83. Janig W, McLachlan EM. Characteristics of function-specific pathways in the sympathetic nervous system. TINS 1992; 15:475-481.

84. Ritter S, Dinh TT. Capsaicin-induced neuronal degeneration: Silver impregnation of cell bodies, axons, and terminals in the central nervous system of the adult rat. J Comp Neurol 1988; 271:79-90.

85. Cui J, Zaror-Behrens G, Himms-Hagen J. Capsaicin desensitization induces atrophy of brown adipose tissue in rats. Am J Physiol 1990; 259:R324-R332.

86. Cui J, Himms-Hagen J. Long-term decrease in body fat and in brown adipose tissue in capsaicin-desensitized rats. Am J Physiol 1992; 262:R568-R573.

87. Cui J, Himms-Hagen J. Rapid but transient atrophy of brown adipose tissue in capsaicin-desensitized rats. Am J Physiol 1992; 262:R562-R567.

88. Osaka T, Kobayashi A, Namba Y et al. Temperature- and capsaicin-sensitive nerve fibers in brown adipose tissue attenuate thermogenesis in the rat. Pflugers Arch 1998; 437:36-42.

89. Seydoux J, Tribollet ER, Bouillaud F. Effectiveness of surgical denervation of interscapular brown adipose tissue in the rat: Further observations. In: Hales JRS, ed. Thermal Physiology. New York: Raven Press, 1984:197-199.

90. Giradier L, Seydoux J. Neural control of brown adipose tissue. In: Trayhurn P, Nicholls DG, editors. Brown Adipose Tissue. London: Edward Arnold, 1986:122-151.

91. Bamshad M, Song CK, Bartness TJ. CNS origins of the sympathetic nervous system outflow to brown adipose tissue. Am J Physiol 1999; 276:R1569-R1578.

92. Strack AM, Sawyer WB, Hughes JH et al. A general pattern of CNS innervation of the sympathetic outflow demonstrated by transneuronal pseudorabies viral infections. Brain Res 1989; 491:156-162.

93. Strack AM, Sawyer WB, Platt KB et al. CNS cell groups regulating the sympathetic outflow to adrenal gland as revealed by transneuronal cell body labeling with pseudorabies virus. Brain Res 1989; 491:274-296.

94. Rotto-Percelay DM, Wheeler JG, Osorio FA et al. Transneuronal labeling of spinal interneurons and sympathetic preganglionic neurons after pseudorabies virus injections in the rat medial gastrocnemius muscle. Brain Res 1992; 574:291-306.

95. Schramm LP, Strack AM, Platt KB et al. Peripheral and central pathways regulating the kidney: a study using pseudorabies virus. Brain Res 1993; 616:251-262.

96. Card JP, Rinaman L, Schwaber JS et al. Neurotropic properties of pseudorabies virus: uptake and transneuronal passage in the rat central nervous system. J Neurosci 1990; 10(6):1974-1994.

97. Strack AM, Loewy AD. Pseudorabies virus: A highly specific transneuronal cell body marker in the sympathetic nervous system. J Neurosci 1990; 10(7):2139-2147.

98. Rinaman L, Card JP, Enquist LW. Spatiotemporal responses of astrocytes, ramified microglia, and brain macrophages to central neuronal infection with pseudorabies virus. J Neurosci 1993; 13(2):685-702.

99. Hoffman GE, Smith MS, Verbalis JG. c-Fos and related immediate early gene products as markers of activity in neuroendocrine systems. Front Neuroendocrinol 1993; 14:173-213.

100. Miyata S, Ishiyama M, Shido O et al. Central mechanism of neural activation with cold acclimation of rats using Fos immunohistochemistry. Neurosci Res 1995; 22:209-218.

101. Joyce MP, Barr GA. The appearance of Fos protein-like immunoreactivity in the hypothalamus of developing rats in response to cold ambient temperatures. Neurosci 1992; 49:163-173.

102. Kiyohara T, Miyata S, Nakamura T et al. Differences in Fos expression in the rat brains between cold and warm ambient exposures. Brain Res Bull 1995; 38:193-201.

103. Bonaz B, Tache Y. Induction of Fos immunoreactivity in the rat brain after cold-restraint induced gastric lesions and fecal excretion. Brain Res 1994; 652:56-64.

104. Ozaki N, Sugiura Y, Yamamoto M et al. Induction of Fos protein expression in spinal cord neurons by herpes simplex virus infections in the mouse. Neurosci Lett 1996; 216:61-64.

105. Egawa M, Yoshimatsu H, Bray GA. Preoptic area injection of corticotropin-releasing hormone stimulates sympathetic activity. Am J Physiol 1990; 259:R799-R806.

106. Sakaguchi T, Bray GA. Effect of norepinephrine, serotonin and tryptophan on the firing rate of sympathetic nerves. Brain Res 1989; 492:271-280.

107. Sakaguchi T, Bray GA, Eddlestone G. Sympathetic activity following paraventricular or ventromedial hypothalamic lesions in rats. Brain Res Bull 1988; 20:461-465.

108. Shiraishi T, Sasaki K, Niijima A et al. Leptin effects on feeding-related hypothalamic and peripheral neuronal activities in normal and obese rats. Nutrition 1999; 15:576-579.

109. Yoshimatsu H, Egawa M, Bray GA. Sympathetic nerve activity after discrete hypothalamic injections of L-glutamate. Brain Res 1993; 601:121-128.

110. Morrison SF. Differential control of sympathetic outflow to brown adipose tissue and to the splanchnic bed by the raphe pallidus (RPa) and the rostral ventrolateral medulla (RVLM). Society for Neuroscience 23, 1997:839-839.

111. Morrison SF, Sved AF, Passerin AM. GABA-mediated inhibition of raphe pallidus neurons regulates sympathetic outflow to brown adipose tissue. Am J Physiol 1999; 276(R290):R297.

112. Morrison SF. RVLM and raphe differentially regulate sympathetic outflows to splanchnic and brown adipose tissue. Am J Physiol 1999; 276:R962-R973.

113. Jansen ASP, Hoffman JL, Loewy AD. CNS sites involved in sympathetic and parasympathetic control of the pancreas: A viral tracing study. Brain Res 1997; 766:29-38.

114. Bamshad M, Aoki VT, Adkison MG et al. Central nervous system origins of the sympathetic nervous system outflow to white adipose tissue. Am J Physiol 1998; 275:R291-R299.

115. Gold RM. Hypothalamic obesity: the myth of the ventromedial nucleus. Science 1973; 182:488-490.

116. Bates MW, Nauss SF, Hagman NC et al. Fat metabolism in three forms of experimental obesity: Body composition. Am J Physiol 1955; 180:301-309.

117. Brobeck JR, Tepperman J, Long CNH. Experimental hypothalamic hyperphagia in the albino rat. Yale J Biol Med 1943; 15:831-853.

118. Luiten PGM, ter Horst GJ, Karst H et al. The course of paraventricular hypothalamic efferents to autonomic structures in medulla and spinal cord. Brain Res 1985; 329:374-378.

119. Kirchgessner AL, Sclafani A. Histochemical identification of a PVN-hindbrain feeding pathway. Physiol Behav 1988; 42:529-543.

120. Holt SJ, Wheal HV, York DA. Hypothalamic control of brown adipose tissue in Zucker lean and obese rats. Effect of electrical stimulation of the ventromedial nucleus and other hypothalamic centres. Brain Res 1987; 405:227-233.

121. Minokoshi Y, Saito M, Shimazu T. Sympathetic denervation impairs responses of brown adipose tissue to VMH stimulation. Am J Physiol 1986; 251:R1005-R1008.

122. Shimazu T, Sudo M, Minokoshi Y et al. Role of the hypothalamus in insulin-independent glucose uptake in peripheral tissues. Brain Research Bulletin 1991; 27(3/4):501-504.

123. Sudo M, Minokoshi Y, Shimazu T. Ventromedial hypothalamic stimulation enhances peripheral glucose uptake in anesthetized rats. Am J Physiol Endocrinol Metab 1991; 261:E298-E303.

124. Woods AJ, Stock MJ. Biphasic brown fat temperature responses to hypothalamic stimulation in the rat. Am J Physiol 1994; 266(R328):R337.

125. Woods AJ, Stock MJ. Inhibition of brown fat activity during hypothalamic stimulation in the rat. Am J Physiol 1996; 270:R605-R613.

126. Thornhill JA, Halvorson I. Electrical stimulation of the posterior and ventromedial hypothalamic nuclei causes specific activation of shivering and nonshivering thermogenesis. Can J Physiol Pharmacol 1994; 72:96.

127. Silverman A-J, Zimmerman EA, Gibson MJ et al. Implantation of normal fetal preoptic area into hypogonadal mutant mice: temporal relationships of the growth of gonadotropin-releasing hormone neurons and the development of the pituitary/testicular axis. Neurosci 1985; 16:69-84.

128. Coscina DV, Chambers JW, Park I et al. Impaired diet-induced thermogenesis in brown adipose tissue from rats made obese with parasagittal hypothalamic knife-cuts. Brain Research Bulletin 1985; 14:585-593.

129. De Luca B, Monda M, Amaro S et al. Lack of diet-induced thermogenesis following lesions of paraventricular nucleus in rats. Physiol Behav 1989; 46(4):685-691.

130. Freeman PH, Wellman PJ. Brown adipose tissue thermogenesis induced by low level electrical stimulation of hypothalamus in rats. Brain Research Bulletin 1987; 18(1):7-11.

131. Amir S. Stimulation of the paraventricular nucleus with glutamate activates interscapular brown adipose tissue thermogenesis in rats. Brain Res 1990; 508:152-155.

132. Holt SJ, Wheal HV, York DA. Response of brown adipose tissue to electrical stimulation of hypothalamus centres in intact and adrenalectomized Zucker rats. Neurosci Lett 1988; 84:63-67.

133. Saito M, Minokoshi Y, Shimazu T. Accelerated norepinephrine turnover in peripheral tissues after ventromedial hypothalamic stimulation in rats. Brain Res 1989; 481:298-303.

134. Amir S. Retinohypothalmic tract stimulation activates thermogenesis in brown adipose tissue in the rat. Brain Res 1989; 503:163-166.

135. Amir S, Shizgal P, Rompre P-P. Glutamate injection into the suprachiasmatic nucleus stimulates brown fat thermogenesis in the rat. Brain Res 1989; 498:140-144.

136. Bamshad M, Demas GE, Bartness TJ. Does white adipose tissue (WAT) receive photic information directly from projections of retinal ganglion cells? Society of Neuroscience Abstracts 24(Pt1), 1998:370.

137. Bamshad M, Aoki VT, Bartness TJ. Sympathetic nervous system (SNS) outflow from the CNS to brown adipose tissue (BAT): A retrograde study using the transneuronal pseudorabies virus (PRV). Society of Neuroscience Abstracts 23(Pt1), 1997:515.

138. Shi H, Bamshad M, Demas GE et al. Neurochemical phenotypes of CNS neurons that participate in the sympathetic nervous system (SNS) innervation of white adipose tissue (WAT). Society of Neuroscience Abstracts 25(Pt1), 1999:79.

139. Card JP, Whealy ME, Robbins AK et al. Two α-herpesvirus strains are transported differentially in the rodent visual system. Neuron 1991; 6:957-969.

140. Larsen PJ, Enquist LW, Card JP. Characterization of the multisynaptic neuronal control of the rat pineal gland using viral transneuronal tracing. Eur J Neurosci 1998; 10:128-145.

141. Ueyama T, Krout KE, Nguyen XV et al. Suprachiasmatic nucleus: A central autonomic clock. Nat Neurosci 1999; 12:1051-1053.

142. Buijs RM, Wortel J, Van Heerikhuize JJ et al. Anatomical and functional demonstration of a multisynaptic suprachiasmatic nucleus adrenal (cortex) pathway. Eur J Neurosci 1999; 11:1535-1544.

143. Grill WM, Erokwu BO, Hadziefendic S et al. Extended survival time following pseudorabies virus injection labels the suprachiasmatic neural network controlling the bladder and urethra in the rat. Neurosci Lett 1999; 270:63-66.

144. Vrang N, Mikkelsen JD, Larsen PJ. Direct link from the suprachiasmatic nucleus to hypothalamic neurons projecting to the spinal cord: A combined tracing study using cholera toxin subunit B and Phaseolus vulgaris-leucoagglutinin. Brain Res Bull 1997; 44:671-680.

145. Warren WS, Champney TH, Cassone VM. The suprachiasmatic nucleus controls the circadian rhythm of heart rate via the sympathetic nervous system. Physiol Behav 1994; 55:1091-1099.

146. Warren WS, Champney TH, Cassone VM. The hypothalamic suprachiasmatic nucleus regulates circadian changes in norepinephrine turnover in sympathetically innervated tissue. Society of Neuroscience Abstracts 21, 1995:178.

147. Ruby NF, Ibuka N, Barnes BM et al. Suprachiasmatic nuclei influence torpor and circadian temperature rhythms in hamsters. Am J Physiol 1989; 257:R210-R215.

148. Thornhill J, Jugnauth A, Halvorson I. Brown adipose tissue thermogenesis evoked by medial preoptic stimulation is mediated via the ventromedial hypothalamic nucleus. Can J Physiol Pharmacol 1994; 72:1042-1048.

149. Shibata M, Benzi RH, Seydoux J et al. Hyperthermia induced by prepontine knife-cut: evidence for a tonic inhibition of nonshivering thermogenesis in anesthetized rat. Brain Res 1987; 436:273-282.

150. Shibata M, Iriki M, Arita J et al. Procaine microinjection into the lower midbrain increases brown fat and body temperatures in anesthetized rats. Brain Res 1996; 716(1-2):171-179.

151. Benzi RH, Shibata M, Seydoux J et al. Prepontine knife cut-induced hyperthermia in the rat. Eur J Physiol 1988; 411:593-599.

152. Woods SC, Seeley RJ, Porte DJr et al. Signals that regulate food intake and energy homeostasis. Science 1998; 280:1378-1383.

153. Abe M, Saito M, Shimazu T. Neuropeptide Y in the specific hypothalamic nuclei of rats treated neonatally with monosodium glutamate. Brain Res Bull 1990; 24:289-291.

154. Bai FL, Yamano M, Shiotani Y et al. An arcuato-paraventricular and -dorsomedial hypothalamic neuropeptide Y-containing neurons in the rat hypothalamus. Brain Res 1985; 331:172-175.

155. Billington CJ, Briggs JE, Grace M et al. Effects of intracerebroventricular injection of neuropeptide Y on energy metabolism. Am J Physiol 1991; 260:R321-R327.

156. Billington CJ, Briggs JE, Harker S et al. Neuropeptide Y in hypothalamic paraventricular nucleus: a center coordinating energy metabolism. Am J Physiol 1994; 266:R1765-R1770.
157. Elias CF, Lee C, Kelly J, Aschkenasi C et al. Leptin activates hypothalamic CART neurons projections to the spinal cord. Neuron 1998; 21:1375-1385.
158. Kanarek RB, Meyers J, Meade RG et al. Juvenile-onset obesity and deficits in caloric regulation in MSG-treated rats. Pharmacol Biochem Behav 1979; 10:717-721.
159. Tsukahara F, Uchida Y, Ohba K et al. The effect of acute cold exposure and norepinephrine on uncoupling protein gene expression in brown adipose tissue of monosodium glutamate-obese mice. Jpn J Physiol 1998; 77:247-249.
160. Rehorek A, Kerecsen L, Muller F. Measurement of tissue catecholamines of obese rats by liquid chromatography and electrochemical detection. Biomed Biochim Acta 1987; 46:823-827.
161. Morris MJ, Tortelli CF, Filippis A et al. Reduced BAT function as a mechanism for obesity in the hypophagic, neuropeptide Y deficient monosodium glutamate-treated rat. Regul Pept 1998; 75-76:441-447.
162. Minokoshi Y, Haque MS, Shimazu T. Microinjection of leptin into the ventromedial hypothalamus increases glucose uptake in peripheral tissues in rats. Diabetes 1999; 48:287-291.
163. Woods AJ, Stock MJ. Leptin activation in hypothalamus. Nature 1996; 381:745-745.
164. Klingenspor M, Ebbinghaus C, Hulshorst G et al. Multiple regulatory steps are involved in the control of lipoprotein lipase activity in brown adipose tissue. J Lipid Res 1996; 37:1685-1695.
165. Klingenspor M, Meywirth A, Stohr S et al. Effect of unilateral surgical denervation of brown adipose tissue on uncoupling protein mRNA level and cytochrome-c-oxidase activity in the Djungarian hamster. J Comp Physiol [B] 1994; 163:664-670.
166. Nedergaard J. Catecholamine sensitivity in brown fat cells from cold-acclimated hamsters and rats. Am J Physiol 1982; 242:C250-C257.
167. Desautels M, Dulos RA. Is adrenergic innervation essential for maintenance of UCP in hamster BAT mitochondria? Am J Physiol 1988; 254:R1035-R1042.
168. Triandafillou J, Hellenbrand W, Himms-Hagen J. Trophic response of hamster brown adipose tissue: Roles of norepinephrine and pineal gland. Am J Physiol 1984; 247:E793-E799.
169. McElroy JF, Wade GN. Short photoperiod stimulates brown adipose tissue growth and thermogenesis but not norepinephrine turnover in Syrian hamsters. Physiol Behav 1986; 37:307-311.
170. Hamilton JM, Mason PW, McElroy JF et al. Dissociation of sympathetic and thermogenic activity in brown fat of Syrian hamsters. Am J Physiol 1986; 250:R389-R395.
171. Trayhurn P, Howe A. Noradenaline turnover in brown adipose tissue and the heart of fed and fasted golden hamsters. Can J Physiol Pharmacol 1989; 67:106-109.
172. Trayhurn P, Wusteman MC. Apparent dissociation between sympathetic activity and brown adipose tissue thermogenesis during pregnancy and lactation in golden hamsters. Can J Physiol Pharmacol 1987; 65:2396-2399.
173. Youngstrom TG, Bartness TJ. Catecholaminergic innervation of white adipose tissue in the Siberian hamster. Am J Physiol 1995; 268:R744-R751.
174. ter Horst GJ, Hautvast RWM, De Jongste MJL et al. Neuroanatomy of cardiac activity-regulating circuitry: A transneuronal retrograde viral labeling study in the rat. Eur J Neurosci 1996; 8(10):2029-2041.
175. Smith JE, Jansen ASP, Gilbey MP et al. CNS cell groups projecting to sympathetic outflow of tail artery: Neural circuits involved in heat loss in the rat. Brain Res 1998; 786:153-164.
176. Jones DD, Ramsay TG, Hausman GJ et al. Norepinephrine inhibits rat preadipocyte proliferation. Int J Obesity 1992; 16:349-354.

CHAPTER 7

Central Nervous System Innervation of White Adipose Tissue

Timothy J. Bartness, Gregory E. Demas, C. Kay Song

The incessant demand for energy by tissues, especially the central nervous system (CNS), is a challenge that must be met for normal physiological functioning and behavior to occur. This is an especially difficult task for small rodents because of their increased energy expenditure as a result of their high metabolic rates and because of their increased heat loss due to their high surface-to-volume ratio.[1] When this demand for energy exceeds the calories available from recently ingested food, and from the relatively small carbohydrate stores (i.e., glycogen), the needed energy is liberated from white adipose tissue (WAT), the principal energy depot, through the process of lipolysis (i.e., breakdown of triglycerides). It now appears that the direct sympathetic nervous system (SNS) innervation of WAT, rather than the indirect SNS control of WAT via the secretion of adrenal medullary catecholamines into the circulation, may be the predominate trigger for lipolysis in adipocytes (for review see 2). Thus, the role of the innervation of WAT by the SNS and sensory nerves is the major focus of this review. In addition, a more recently discovered role of the SNS innervation of WAT, the control of adipocyte proliferation (i.e., fat cell number)[3,4] also will be reviewed. Because we are unaware of any convincing data for the parasympathetic innervation of WAT, this topic will not be discussed.[5,6]

Note that the neuroanatomy of the innervation of WAT, as well as the SNS control of leptin secretion by WAT will be discussed here, but also will be discussed in greater detail in the chapters by S. Cinti and P. Trayhurn in this volume. In addition, there will be little discussion of adrenergic receptor and function in this review, but the interested reader should see reviews of this area by M. Lafontan (see e.g., 7-9). Finally, this review of the innervation of WAT will not be exhaustively comprehensive, but such a review that spans the literature from 1898 to 1998 has been published recently.[2]

Innervation of WAT

Free fatty acids (FFAs) are liberated from WAT adipocytes via lipolysis under conditions where energy needs cannot be met by circulating fuels or stored carbohydrate. Lipolysis occurs in WAT when lipolytic neurotransmitters such as norepinephrine (NE) or epinephrine (EPI), or hormones (e.g., glucagon[10,11]) bind to their receptors, ultimately activating hormone-sensitive lipase that breaks triglycerides down to glycerol and FFAs.[12] Traditionally, then, the focus has been on the mediation of lipolysis by humoral substances.

There has been a growing amount of support that the SNS innervation of WAT, rather than adrenal medullary catecholamine secretion, is the primary stimulus for WAT lipolysis. For example, cold exposure increases WAT norepinephrine turnover (i.e., NETO or sympathetic

Adipose Tissues, edited by Susanne Klaus. ©2001 Eurekah.com.

drive) and circulating FFA concentrations that still occur despite adrenal demedullation.[13, 14] The results of this and other findings (for review see ref. 2) have been bolstered by more recent discoveries. For example, fasting increases WAT, but not BAT, NETO.[15] Despite these and other findings (for review see ref. 2), the significance of the role of the SNS innervation of WAT has been debated for over a hundred years.[2] The principal reason for the continuance of this debate was that, until recently (see below, ref. 16), there was no direct neuroanatomical evidence for the innervation of WAT by SNS postganglionic or more rostral sites (e.g., brainstem or hypothalamus).

Neuroanatomy at the Level of the Fat Pad

The innervation of WAT at the level of the fat pad also is discussed in this volume (see Chapter 2). Dogiel reported the first histological evidence of the SNS innervation of WAT over 100 years ago staining the nerves with methylene blue.[17] Despite that early beginning, we still have little specific information regarding the exact types of nerve contacts in this tissue (unlike that for BAT; see Chapter 6 of this volume). That is, the innervation of WAT could take one, all or a subset of three forms: 1) innervation of adipocytes by neural processes forming presynaptic units, not unlike the motor innervation of skeletal muscle, 2) en passant innervation of adipocytes, with a lesser developed presynaptic structure, or 3) innervation of the vasculature of WAT only. Previous studies of WAT using electronmicroscopy yielded mixed results with some indication of direct or en passant innervation, whereas only vascular innervation is seen in other studies.[19, 20, 21] Clearly, this remains an unanswered question and may require a means of making the nerve endings more salient at the electronmicroscopic level (e.g., application of the anterograde tract tracer *Phaseolus vulgaris-leucoagglutinin* to the sympathetic chain combined with electronmicroscopic analysis of WAT).

Is it important to know the exact type of SNS innervation to know whether it is an important physiological controller of lipolysis? We think not. Each of the three classes of innervation could result in enhanced WAT lipolysis as the result of an increased SNS drive on the tissue. The first two possibilities seem relatively straightforward, whereas only SNS innervation of the vasculature requires some explanation. Rosell suggests that SNS neurons innervating the epithelial lining of the vasculature affect the SNS-induced increases in lipolysis indirectly.[22] The capillary filtration coefficient increases with electrical stimulation of the nerves innervating the inguinal subcutaneous WAT pad in dogs in situ, due to increases in capillary membrane permeability.[23] Thus, although whole organ studies show that SNS stimulation increases peripheral resistance due to constriction of the arterioles in the vascular bed of WAT, the blood pressure within the pads does not change because vascular permeability increases through SNS-induced increases in pore size.[23] These increases in pore size may allow large molecules, such as albumin, partially to penetrate or become accessible to the intracellular space. FFAs are insoluble in water and therefore require a carrier, such as albumin, for transport in blood after their release from white adipocytes due to lipolysis.[12] Therefore, the ready acceptance of FFAs by albumin facilitates their removal thus permitting high rates of lipolysis by minimizing end product inhibition due to build-up of released FFAs.[22]

With the advent of the histofluorescence technique[24] the nerves innervating WAT were identified as catecholaminergic and thus of SNS origin.[6,24-29] Using this technique, it initially appeared that WAT and BAT only had vascular SNS innervation,[25] but delicate fibers making apparent contact with BAT, but not WAT, adipocytes soon were reported.[21] The catecholaminergic innervation of WAT vasculature was soon confirmed,[27] with both vascular and parenchymal innervation identified next.[6,19,28-30] The most extensive of these early studies using histofluorescence to define the SNS innervation of WAT[19] found catecholaminergic innervation of the vasculature and parenchyma of the mesenteric, epididymal (EWAT) and subcutaneous

WAT depots. Specifically, the arteries and arterioles were encapsulated by catecholaminergic neurons and the innervation was most extensive for the mesenteric WAT and EWAT pads.[19] Capillary catecholaminergic innervation was best seen in mesenteric and EWAT from fasted rats that were then refed, whereas parenchymal innervation was best seen in these same pads from rats that only were fasted. The relative ease with which parenchymal innervation could be seen in fasted animals most likely is due to decreases in adipocyte size caused by lipid mobilization and consequently a larger visible parenchymal area, as well as due to increases in NE content in the terminals of the fasted rats. At the electron microscopic level, the innervation of the vasculature, including the capillaries, was readily apparent, with only ~2-3 % of the adipocytes receiving direct innervation.[19] More recently, the most convincing evidence of the vascular and nonvascular catecholaminergic innervation of WAT was obtained by combining the histofluorescence technique with confocal microscopy.[30] Unfortunately, this is a preliminary report that never appeared in full form; however the verbal descriptions of the histology are striking. Specifically, direct catecholaminergic neural fluorescence of adipocytes was seen in EWAT, perirenal, mesenteric and inguinal (IWAT) WAT from laboratory rats, in addition to vascular innervation. Moreover, mesenteric WAT had the highest and IWAT the lowest, degree of direct catecholaminergic innervation of adipocytes,[30] a finding that correlates well with the highest and lowest rates of NE-stimulated lipolysis in vitro by isolated adipocytes from these pads.[31] The first direct neuroanatomical evidence of the innervation of WAT by the postganglionic neurons of the SNS in any species was obtained using fluorescent anterograde and retrograde neuronal tract tracers.[32] Injections of the fluorescent retrograde tract tracer FluoroGold into IWAT or EWAT pads of Siberian hamsters and laboratory rats yielded relatively separate distributions of postganglionic neurons in the sympathetic chain with the EWAT pad having a more rostral pattern of labeling compared with the IWAT pad.[32] These connections were confirmed bidirectionally by injecting the sympathetic chain with the fluorescent anterograde tract tracer, iodocarbocyanine perchlorate (DiI) yielding rings of fluorescence around each adipocyte in both pads.[32] These results supported and extended the histofluorescence evidence of SNS innervation of WAT and also offered a possible neuroanatomical basis for the differential lipid mobilization seen by WAT pads across a variety of conditions including age and anatomical location of the pads,[33] exposure to short photoperiods,[34] estrogen treatment,[35] and food deprivation,[35] as well as other manipulations. Support for the view that a relatively separate postganglionic SNS innervation of WAT underlies the differential mobilization of lipid from fat pads comes from studies of the naturally occurring decreases in body fat exhibited by short 'winter-like' photoperiod exposed Siberian hamsters.[34,36,37] Specifically, the short day-induced decrease in body fat is not uniform with the more internally located EWAT and retroperitoneal WAT (RWAT) pads showing a more rapid and pronounced decrease in lipid mass compared with the more externally located IWAT pad. Moreover, these decreases in WAT pad are accompanied by greater NETO in the more internally located pads (EWAT and RWAT), than the more externally located (IWAT) pad,[32] and thus corresponding to the faster and greater rates of lipid mobilization by the former than latter pads. Thus, this relatively separate neuroanatomical innervation of WAT pads at this level of the neuroaxis, and accompanying differential SNS drives to these adipose depots' most likely accounts for the fat pad-specific differences observed in lipid mobilization many mammals.

Sensory Innervation of WAT

An Introduction

The sensory innervation of WAT was initially suggested by the discovery of substance P in WAT, a neurotransmitter typically associated with sensory innervation.[38] The sensory innervation of WAT was first shown neuroanatomically when the anterograde tract tracer true blue labeled cells in the dorsal root ganglia after placement in the dorsal subcutaneous WAT and IWAT pads in rats.[39] These data beg the question of what is being sensed? Perhaps the sensory innervation of WAT informs the brain of the size of the peripheral lipid stores, a putative role recently given to leptin (see e.g., ref. 40), and postulated previously for pancreatic insulin (see e.g., ref. 40, 42). At this point, we can only speculate with possibilities ranging from mechanoception of fat pad expansion and contraction to monitoring of lipolytic rate via chemoception. Regardless of what is being sensed there are at least four possibilities for the role of WAT sensory nerves in the control of lipolysis. Sensory innervation could be involved in controlling lipolytic rate via:

1. the feedback control of vascular perfusion rate affecting access of lipolytic humoral substances to WAT adipocytes,
2. alterations in capillary permeability to albumin (see above),
3. a 'long' neural negative feedback loop that monitors a neurochemical corollary of lipolysis with sensory afferents terminating in the brain at loci controlling the SNS outflow from the CNS to WAT,[16] and
4. a 'short' neural feedback loop sending information about lipolytic rate to the sympathetic ganglia or the intermediolateral (IML) horn of the spinal cord to control WAT sympathetic drive relatively locally.

Support for each of these possibilities is scant at best, but the notion of a short feedback loop may have some credibility in that in another system (i.e., the SNS innervation of laboratory rat mesenteric arteries), sensory denervation produced by capsaicin results in enhanced SNS-driven vasoconstriction responses.[43] Conversely, chemical sympathectomy (guanethidine) increases substance P and calcitonin-gene related peptide (CGRP; another sensory nerve associated neurotransmitter) fiber densities in all peripheral tissues examined.

Role of Leptin

Recently, another possibility has emerged for a role of sensory nerves in WAT related to the apparent regulation of lipid stores. WAT possesses leptin receptors.[44] When these receptors are purportedly stimulated via injections of the leptin into WAT, an increase in sensory afferent nerve activity results.[45,46] Moreover, this increase in sensory input to the CNS, in turn, triggers increases in SNS efferent nerve firing rates back to WAT.[45,46] Thus, these data suggest that, inasmuch as leptin participates in the regulation of lipid stores (for review see ref. 47), which is not without debate, a negative feedback system regulates lipid stores involving leptin and the sensory and SNS innervation of WAT. If this system works properly, then obesity should not exist given the self-correcting nature of the hypothesized feedback system—of course that is not the case. This feedback system can be disrupted (e.g., lack of leptin secretion or leptin receptors, decreases SNS innervation or sympathetic drive) and results in increased adiposity. Explicit tests of the regulation of leptin secretion via itself through the sensory and SNS innervation of WAT remain to be conducted.

Trophic Influences

Finally, with regard to the sensory innervation of WAT, there may be a trophic influence of these nerves on the growth or maintenance of WAT mass. This notion comes from the results of global sensory denervation produced by systemic injections of capsaicin, the principal piquant of chili peppers that destroys unmyelnated sensory nerves.[48,49] In addition to causing an atrophy of BAT mass and impairing BAT function,[50] long-term capsaicin desensitized rats have reduced adiposity reflected as decreases in EWAT and RWAT masses and, moreover, decreases in RWAT but not EWAT fat cell number.[51] The current inability to create a sensory denervated WAT pad on one side, with the contralateral side serving as a within animal neurally intact control, has hampered our understanding of the role of the sensory innervation of WAT.

Peripheral Denervation of WAT

Effects on Lipolysis

Because each WAT pad is unilaterally innervated, one of a pair of WAT pads can be denervated with the contralateral pad serving as a within-animal neurally-intact control. Ascribing the effects of surgical denervation of WAT to the absence of SNS innervation should be done cautiously because it is impossible to selectively sever only the SNS innervation of the tissue without also cutting the sensory nerves. This caveat aside, the surgical denervation of WAT is at least permanent, whereas chemical sympathectomy (e.g., 6-hydroxydopamine [6-OHDA] or guanethidine), although selective for the SNS innervation, has only been done via systemic treatment. This results in a global sympathetic denervation that also is not permanent (e.g., ref. 52,53).

The first suggestions that the SNS innervation of WAT was important for lipid mobilization came from observations that human patients had increased fat deposition in areas without sympathetic innervation.[54] Animals with surgically denervated WAT were fasted to trigger lipolysis resulting in fat pad masses that were larger in denervated than nondenervated depots.[55-57] Because denervation of IWAT can decrease blood flow to WAT, the interpretation of these results were more complicated than they first seemed because of the possibility that access to circulating stimulators of lipolysis, such as glucagon and adrenal medullary-released catecholamines, also were affected.[56] Thus, although the results of these denervation studies suggest that lipolysis is impaired in denervated tissue the extent to which this is due solely to decreased SNS neurally stimulated lipolysis is not known.

Trophic Effects

In addition to the potentially vital role of the SNS innervation of WAT in lipolysis, sympathetic innervation also may affect WAT mass through a trophic function by affecting adipocyte proliferation. For example, fat cell number is markedly increased after surgical denervation of the RWAT pad in laboratory rats,[4] and we found a similar and even more impressive increase in fat cell number.[3] Specifically, long day-housed Siberian hamsters, that are at their annual body fat peak,[58] had their IWAT pads unilaterally denervated (contralateral pad served as intact within animal control) and then half these animals were moved to short days, where body fat reaches its annual nadir.[58] In long day- and short day-housed hamsters the denervated pad mass increased 200 % in both photoperiods and fat cell number increased ~250 and ~180 % in LDs and SDs, respectively compared with the contralateral intact pads.[3] Remarkably, this denervation-induced increase in fat cell number occurred in short days despite a naturally occurring decrease in food intake of ~30 %.[58] These findings suggest that the SNS normally inhibits the growth of fat pads, not only through changes in fat cell size via alterations in lipolysis, but also through an as yet to be determined mechanism affecting fat cell number. Support for this notion also comes from the inhibition of fat cell proliferation by NE in an in vitro cell culture system of stromal-vascular cells.[59] This trophic function of the SNS innervation of WAT may be important in the etiology of human obesity because a variety of obesities have been associated with decreases in SNS drive (for review see ref. 60). Moreover, obesity, including human obesity, is often, if not usually, associated with increases in fat cell number ('hypercellular obesity'; for review see ref. 61). This suggests that decreases in SNS drive, exaggerated to no SNS drive with denervation, may be the principal factor stimulating white adipocyte proliferation in a variety of types of obesity.

SNS Outflow from Brain to WAT

A schematic diagram of the origins of SNS innervation of WAT is presented in Figure 7.1. Although we defined the distribution of postganglionic SNS neurons projecting from the spinal cord to IWAT and EWAT, the rostral brain origins of the innervation of the preganglionic sympathetic neurons in the spinal cord that project to these postganglionic neurons, remained to be defined neuroanatomically until recently.[32.16] This was made possible by the development of a transneuronal viral tract tracer, the Bartha's K strain of the pseudorabies virus (PRV) that was used to trace the SNS outflow from brain to peripheral sympathetic targets (see e.g., 62-65). Specifically, some viruses, like the PRV, are taken up into neurons following binding to viral attachment protein molecules located on the surface of neuronal membranes. These protein surface molecules act as 'viral receptors'. Neurons synapsing on infected cells become exposed to relatively high concentrations of the virus particles that have been exocytosed. The virus particles are then taken up at synaptic contact sites and this process continues causing an infection along the neuronal chain from the periphery to higher CNS sites.[66-68] The infected neurons can be visualized easily using standard immunocytochemistry. The major advantage of using viruses as transneuronal markers over some of the more traditionally used tracers, is their ability to replicate within the neuron and thus act as a self-amplifying cell marker. Furthermore, because the transfer of the virus only is by a transsynaptic mechanism, rather than by lateral spread to adjacent but unrelated neurons or by a nonsynaptic mechanism,[66-68] the transsynaptic transfer of the PRV after injection into WAT yields a hierarchical chain of functionally connected neurons from brain to WAT. Therefore, we will review the functional studies of CNS-SNS-WAT connections after we describe the results of the only neuroanatomical description of the SNS outflow from brain to WAT.[16]

We retrogradely labeled the SNS outflow from brain to WAT (both IWAT and EWAT) in laboratory rats and Siberian hamsters.[16] Briefly, neurons comprising the CNS-SNS-WAT connections were identified throughout the neural axis. In the spinal cord sympathetic preganglionic cells were found in a well-defined cluster located in the intermediolateral cell group and the central autonomic nucleus of the spinal cord ipsilateral to the unilateral IWAT or EWAT injection site. In the brainstem, most of the classically-defined autonomic areas had PRV immunoreactivity and these were bilateral and uniform including the reticular area (gigantocellular reticular nucleus [ventral and alpha parts] lateral reticular nucleus), C1 adrenaline and A5 noradrenaline cell groups, the rostroventrolateral and rostroventromedial area, the caudal raphe area ([raphe obscurus]) and the nucleus of the solitary tract. In the midbrain the infection also was bilateral and uniform with the central gray being the most heavily infected area. In the forebrain, infections were bilateral, but a significantly greater density of labeling was seen ipsilateral to the injection site. PRV immunoreactivity was seen in hypothalamic (arcuate nucleus, dorsal and lateral hypothalamic areas, PVN, SCN, ventral premammillary nucleus, dorsomedial nuclei, and nonhypothalamic (zona incerta, medial amygdala, medial preoptic area, septum, and bed nucleus of the stria terminalis) forebrain areas. Despite the large literature suggesting that VMH manipulations affect WAT via the SNS (see below for discussion), only a few scattered neurons within the VMH, mostly on the lateral edges, became infected. The general patterns of infections for the IWAT and EWAT pads of the Siberian hamsters were more similar than different, as were the patterns of infection for the IWAT pads between Siberian hamsters and laboratory rats (EWAT was not injected in rats). Some noteworthy differences did exist in Siberian hamsters, however. Specifically, higher percentages of labeled cells within

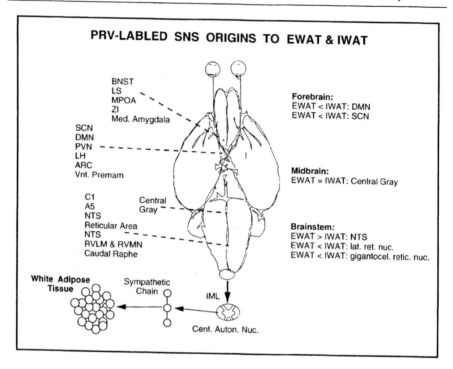

Fig. 7.1. Schematic diagram of origins of the SNS innervation of epididymal and inguinal white adipose tissue (EWAT and IWAT, respectively). Summary of major structures showing infections after injection of the transneural tract tracer, the pseudorabies virus (PRV), into EWAT and IWAT (left side). Significant differences in major structures infected with the PRV after injection into either EWAT or IWAT (right side). BNST, bed nucleus of the stria terminalis; LS, lateral septal area; MPOA, medial preoptic area; ZI, zona incerta; Med Amygdala, medial amygdaloid nucleus; SCN, suprachiasmatic nucleus; DMN, dorsomedial hypothalamic nucleus; PVN, hypothalamic paraventricular nucleus; LH, lateral hypothalamic nucleus; ARC, arcuate nucleus; Vnt. Premam, ventral premammillary; Central Gray, midbrain central gray; C1, adrenergic area; A5, noradrenergic area; Reticular Area; includes: gigantocellular reticular nucleus [ventral and alpha parts] lateral reticular nucleus; NTS, nucleus of the solitary tract; RVLM, rostral ventrolateral medullary region; RVMM, rostral ventromedial medullary area; Caudal Raphe, includes raphe obscurus; IML, intermediolateral horn of the spinal cord, Cent. auton. nuc., central autonomic nucleus of the spinal cord; Cent. Gray, central gray; lat. retic. nucleus, lateral reticular nucleus; gigantocel. retic. nuc., gigantocellular reticular nucleus [ventral and alpha parts].

brainstem occurred after EWAT virus injections than IWAT for the nucleus of the solitary tract, and were greater after IWAT virus injection compared with EWAT for the lateral reticular and gigantocellular reticular nucleus. No differences were found in the midbrain. Finally, in the forebrain, higher percentages of PRV-infected neurons occurred in the dorsomedial hypothalamic nucleus and SCN after virus injections into IWAT than EWAT. Collectively, it appears that WAT receives input from CNS cell groups that are part of the general SNS outflow from the brain to the periphery.

Electrophysiological Studies of WAT

We only are aware of three studies, all by Niijima,[45,46,69] where the electrophysiological activity of WAT nerves was measured and two of these studies were discussed above with regard to leptin. In the other study, the SNS efferents firing rates to WAT and BAT are opposite one another with several metabolic manipulations.[69] Specifically, the firing rates of the SNS nerves innervating rat EWAT decrease after added carbohydrate fuel (intravenous glucose infusions), but increase when glucose utilization is blocked (intravenous 2-deoxy-D-glucose [2DG]) or when metabolic fuel storage is enhanced (intravenous insulin infusions); moreover, opposite responses to the same stimuli are found in the SNS nerves innervating BAT.[69] It seems likely that these increases in SNS WAT nerve firing rates originate in brain sites that comprise the general SNS outflow from the brain to periphery because 2DG causes a state of energetic emergency in the CNS.[62,63,70] Furthermore, because 2DG-induced glucoprivation selectively stimulates the SNS innervation of the adrenal medulla, rather than WAT, the interpretation of these data in terms of sympathetic neural stimulation of WAT is unwarranted.[71]

CNS Manipulations Affecting WAT

An Introduction

An extensive review of the electrical and chemical stimulation, and damage of brain sites thought to affect WAT through its SNS innervation has been presented recently in light of the SNS outflow from brain to WAT.[16] Highlights of these studies, as well as additional new results will be presented briefly. Together, the results of these studies implicate a relatively small number of brain structures in the control of the SNS drive on WAT, with the VMH receiving the most attention as it was the case for BAT (see Chapter 6 in this volume).

Role of the VMH/PVN

Recall that we found few or no infected cells in the VMH after injections of PRV into WAT; therefore there appears to be little neuroanatomical support for the notion of a VMH-SNS-WAT circuitry. A similar lack of neuroanatomical support exists for previously purported VMH-SNS-peripheral tissue connections including the pancreas and adrenal medulla also demonstrated using the PRV technique.[72,62,63] Thus, we conclude that the VMH is not part of the SNS outflow to the periphery generally, or WAT specifically.[2,16,73]

One likely reason for this discrepancy between the results of the nonneuroanatomical studies (for review see ref. 2), and those of our PRV neuroanatomical study is that the targeted stimulation or destruction of the VMH most likely affected nearby brain sites known to be involved in the SNS outflow from the brain to WAT.[16] These include the PVN, retrochiasmatic area, as well as the dorsomedial, arcuate and periventricular nuclei.[62] This misplaced focus on the VMH is reminiscent of the 'myth of the VMH' that resulted originally from the misguided emphasis on these nuclei in relation to the control of body fat and food intake (see e.g., 75, 76).[74] This focus on the VMH has been subsequently redirected to other brain regions, especially the PVN and its descending projections that course just lateral to the VMH and that make direct and indirect connections with the spinal preganglionics (see e.g., 77, 78). Indeed, using the asymmetrical lesion technique where the PVN on one side and the VMH on the other is destroyed, it appears that the post lesion-induced increases in food intake and body weight involves a longitudinally projecting system consisting of cell bodies within the PVN projecting through or on the edges of the VMH nucleus.[79] Note, however, that apparent bona fide differences between the effects of PVN and VMH lesions on adiposity and food intake exist.[80-82] Nevertheless, despite the appeal of interpreting the effects of VMH manipulations to the descending PVN pathways that comprise part of the SNS outflow to WAT, we clearly have an inadequate understanding of this neurology as a forebrain pathway modulating SNS neural mediation of lipolysis in WAT. For example, unilateral parasaggital microknife cuts that transect this presumed critical descending longitudinal pathway from the PVN and other rostral forebrain structures[78] do not result in unilateral sparing of lipid mobilization in the retroperitoneal WAT pads following fasting,[83] as do unilateral VMH lesions.[84,85]

Electrical or chemical stimulation of the VMH, but not the LH, increases plasma FFA concentrations as does electrical stimulation of other brain sites including the premammillary area, SCN (see below) and the zona incerta.[86-92] This latter finding is interesting in light of the close proximity of the zona incerta to the dorsomedial nucleus of the hypothalamus, and the participation of these nuclei in the SNS outflow from brain to WAT.[16]

Role of the MPOA/POA

Injections of 2DG into the preoptic area of laboratory rats also increases plasma FFA concentrations, a finding consistent with the presence of infected neurons in the MPOA after PRV injections into WAT.[93,16] Further support for participation of the preoptic area in mobilization of lipid from WAT is that the MPOA is part of a noradrenergic circuit that is activated by the cold and then increases plasma FFA concentration.[94] This effect is blocked by precold administration of the noradrenergic beta receptor antagonist, propranolol into the area.[95]

Involvement of the Adrenal Medulla

Involvement of the adrenal medulla always is a potentially confounding influence in CNS stimulation experiments interested in WAT lipolysis. The adrenal medulla is not involved in at least some of these responses, because stimulation of some of these sites primarily increase plasma FFA, but not glucose concentrations,[96,97] and because the VMH electrical stimulation-induced increase in plasma FFA concentrations is either slightly or not diminished in adrenalectomized or adrenal demedullated animals.[96-98,87,88] Collectively, these electrical and chemical stimulation data indicate that lipolysis can be selectively triggered by stimulation of several CNS sites and that these lipid mobilization effects are most likely through the SNS innervation of WAT, rather than via adrenal medullary-released catecholamines.

Role of the SCN

Increases in plasma FFA concentration triggered by intracerebroventricular (icv) administration of 2DG are blocked by coronal microknife cuts just behind the SCN.[97] These cuts also block increases in lipid mobilization associated with 24h fasts, forced exercise (swimming), cold exposure and insulin-induced hypoglycemia.[99] Furthermore, electrical stimulation of the SCN decreases the respiratory quotient, indicative of the metabolism of lipid-derived fuel.[100] Recall that the SCN is part of the SNS outflow to peripheral tissues such as WAT, BAT and the pineal gland, a finding confirmed and extended to other tissues recently.[2,16,73,101-107] Thus, the SCN appears to be involved in triggering SNS neurally-induced lipolysis in WAT. Indeed, bilateral SCN lesions block the increases in plasma FFA and glucose concentrations, and in food intake, elicited by icv administered 2DG that normally occur during the day only.[108] This SCN-SNS-WAT circuitry may serve as a modulator of lipid mobilization, perhaps helping to account for the 24h rhythms of lipolysis/lipogenesis shown by laboratory rats and mice, given the role of the SCN in the generation of circadian rhythms and their entrainment by the photocycle.[109-111] Perhaps the SCN also is involved in the rhythmic pulsatile release of leptin, presumably from WAT, in obese and normal weight humans via this SCN-SNS-WAT circuitry, given the control of leptin secretion by the SNS.[112-115] Collectively, it may be that the SCN controls the daily cycles of lipid mobilization by rhythmically varying the SNS drive on WAT.[116]

Effects of Leptin

Note that an extensive discussion of the role of leptin and control of its expression and secretion by WAT and BAT is found in the Chapter 9 of this volume. In the context of the present discussion, it is intriguing to suggest that the increases in lipid mobilization by leptin may be mediated by the SNS innervation that is stimulated by central sites, although centrally administered leptin appears to increase NETO in BAT, but not WAT, and leptin elicits lipolysis in vitro.[117-119,44]

Summary

In summary, there is overwhelming evidence for the SNS innervation of WAT, but less extensive information on its sensory innervation. The exact contribution of the SNS innervation of WAT versus the contribution of adrenal medullary-released catecholamines in lipolysis remains to be defined. Moreover, the roles of both types of WAT innervation, as well as of adrenal medullary-released catecholamines, most likely vary across conditions of lipid mobilization. Furthermore, the CNS components of the SNS outflow from brain to WAT involved in cold-, fasting-, and exercise-induced lipolysis, as well as other conditions promoting lipid mobilization, are unexplored. Finally, the SNS innervation of WAT appears important not only in the control of lipid mobilization, but also for maintenance and growth of WAT pad mass and for secretion of leptin as well.

Acknowledgments

This research is supported, in part, from NIH research grant RO1 DK35254 and NIMH Research Scientist Development Award KO2 MH00841 to TJB.

References

1. Kleiber M. Metabolic turnover rate: A physiological meaning of the metabolic rate per unit body weight. J Theor Biol 1975; 53:199-204.
2. Bartness TJ, Bamshad M. Innervation of mammalian white adipose tissue: Implications for the regulation of total body fat. Am J Physiol 1998; 275:R1399-R1411.
3. Youngstrom TG, Bartness TJ. White adipose tissue sympathetic nervous system denervation increases fat pad mass and fat cell number. Am J Physiol 1998; 275:R1488-R1493.
4. Cousin B, Casteilla L, Lafontan M et al. Local sympathetic denervation of white adipose tissue in rats induces preadipocyte proliferation without noticeable changes in metabolism. Endocrinology 1993; 133:2255-2262.
5. Ballantyne B. Histochemical and biochemical aspects of cholinesterase activity of adipose tissue. Arc It Pharmacodyn 1968; 173:343-350.
6. Ballantyne B, Raftery AT. The intrinsic autonomic innervation of white adipose tissue. Cytobios 1974; 10:187-197.
7. Carpene C, Bousquet-Melou A, Galitzky J et al. Lipolytic effects of beta 1-, beta 2-, and beta 3-adrenergic agonists in white adipose tissue of mammals. Ann NY Acad Sci 1998; 839:186-189.
8. Lafontan M, Berlan M. Fat cell a2-adrenoceptors: The regulation of fat cell function and lipolysis. Endocrine Rev 1995; 16(6):716-738.
9. Lafontan M, Bousquet-Melou A, Galitzky J et al. Adrenergic receptors and fat cells: Differential recruitment by physiological amines and homologous regulation. Obesity Res 1995; 3(Suppl. 4):507S-514S.
10. Lefebvre P, Luyckx A, Bacq ZM. Effects of denervation on the metabolism and the response to glucagon of white adipose tissue of rats. Horm Metab Res 1973; 5:245-250.
11. Luyckx AS, Dresse A, Cession-Fossion A et al. Catecholamines and exercise-induced glucagon and fatty acid mobilization in the rat. Am J Physiol 1975; 229:376-383.

12. Newsholme EA, Leech AR. Biochemistry for the Medical Sciences. Chichester: John Wiley, 1983.
13. Garofalo MAR, Kettelhut IC, Roselino JES et al. Effect of acute cold exposure on norepinephrine turnover rates in rat white adipose tissue. J Auton Nerv Syst 1996; 60(3):206-208.
14. Gilgen A, Maickel RP. Essential role of catecholamines in the mobilization of free fatty acids and glucose after exposure to cold. Life Sci 1962; 12:709-715.
15. Migliorini RH, Garofalo MAR, Kettelhut IC. Increased sympathetic activity in rat white adipose tissue during prolonged fasting. Am J Physiol 1997; 272:R656-R661.
16. Bamshad M, Aoki VT, Adkison MG et al. Central nervous system origins of the sympathetic nervous system outflow to white adipose tissue. Am J Physiol 1998; 275:R291-R299.
17. Dogiel AS. Die sensiblen Nervenendigungen im Herzen und in den Blutgefassen der Saugethiere. Arch Mikr Anat 1898; 52:44-70.
18. Bartness TJ, Song CK, Demas GE. Central nervous system innervation of brown adipose tissue. In: Klaus S, editor. Adipose Tissue. Georgetown: Landes Bioscience, 2000.
19. Slavin BG, Ballard K. Morphological studies on the adrenergic innervation of white adipose tissue. Anat Rec 1978; 191:377-390.
20. Slavin BG. The morphology of adipose tissue. In: Cryer A, Van RLR, eds. New Perspectives in Adipose Tissue: Structure, Function and Development. London: Butterworth, 1985:23-43.
21. Wirsen C. Studies in lipid mobilization with special reference to morphological and histochemical aspects. Acta Physiol Scand 1965; 65(Suppl. 252):1-46.
22. Rosell S. Neuronal control of microvessels. Ann Rev Physiol 1980; 42:359-371.
23. Fredholm BB, Oberg B, Rosell S. Effects of vasoactive drugs on circulation in canine subcutaneous adipose tissue. Acta Physiol Scand 1970; 79:564-574.
24. Falck B, Hillarp NA, Thieme G et al.. Fluorescence of catecholamines and related compounds condensed with formaldehyde. J Histochem Cytochem 1962; 10:348-354.
25. Wirsen C. Adrenergic innervation of adipose tissue examined by fluorescence microscopy. Nature 1964; 202:913.
26. Wirsen C. Distribution of adrenergic nerve fibers in brown and white adipose tissue. In: Renold AE, Cahill GFJr, eds. Adipose Tissue. Washington,D.C.: Am Physiol Soc 1965:197-199.
27. Daniel H, Derry DM. Criteria for differentiation of brown and white fat in the rat. Can J Physiol Pharmacol 1969; 47:941-945.
28. Diculescu I, Stoica M. Fluorescence histochemical investigations on the adrenergic innervation of the white adipose tissue in the rat. J Neuro-Visceral Relations 1970; 32:25-36.
29. Ballard K, Malmfors T, Rosell S. Adrenergic innervation and vascular patterns in canine adipose tissue. Microvasc Res 1974; 8:164-171.
30. Rebuffe-Scrive M. Neuroregulation of adipose tissue: Molecular and hormonal mechanisms. Int J Obesity 1991; 15:83-86.
31. Storck R, Spitzer JA. Metabolism of isolated fat cells from various tissue sites in the rat: Influence of hemorrhagic hypotension. J Lipid Res 1974; 15:200-205.
32. Youngstrom TG, Bartness TJ. Catecholaminergic innervation of white adipose tissue in the Siberian hamster. Am J Physiol 1995; 268:R744-R751.
33. Hartman AD, Christ DW. Effect of cell size, age and anatomical location on the lipolytic response of adipocytes. Life Sci 1978; 22:1087-1096.
34. Bartness TJ, Hamilton JM, Wade GN et al. Regional differences in fat pad responses to short days in Siberian hamsters. Am J Physiol 1989; 257:R1533-R1540.
35. Krotkiewski M. The effects of estrogens on regional adipose tissue cellularity in the rat. Acta Physiol Scand 1976; 96:128-133.
36. Bartness TJ. Photoperiod, sex, gonadal steroids and housing density affect body fat in hamsters. Physiol Behav 1996; 60:517-529.
37. Bartness TJ. Short day-induced depletion of lipid stores is fat pad- and gender-specific in Siberian hamsters. Physiol Behav 1995; 58:539-550.
38. Fredholm BB. Nervous control of circulation and metabolism in white adipose tissue. In: Cryer A, Van RLR, eds. New Perspectives in Adipose Tissue: Structure, Function and Development. Boston: Butterworth, 1985:45-64.
39. Fishman RB, Dark J. Sensory innervation of white adipose tissue. Am J Physiol 1987; 253:R942-R944.

40. Campfield LA, Smith FJ, Guisez Y et al. Recombinant mouse OB protein: Evidence for a peripheral signal linking adiposity and central neural networks. Science 1995; 269:546-549.

41. Woods SC, Seeley RJ, Porte DJr et al. Signals that regulate food intake and energy homeostasis. Science 1998; 280:1378-1383.

42. Kaiyala K, Woods SC, Schwartz MW. New model for the regulation of energy balance and adiposity by the central nervous system. Am J Clin Nutr 1995; 62(Suppl.):1123S-1134S.

43. Ralevic V, Karoon P, Burnstock G. Long-term sensory denervation by neonatal capsaicin treatment augments sympathetic neurotransmission in rat mesenteric arteries by increasing levels of norepinephrine and selectively enhancing postjunctional actions. J Pharm Exp Ther 1995; 274(1):64-71.

44. Siegrist-Kaiser CA, Pauli V, Juge-Aubry CE et al. Direct effects of leptin on brown and white adipose tissue. J Clin Invest 1997; 100:2858-2864.

45. Niijima A. Afferent signals from leptin sensors in the white adipose tissue of the epididymis, and their reflex effect in the rat. J Auton Nerv Syst 1998; 73:19-25.

46. Niijima A. Reflex effects from leptin sensors in the white adipose tissue of the epididymis to the efferent activity of the sympathetic and vagus nerve in the rat. Neurosci Lett 1999; 262:125-128.

47. Campfield LA, Smith FJ, Burn P. The OB protein (leptin) pathway-a link between adipose tissue mass and central neural networks. Horm Metab Res 1996; 28:619-632.

48. Jansco G, Kiraly E, Jansco-Gabor A. Direct evidence for an axonal site of action of capsaicin. Nauyn-Schmiedeberg's Arch Pharmacol 1980; 31:91-94.

49. Jansco G, Kiraly E, Joo F et al. Selective degeneration by capsaicin of a subpopulation of primary sensory neurons in the adult rat. Neurosci Lett 1985; 59:209-214.

50. Cui J, Zaror-Behrens G, Himms-Hagen J. Capsaicin desensitization induces atrophy of brown adipose tissue in rats. Am J Physiol 1990; 259:R324-R332.

51. Cui J, Himms-Hagen J. Long-term decrease in body fat and in brown adipose tissue in capsaicin-desensitized rats. Am J Physiol 1992; 262:R568-R573.

52. Saulnier-Blache J-S, Atgie C, Carpene C et al. Hamster adipocyte alpha2-adrenoceptor changes during fat mass modifications are not directly dependent on adipose tissue norepinephrine content. Endocrinology 1990; 126:2425-2434.

53. Powley TL, Walgren MC, Laughton WB. Effects of guanethidine sympathectomy on ventromedial hypothalamic obesity. Am J Physiol 1983; 245:R408-R420.

54. Mansfeld G, Muller F. Der Einfluss der Nervensystem auf die Mobilisierung von Fett. Arch Physiol 1913; 152:61-67.

55. Beznak ABL, Hasch Z. The effect of sympathectomy on the fatty deposit in connective tissue. Quart J Exptl Physiol 1937; 27:1-15.

56. Cantu RC, Goodman HM. Effects of denervation and fasting on white adipose tissue. Am J Physiol 1967; 212:207-212.

57. Clement G. La mobilisation des glycerides de reserve chez le rat. II. Influence due systeme nerveux sympathique. Etablissement d'un procede de mesure de l'intensite de la mobilisation. Arch Sci Physiol 1950; 4:13-29.

58. Wade GN, Bartness TJ. Effects of photoperiod and gonadectomy on food intake, body weight and body composition in Siberian hamsters. Am J Physiol 1984; 246:R26-R30.

59. Jones DD, Ramsay TG, Hausman GJ et al. Norepinephrine inhibits rat preadipocyte proliferation. Int J Obesity 1992; 16:349-354.

60. Bray GA. Obesity—a state of reduced sympathetic activity and normal or high adrenal activity (the autonomic and adrenal hypothesis revisited). Int J Obesity 1990; 14(Suppl. 3):77-91.

61. Faust IM. Role of the fat cell in energy balance physiology. In: Stunkard AJ, Stellar E, editors. Eating and Its Disorders. New York: Raven Press, 1984:97-107.

62. Strack AM, Sawyer WB, Hughes JH et al. A general pattern of CNS innervation of the sympathetic outflow demonstrated by transneuronal pseudorabies viral infections. Brain Res 1989; 491:156-162.

63. Strack AM, Sawyer WB, Platt KB et al. CNS cell groups regulating the sympathetic outflow to adrenal gland as revealed by transneuronal cell body labeling with pseudorabies virus. Brain Res 1989; 491:274-296.

64. Rotto-Percelay DM, Wheeler JG, Osorio FA et al. Transneuronal labeling of spinal interneurons and sympathetic preganglionic neurons after pseudorabies virus injections in the rat medial gastrocnemius muscle. Brain Res 1992; 574:291-306.

65. Schramm LP, Strack AM, Platt KB et al. Peripheral and central pathways regulating the kidney: A study using pseudorabies virus. Brain Res 1993; 616:251-262.

66. Card JP, Rinaman L, Schwaber JS et al. Neurotropic properties of pseudorabies virus: Uptake and transneuronal passage in the rat central nervous system. J Neurosci 1990; 10(6):1974-1994.

67. Strack AM, Loewy AD. Pseudorabies virus: a highly specific transneuronal cell body marker in the sympathetic nervous system. J Neurosci 1990; 10(7):2139-2147.

68. Rinaman L, Card JP, Enquist LW. Spatiotemporal responses of astrocytes, ramified microglia, and brain macrophages to central neuronal infection with pseudorabies virus. J Neurosci 1993; 13(2):685-702.

69. Niijima A. The effect of glucose and other sugars on the efferent activity of the sympathetic nerves innervating the fatty tissue. Jpn J Physiol 1997; 47:S40-S42.

70. Stricker EM, Rowland NE, Saller CF et al. Homeostasis during hypoglycemia: central control of adrenal secretion and peripheral control of feeding. Science 1977; 196:79-81.

71. Scheurink AJW, Ritter S. Sympathoadrenal responses to glucoprivation and lipoprivation in rats. Physiol Behav 1993; 53:995-1000.

72. Jansen ASP, Hoffman JL, Loewy AD. CNS sites involved in sympathetic and parasympathetic control of the pancreas: A viral tracing study. Brain Res 1997; 766:29-38.

73. Bamshad M, Song CK, Bartness TJ. CNS origins of the sympathetic nervous system outflow to brown adipose tissue. Am J Physiol 1999; 276:R1569-R1578.

74. Gold RM. Hypothalamic obesity: The myth of the ventromedial nucleus. Science 1973; 182:488-490.

75. Bates MW, Nauss SF, Hagman NC, Mayer J. Fat metabolism in three forms of experimental obesity: Body composition. Am J Physiol 1955; 180:301-309.

76. Brobeck JR, Tepperman J, Long CNH. Experimental hypothalamic hyperphagia in the albino rat. Yale J Biol Med 1943; 15:831-853.

77. Luiten PGM, ter Horst GJ, Karst H et al. The course of paraventricular hypothalamic efferents to autonomic structures in medulla and spinal cord. Brain Res 1985; 329:374-378.

78. Kirchgessner AL, Sclafani A. Histochemical identification of a PVN-hindbrain feeding pathway. Physiol Behav 1988; 42:529-543.

79. Cox JE, Sims JS. Ventromedial hypothalamic and paraventricular nucleus lesions damage a common system to produce hyperphagia. Behav Brain Res 1988; 28:297-308.

80. Fukushima M, Tokunaga K, Lupien J et al. Dynamic and static phases of obesity following lesions in PVN and VMH. Am J Physiol 1987; 253:R523-R529.

81. Tokunaga K, Fukushima M, Kemnitz JW et al. Comparison of ventromedial and paraventricular lesions in rats that become obese. Am J Physiol 1986; 251:R1221-R1227.

82. Weingarten HP, Chang P, McDonald TJ. Comparison of the metabolic and behavioral disturbances following paraventricular- and ventromedial-hypothalamic lesions. Brain Res Bull 1985; 14:551-559.

83. Jones AP, Assimon SA, Gold RM et al. Adipose tissue mobilization is unaffected by obesifying hypothalamic knife cuts. Physiol Behav 1985; 34:29-31.

84. Nishizawa Y, Bray GA. Ventromedial hypothalamic lesions and the mobilization of fatty acids. J Clin Invest 1978; 61:714-721.

85. Bray GA, Nishizawa Y. Ventromedial hypothalamus modulates fat mobilization during fasting. Nature 1978; 274:900-901.

86. Steffens AB, Damsma G, van der Gugten J et al. Circulating free fatty acids, insulin, and glucose during chemical stimulation of hypothalamus in rats. Am J Physiol 1984; 247:E765-E771.

87. Takahashi A, Shimazu T. Hypothalamic regulation of lipid metabolism in the rat: effect of hypothalamic stimulation on lipolysis. J Auton Nerv Syst 1981; 4:195-205.

88. Kumon A, Takahashi A, Hara T et al. Mechanism of lipolysis induced by electrical stimulation of the hypothalamus in the rabbit. J Lipid Res 1976; 17:551-558.

89. Correll JW. Central neural structures and pathways important for free fatty acid mobilization demonstrated in chronic animals. Fed Proc 1963; 22:547.

90. Barkai A, Allweis C. Effect of electrical stimulation of the hypothalamus on plasma levels of free fatty acids and glucose in rats. Metabolism 1972; 21(10):921-927.

91. Barkai A, Allweis C. Effect of electrical stimulation of the hypothalamus on plasma free fatty acid concentration in cats. J Lipid Res 1972; 13:725-732.

92. Oro L, Wallenberg LR, Bolme P. Influence of electrical supramedullary stimulation on the plasma level of free fatty acids, blood pressure and heart rate in the dog. Acta Med Scand 1965; 178:697.

93. Coimbra CC, Migliorini RH. Insulin-sensitive glucoreceptors in rat preoptic area that regulate FFA mobilization. Am J Physiol 1986; 251:E703-E706.

94. Coimbra CC, Migliorini RH. Cold-induced free fatty acid mobilization is impaired in rats with lesions in preoptic area. Neurosci Lett 1988; 88:1-5.

95. Ferreira ML, Marubayashi U, Coimbra CC. The medial preoptic area modulates the increase in plasma glucose and free fatty acid mobilization induced by acute cold exposure. Brain Res Bull 1999; 49:189-193.

96. Robinson RL, Culberson JL, Carmichael SW. Influence of hypothalamic stimulation on the secretion of adrenal medullary catecholamines. J Auton Nerv Syst 1983; 8:89-96.

97. Teixeira VL, Antunes-Rodrigues J, Migliorini RH. Evidence for centers in the central nervous system that selectively regulate fat mobilization in the rat. J Lipid Res 1973; 14:672-677.

98. Correll JW. Adipose tissue: Ability to respond to nerve stimulation in vitro. Science 1963; 140:387-388.

99. Coimbra CC, Migliorini RH. Evidence for a longitudinal pathway in rat hypothalamus that controls FFA mobilization. Am J Physiol 1983; 245:E332-E337.

100. Caulliez R, Viarouge C, Nicolaidis S. Electrical stimulation of the suprachiasmatic nucleus region modifies energy metabolism in the rat. In: Nakagawa H, Oomura Y, Nagai K, eds. New Functional Aspects of the Suprachiasmatic Nucleus of the Hypothalamus. London: John Libbey, 1991:75-84.

101. Bamshad M, Aoki VT, Bartness TJ. Sympathetic nervous system (SNS) outflow from the CNS to brown adipose tissue (BAT): A retrograde study using the transneuronal pseudorabies virus (PRV). Society of Neuroscience Abstracts 23(Part 1), 1997:515.

102. Bamshad M, Demas GE, Bartness TJ. Does white adipose tissue (WAT) receive photic information directly from projections of retinal ganglion cells? Society of Neuroscience Abstracts 24 (Part 1), 1998:370.

103. Larsen PJ, Enquist LW, Card JP. Characterization of the multisynaptic neuronal control of the rat pineal gland using viral transneuronal tracing. Eur J Neurosci 1998; 10:128-145.

104. Buijs RM, Wortel J, Van Heerikhuize JJ et al. Anatomical and functional demonstration of a multisynaptic suprachiasmatic nucleus adrenal (cortex) pathway. Eur J Neurosci 1999; 11:1535-1544.

105. Grill WM, Erokwu BO, Hadziefendic S et al. Extended survival time following pseudorabies virus injection labels the suprachiasmatic neural network controlling the bladder and urethra in the rat. Neurosci Lett 1999; 270:63-66.

106. Sly DJ, Colvill L, McKinley MJ, Oldfield BJ. Identification of neural projections from the forebrain to the kidney, using the virus pseudorabies. J Auton Nerv Syst 1999; 77(73):82.

107. Ueyama T, Krout KE, Nguyen XV et al. Suprachiasmatic nucleus: A central autonomic clock. Nat Neurosci 1999; 12:1051-1053.

108. Yamamoto H, Nagai K, Nakagawa H. Bilateral lesions of the suprachiasmatic nucleus enhance glucose tolerance in rats. Biomed Res 1984; 5:47-54.

109. LeMagnen J, Devos M. Metabolic correlates of the meal onset in the free food intake of rats. Physiol Behav 1970; 5:805-814.

110. Cornich S, Cattene C. Fatty acid synthesis in mice during the 24 hr cycle and during meal feeding. Horm Metab Res 1978; 10:276-290.

111. Rusak B, Zucker I. Neural regulation of circadian rhythms. Physiol Rev 1979; 59:449-526.

112. Licinio J, Mantzoros C, Negrao AB et al. Human leptin levels are pulsatile and inversely related to pituitary-adrenal function. Nat Med 1997; 5:575-579.

113. Trayhurn P, Duncan JS, Rayner DV. Acute cold-induced suppression of *ob* (obese) gene expression in white adipose tissue of mice: mediation by the sympathetic nervous system. Biochm J 1995; 311:729-733.

114. Rayner DV, Simon E, Duncan JS et al. Hyperleptinaemia in mice induced by administration of the tyrosine hydroxylase inhibitor alpha-methyl-p-tyrosine. FEBS Lett 1998; 429:395-398.

115. Sivitz WI, Fink BD, Morgan DA et al.Sympathetic inhibition, leptin, and uncoupling protein subtype expression in normal and fasting rats. Am J Physiol 1999; 277:E668-E677.

116. Nagai K, Nagai N, Sugahara K et al. Circadian rhythms and energy metabolism with special reference to the suprachiasmatic nucleus. Neurosci Biobehav Rev 1994; 18:579-584.

117. Hwa JJ, Ghibaudi L, Compton D et al. Intracerebroventricular injection of leptin increases thermogenesis and mobilizes fat metabolism in *ob/ob* mice. Horm Metab Res 1996; 28(12):659-663.

118. Hwa JJ, Fawzi AB, Graziano MP et al. Leptin increases energy expenditure and selectively promotes fat metabolism in *ob/ob* mice. Am J Physiol 1997; 272:R1204-R1209.

119. Collins S, Kuhn CM, Petro AE et al. Role of leptin in fat regulation. Nature 1996; 380:677-677.

CHAPTER 8

Heterogeneity of Adipose Tissue Metabolism

Michael Boschmann

Adipose tissue is a highly specialized organ dealing mainly with processes for storage and release of energy. The underlying metabolic routes of these processes are regulated by a complex network to meet the energy requirements of the body. In times of a positive energy balance, i.e., energy (food) intake exceeding energy requirements, excess energy is stored as fat. In times of a negative energy balance, i.e., energy requirements exceeding energy intake, including the extreme situation of long-term fasting, fat resources are mobilized. Both processes operate in a highly efficient manner to prevent wasting of energy and to ensure survival of the organism. Adipose tissue has a highly adaptive capability for storing large amounts of fat. This can be achieved by increasing fat cell size or fat cell number (see also chapter 10 of this volume). Normally, a human subject has about 10 to 20 kg of body fat. Taking into account a caloric value of about 7,000 kcal per kg fat tissue, an energy reserve of 70,000 to 140,000 kcal (that is 300–600 MJ) is stored in adipose tissue. The caloric value for adipose tissue is lower than that for pure fat, i.e., triglycerides (about 9,000 kcal per kg) because the fat-storing adipocytes are normally not completely catabolized for generation of energy. Body fat mass can increase tremendously (between 40 to 100 kg or more) and this accumulation of excess body fat is called obesity. Obesity is by no means an appearance of the late 20th and the early 21st century. One of the first reports of obesity is the "Venus of Willendorf", a small statuette showing clear signs of abdominal obesity.[1] Obesity has been found in almost all old medical traditions including Egyptian, Chinese, Indian, Meso-American, Greco-Roman and Arabic medicine and there have been already single reports about treating obesity as a disease.[1] However, in all these traditions occurrence of obesity was limited to the "higher classes".[1]

It was the merit of Vague et al[2] to show that not obesity per se but abdominal rather than gluteo-femoral body fat accumulation predisposes to obesity-associated diseases such as dyslipidemia, hypertension, diabetes mellitus, or arteriosclerosis—a complex known also as metabolic syndrome.[2] This was the first indication of an obvious functional heterogeneity of adipose tissue depending on the location within the human body. With respect to anatomic location, different regions of white adipose tissue can be distinguished as listed in Table 8.1.

The recent discovery that adipose tissue functions also as an endocrine organ has pushed adipose tissue research into new areas and today adipose tissue is a widely-used model for investigating interactions between "Joules" and "Genes".

This chapter focuses on basic and regulatory aspects of adipose tissue metabolism. Special attention is paid, in particular, to site- and sex-specific differences within the regulatory processes. Additionally, the impact of nutrition, obesity, exercise and aging on physiological adaptation of adipose tissue metabolism is covered. Adaptation to pregnancy and lactation is beyond the view of this chapter and is, therefore, not touched.

Adipose Tissues, edited by Susanne Klaus. ©2001 Eurekah.com.

Table 8.1. White adipose tissue classified according to anatomical distribution

Subcutaneous adipose tissue	abdominal subcutaneous adipose tissue
	gluteo-femoral subcutaneous adipose tissue
	mammary subcutaneous adipose tissue
Visceral adipose tissue	omental adipose tissue
	mesenteric adipose tissue
	retroperitoneal adipose tissue
Intraorgan adipose tissue	intrahepatic adipose tissue
	intramuscular adipose tissue

Techniques for Investigating Adipose Tissue Metabolism

In Vitro Techniques

The isolated adipocyte is the most simple physiological system to study adipose tissue metabolism allowing also the control of a number of variables in the fat cell environment. Isolated adipocytes can be prepared by treating tissue samples with collagenase as originally described by Rodbell et al.[3] After isolation, adipocytes can be incubated in silicon-coated Erlenmeyer flasks or plastic tubes containing a basic incubation medium such as Ringer's solution or Krebs/Henseleit-bicarbonate buffer which is supplemented with albumin and various substrates such as glucose or fatty acids and/or different metabolic effectors such as hormones or catecholamines. In order to quantify changes in lipolytic rates, the release of glycerol and free fatty acids (FFA) into the incubation medium is measured. Using labeled substrates such as ^{14}C-U-glucose or ^{14}C-3-O-methylglucose, turnover rates of major metabolic pathways or characteristics of various transporters such as the glucose transporter can be estimated.[4] In any case, it is important to relate the metabolic data to the mean cell size and/or correct for differences in surface area. That is because fat cell size markedly influences metabolic rates of adipocytes.[5]

The perifusion technique is a more sophisticated method to investigate adipocytes in a more physiological situation. First reports describing this technique for the metabolic characterization of isolated adipocytes or hepatocytes were published in the early seventies.[6-9] Compared to the incubation technique, the perifusion technique offers some major advantages:

1. metabolic studies under flow conditions,
2. continuous supply with substrates and/or metabolic effectors,
3. continuous removal of products formed by the adipocytes (no product accumulation!),
4. easy sampling,
5. easy switching between "on" and "off" load when different substrate or metabolic effectors are added,
6. possibility of running different experimental protocols once adipocytes are filled into the perifusion system.

The main part of that technique is the perifusion chamber connected to a peristaltic pump.[10] In contrast to hepatocytes, adipocytes are floating to the surface of the chamber due to their lower density when compared to water. To avoid an escape of the cells, the chamber is closed by a membrane filter and a stainless steel mash. Cells are normally perfused at flow rates of about

0.5 ml/min and flow direction is upside down allowing an optimal substrate supply of the cells. A cell concentration of about 100.000 per ml should be appropriate.[11] When using larger cell numbers, oxidative glucose-metabolism is increased whereas non-oxidative glucose metabolism is decreased.[11] Unfortunately, this technique has not widely been used, possibly due to the higher technical requirements for running and maintaining such a system.

The use of preadipocyte cell lines such as 3T3-L1, 3T3-F442A or ob17 is another valuable tool for investigating cell growth and differentiation or performing long-term metabolic studies.

In Vivo Techniques

Over years, there were only very limited approaches to study the regulation of adipose tissue metabolism in vivo during different physiological and pharmacological manipulations. Measuring changes in plasma FFA and glycerol or the appearance and disappearance rates of either radio- or stable isotopes in easily accessible compartments gave more or less specific insights into regulatory aspects of adipose tissue metabolism. The experimental outcome could be improved considerably by analyzing biopsies taken from different adipose tissue sites. However, an integral investigation of adipose tissue metabolism covering different regulatory levels was almost impossible.

A deeper insight into in vivo regulation of adipose tissue metabolism was achieved by the introduction of two new techniques, i.e., the arterio-venous difference technique and the microdialysis technique. The arterio-venous difference technique implies catheterization of a major artery in the forearm or leg and a superficial vein draining the subcutaneous adipose depot of the anterior abdominal wall.[12] Measuring the arterio-venous difference of various metabolites across that adipose tissue depot, information is obtained about metabolites accumulated in or released from adipose tissue under different physiological or pathophysiological situations.[12]

A landmark in adipose tissue research was the introduction of the microdialysis technique for in situ studies of both basic aspects of physiological/pathophysiological processes as well as effects of pharmaco-toxicological interventions.[13,14] Originally developed for pharmacokinetic studies in rat brain almost 25 years ago,[15] this technique has meanwhile found broad acceptance in patient-oriented research for investigating regulatory aspects of the metabolism in easily accessible organs such as adipose tissue, muscle and skin. Moreover, insertion of a catheter into the subcutaneous adipose tissue allows continuous long-term glucose monitoring in ambulatory insulin-dependent diabetic patients or controlling the metabolic adaptation of critically ill neonates.[16,17] The main advantage of the microdialysis technique is the possibility to study different regulatory levels (metabolic, hormonal, neuronal, hemodynamic) simultaneously at the tissue level. The core part is the microdialysis catheter with the semipermeable dialysis tubing. In principal, there are two types of catheters currently used, a) the older, linear one and b) the newer, concentric one.[13,14] In both cases, the perfusate (inflow) supplemented with substrates and/or various biological effectors enters the tissue at a flow rate between 1 and 5 μl/min and, after exchanging substrates for products, leaves the tissue as dialysate (outflow). In all non-ethanol metabolizing organs such as adipose tissue blood flow changes can be monitored very easily by the so-called ethanol dilution technique.[18,19] Briefly, ethanol is added to the perifusion medium. Due to the concentration gradient, ethanol diffuses into the intercellular water space and also into the surrounding capillaries where it is removed. The higher the capillary blood flow the higher the removal of ethanol out of the tissue. Measuring the ethanol concentration in the dialysate (outflow) and the perfusate (inflow) and calculating the $[Ethanol]_{outflow} / [Ethanol]_{inflow}$ ratio, changes in blood flow can be estimated qualitatively. Recently, this technique has been validated against the ^{133}Xe-clearance technique.[20]

The arterio-venous difference technique and the microdialysis technique should be seen as complementary.

Basic Aspects of Adipocyte Metabolism

Compared to tissues such as liver and muscle, the metabolism of adipose tissue is rather simple. The two major metabolic processes of white adipose tissue are lipogenesis and lipolysis. Two key enzymes are involved in the regulation of these processes, i.e., lipoprotein lipase (LPL) located extracellularly and hormone sensitive lipase (HSL) located intracellularly for lipogenesis and lipolysis, respectively.

The term "lipogenesis" is not sharply defined. In a restricted sense, "lipogenesis" can be applied to de novo synthesis of fatty acids. For this process the term "lipacidogenesis" has been proposed but it has not been widely accepted in the scientific literature. In a broader sense, "lipogenesis" can be applied to acylation of glycerol with free fatty acids to form triacylglycerols (TAG). After synthesis, TAG can be either stored inside the cell (adipocyte) or incorporated into more complex structures, i.e., lipoproteins, for example, chylomicrons (small intestine) or very low density lipoproteins (VLDL) (liver). In this chapter, the term "lipogenesis" is used in that broader sense, only. In humans, de novo synthesis of FFA from carbohydrates plays normally a minor role.[21,22] FFA for TAG synthesis in adipose tissue are mainly derived from the diet.[23] Nevertheless, diets very high in carbohydrate, but low in fat can lead to enhanced de novo FFA synthesis.[24]

Glucose is an essential component for adipose tissue metabolism, primarily for TAG synthesis. Glucose is taken up into adipocytes via specific glucose transporters which are regulated by various hormones.[25,26] Until now five different glucose transporters have been described whereof GLUT4 is of major importance and GLUT1 is less important in WAT.[27,28]

Glucose provides the adipocyte with glycerol-3-phosphate via glycolysis. Glycerol-3-phosphate forms the backbone of TAG because it can be used directly for esterification with FFA. Lipolysis-derived glycerol can not be used for TAG synthesis because adipose tissue is lacking the enzyme glycerol-kinase.[29,30] Measuring the incorporation of a pulse of ^{14}C-glucose into triglycerides in humans showed that only 1 to 2% of glucose normally ingested is used for TAG synthesis.[31] Therefore, adipose tissue seems to be of minor importance in glucose homeostasis. However, recently it has been shown in vitro and in vivo that human adipose tissue can produce substantial amounts of lactate.[32-34] Lactate production can account for up to 50% of glucose metabolism in adipose tissue and this might be of importance if fat mass is considerably increased.[32,35]

In order to match lipogenesis and lipolysis with the existing energy balance, regulation of the corresponding key enzymes, i.e., lipoprotein lipase (LPL) and hormone sensitive lipase (HSL), respectively, is highly coordinated.[36] Regulatory components include substrates, hormones, and neurotransmitters. Drugs can also exert significant interactions with lipid metabolism (for an overview see Table 8.2). However, the role of a single regulatory component can differ considerably between various adipose tissue sites.[37-39]

Lipogenesis and the Role of Lipoprotein Lipase (LPL)

In the situation of a positive energy balance, i.e., energy (food) intake exceeding energy consumption, excess energy is stored mainly as fat in adipose tissue. In all highly developed, industrialized countries excess energy is mainly taken up as animal-derived saturated fat, i.e., triacylglycerol. TAGs enter the adipose tissue mainly via circulating chylomicrons or VLDL. To get inside the adipocyte, triglycerides are hydrolyzed into free fatty acids (FFA) and monoacylglycerol. FFA are taken up into the adipocytes followed by re-esterification to TAG.

LPL is synthesized in various tissues including adipose tissue. In adipose tissue, LPL is the rate-limiting enzyme for the import of triglyceride-derived fatty acids into the adipocytes for storage.[40] After synthesis, LPL becomes glycosylated in the Golgi apparatus where it is also packed into secretory vesicles.[41] Glycosylation of LPL is necessary in order to get secreted out of the cell and to get activated.[42] After secretion, LPL is attached to the capillary endothelium by glycosaminoglycans.[41] Binding of apolipoprotein CII to LPL is necessary to achieve its full activity. The human LPL-gene has been cloned, the protein has been sequenced and a molecular mass of about 53 kD has been determined using Western blot analysis.[43,44] It is important to note that adipose tissue competes with other tissue, for example muscle, for TAG-rich lipoproteins. Under various physiological and pathophysiological situations LPL in adipose tissue is reciprocally regulated to that in muscle.

There are rare cases of familial LPL deficiency.[45] Surprisingly, these patients can also accumulate adipose tissue indicating the existence of compensatory mechanisms.[45] In mutant mice deficient in adipose tissue LPL, fat mass is preserved by endogenous synthesis which is achieved by up-regulation of lipogenic enzymes such as fatty acid synthetase and acetyl CoA carboxylase.[46] Although transgenic mice overexpressing GLUT4 in adipose tissue showed an increase in lipogenesis from glucose such a mechanism seems to be unlikely in men.[47] It is rather possible that free fatty acids are provided from plasma TAGs to adipose tissue by a hepatic or a adipose tissue-specific lipase.[45]

Alternatively, fatty acids bound to albumin can be also taken up from the circulation when LPL is absent. Additionally, a VLDL-receptor present in adipose tissue concomitantly with LPL could also provide the tissue with free fatty acids.

LPL activity shows gender- and site-specific differences in normal-weight human subjects. Generally, women show higher levels of LPL in adipose tissue when compared to men.[40] This is in line with the greater body fat content of women vs. men. In women, however, LPL activity is higher in the gluteo-femoral vs. abdominal region of subcutaneous adipose tissue whereas in men such difference could not be demonstrated.[48-50] The higher LPL activity in the gluteo-femoral vs. abdominal region of women fits with the more peripheral fat distribution pattern in women.

Insulin

Insulin is the most important anabolic hormone. Lipogenesis is promoted directly by insulin via increasing LPL activity in adipose tissue.[51,52] This effect is achieved already at insulin concentrations within the physiological range.[40] Insulin also stimulates glucose transport into the adipocytes by increasing the number of GLUT4 in the plasma membrane.[28]

Insulin binds to a specific cellular receptor located at the surface of the membrane. This receptor is a heterodimer constituted of two glycoproteins, i.e., α- and β-subunits.[53] After binding of insulin to its receptor and internalization of the hormone-receptor complex by endocytosis, β-subunits become autophosphorylated leading to expression of a tyrosine kinase activity.[53] Therefore, the insulin receptor represents both a phosphoprotein and a phosphokinase.[53] In the following, two large docking proteins, the insulin receptor substrates 1 and 2 (IRS-1, IRS-2) become phosphorylated at their tyrosine residues by the receptor tyrosine kinase. Phosphorylated IRS-1, for example, binds to the p85 regulatory subunit of phosphatidyl inositol 3-kinase (PI3-kinase) and this is followed by binding of the p110 catalytic subunit to that complex.[54] Activation of PI3-kinase is a crucial step in the phosphorylation cascade initiated by insulin which finally leads to phosphorylation, i.e., activation of protein kinase B (PKB).[55] PKB does not only activate protein synthesis and cell growth but also the translocation of GLUT4 to the cell surface.[54] Both PKB and PI3-kinase can also activate cGMP-inhibitable phosphodiesterase (PDE3B).[54] This is followed by a decrease in cytosolic cAMP levels and, consequently, decreases in PKA and HSL activity and hence inhibition of lipolysis.[54] This action is called the antilipolytic effect of insulin. Binding of insulin to its receptor also leads to stimulation of phospholipase C which hydrolyzes phosphatidylinositol-4,5-bisphosphate into diacylglycerol (DAG) and inositol-1,4,5-triphosphate (IP3).[56,57] IP3 serves as substrate for PI3-kinase. The signaling cascade for insulin and α_1-receptors is depicted in Figure 8.1

It has been shown that the effect of insulin on LPL is very similar in the abdominal and gluteo-femoral regions of subcutaneous adipose tissue.[58] However, in omental adipose tissue, this effect is rather weak, possibly because of the diminished insulin sensitivity in that region.[40,52] It is important to note, however, that dexamethasone is needed in omental adipocyte tissue cultures to facilitate the action of insulin.[40]

Glucocorticoids, Growth Hormone

Glucocorticoids such as cortisol have a marked stimulatory effect on LPL activity in vitro, especially in the presence of insulin.[59] This effect of cortisol involves both an increased expression of LPL mRNA and additional post-translational regulation.[59] Cortisol acts via a specific glucocorticoid receptor (GR) and the density of this receptor ranks in the order visceral > abdominal subcutaneous > gluteo-femoral subcutaneous adipose tissue.[60-62] Therefore, chronically elevated cortisol-concentrations lead to a higher accumulation of fat in visceral adipose tissue.[61,62] However, growth hormone (GH) can totally inhibit the effect of insulin + cortisol on LPL in vitro and this effect is achieved mainly at the post-translational level.[63,64] The inhibitory effect of GH on LPL is accompanied by an increased lipid mobilization[63,64] Treatment of GH-deficient, abdominally obese men reduces abdominal fat mass and improves glucose and lipoprotein metabolism.[65]

Adipsin / Acylation-Stimulating Protein

In 1989, another protein important for lipogenesis was purified and characterized as acylation-stimulating protein, ASP.[66] ASP is a small protein identical to a fragment of the third component of complement, C3adesArg.[67] C3a is formed by the action of the specific serine protease adipsin (also called factor D) on a complex formed of factor B and the precursor protein C3.[67] Arginine is removed at the carboxy terminal of C3a by a carboxypeptidase forming C3adesArg.[67] ASP can increase triglyceride synthesis in fibroblasts and cultured human adipocytes mainly by increasing the activity of diacylglycerol acyltransferase and by stimulating glucose transport into the adipocyte.[67]

Nutrition, Obesity, Exercise, Aging

Nutrition

Adipose tissue LPL activity normally increases after a meal.[40,68,69] However, LPL acts only as one member of a series of metabolic steps which are regulated in a highly coordinated manner. A meal high in carbohydrate content leads to a higher increase in LPL than a meal high in fat content.[70] This occurs not only after a single carbohydrate-rich meal but also during longer-term feeding of such a diet.[70] The adipose tissue LPL is increased with both oral and intravenous fat loads.[71] However, there is an increased escape of LPL-derived fatty acids into the circulation from adipose tissue when compared to muscle indicating that insulin is important

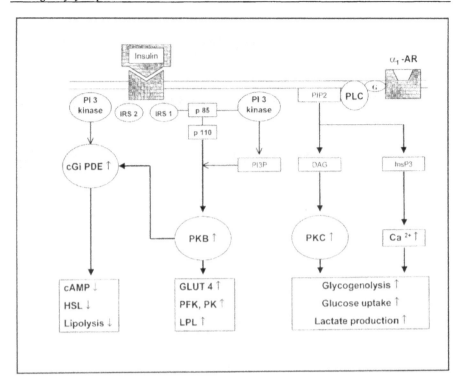

Fig. 8.1 Signaling pathway of the insulin and alpha-1 adrenergic receptor pathways. IRS 1 / IRS 2, insulin receptor substrate 1 and 2; PI 3 kinase, phosphatidylinositol-3-kinase; PI 3P, phosphatidylinositol-3-phosphate; PKB, protein kinase B; GLUT 4, glucose transporter 4; PFK, 6-phosphofructo-1-kinase; PK, pyruvate kinase; LPL, lipoprotein lipase; cGi-PDE, cGMP-inhibitable phosphodiesterase; HSL, hormone-sensitive lipase; G, G-protein(s); PLC, phospholipase C; PIP2, phosphatidylinositol-4,5-bisphosphate; diacylglycerol; InsP3, inositol-1,4,5-triphosphate; PKC, protein kinase C. For explanation of details see text.

for re-esterification of fatty acids in adipose tissue.[71] In rats, high-fat diets decrease LPL expression in epididymal but not in perirenal white adipose tissue.[72] The degree of saturation of dietary fats does not show any site-specific effects on LPL activity in adipose tissue.[72,73]

Obesity

During developing obesity, LPL activity per cell is increased.[74] This can partially be attributed to the hyperinsulineamia often associated with the obese state. There is no difference in expression and activity of LPL between omental and subcutaneous adipose tissue.[75,76] However, total amount of insulin receptor expression is significantly higher in omental adipose tissue and most of this increase is accounted for by expression of the differentially spliced insulin receptor lacking exon 11.[76] This splice variant is considered to transmit the insulin signal less efficiently than the insulin receptor with exon 11.[76] Massive accumulation of adipose tissue as seen in severely obese patients leads to a decrease in LPL activity in adipose tissue but an increase of it in muscle.[77] This is possibly a consequence of increasing insulin resistance because higher insulin amounts are necessary in obese vs. normal-weight subjects in order to obtain the same LPL activity. [77] Developing insulin resistance and the shift in LPL activity from adipose tissue to muscle serves obviously for an "emergency brake" used to prevent the body from further body fat accumulation.[77] There are controversial reports about the effect of fasting and body weight reduction on adipose tissue LPL activity. In two studies, fasting or rather body weight reduction was accompanied by reduced adipose tissue LPL followed by an re-increase during the weight stabilizing phase.[78,79] In another study, LPL activity of both abdominal and gluteo-femoral subcutaneous adipose tissue did not change in obese men, whereas it decreased in obese women in response to 15-week weight reducing program.[80] No changes in LPL gene expression were observed.[80] The increase in LPL activity during weight stability after a weight loss is likely to be a secondary effect of partial refeeding; however, the individual sensitivity of adipose tissue LPL to nutritional induction could be of critical importance [81]

Exercise

In normal untrained human subjects, exercise results in an increase in adipose LPL activity.[82,83] This increase possibly aims to restore the amount of TAG that was mobilized during exercise. In trained human subjects, however, adipose tissue LPL is unchanged or lower whereas muscle LPL is increased after prolonged exercise.[84,85] Changing from a trained to a detrained state is followed by an increase in adipose tissue LPL and a decrease in muscle LPL.[85] Compared to their sedentary controls, endurance-trained women show lower LPL activities in abdominal subcutaneous tissue, but no differences in the femoral adipose tissue region.[86] LPL activity is decreased in abdominal subcutaneous adipose tissue of obese women in response to exercise training and this is related to an improved insulin sensitivity.[87]

Aging

There are only few studies available investigating the effect of aging on adipose tissue LPL activity. Aging is positively associated with insulin resistance.[88] Therefore, glucose metabolism and anti-lipolysis should be decreased in older subjects. Comparing subcutaneous abdominal and gluteal heparin-elutable LPL activity in master athletes and lean sedentary men vs. obese sedentary men, LPL activity was significantly higher in adipose tissue of the obese sedentary men.[89] No effect of aging and cardiovascular fitness on adipose tissue LPL was observed. In young rats, adipose tissue LPL activity decreased during fasting but this downregulation of LPL was blunted in older rats.[90]

Lipolysis and the Role of Hormone Sensitive Lipase

If the body is confronted with a negative energy balance, i.e., if energy expenditure is exceeding energy intake, TAG stores are mobilized in order to meet the energy deficit. TAG mobilization is achieved mainly by the hormone sensitive lipase (HSL), an enzyme located inside the adipocyte. HSL hydrolyzes TAGs in a stepwise manner into diacylglycerol, monoacylglycerol and, finally glycerol and the respective free fatty acids (FFA). Following complete hydrolysis, a molar FFA / glycerol ratio of 3:1 is achieved. FFA are released into circulation and are transported via blood circulation to the organs of respective destination, i.e., mainly liver and muscle. Glycerol, once formed can not be used for re-esterification with FFA in adipocytes because of the lack of glycerokinase in adipose tissue.[29,30] Instead, glycerol is transported to the liver where it is used as gluconeogenic precursor. [30]

Table 8.2. Regulating effectors of lipolysis

	Stimulation	Inhibition
Substrates, Metabolites	prostaglandins (PGI$_2$)	prostaglandins (PGE) adenosine, free fatty acids, ketone bodies, lactate
Hormones	TSH, ACTH, thyroxine, parathyroid hormone, growth hormone, vasopressin, glucagon, secretin, cholecystokinin	insulin, insulin-like growth factor 1 (IGF-1) and 2 (IGF-2), glucocorticoids, somatostatin, neuropeptide Y (NPY)
Neurohormones, Neurotransmitters	epinephrine (high concentration), norepinephrine	epinephrine (low concentration), neuropeptide Y (NPY)
Drugs	ß-adrenergic agonists, heparin, caffeine	ß-blockers, α_2-adrenergic agonists nicotinic acid

Catecholamines

The sympathetic nervous system (SNS) is the most powerful system for controlling lipolysis in adipose tissue. Neuronal signals derived from the SNS are mediated to the adipocyte and the surrounding blood vessels via specific receptors. There are two natural occurring catecholamines, i.e., epinephrine and norepinephrine acting as physiological receptor-binding ligands. Whereas epinephrine is rather a neurohormone released from the adrenal medulla, norepinephrine is a classical neurotransmitter released from the synaptic vesicles of the SNS. There are two types of catecholamine-binding adrenergic receptors (also called adrenoreceptors or adrenoceptors, AR), i.e., α- and β-adrenergic receptors. Both receptor types can be subdivided: α AR into α_1 and α_2-AR, β-AR into β_1, β_2 and β_3-AR. There is a huge number of articles describing molecular, biochemical, physiological and pharmacological characteristics of the different adrenoceptor subtypes and it is impossible to cover all those aspects in this chapter. Where convenient and necessary, the reader is referred to some relevant reviews published during the last years.[38,39,91,92]

All adrenoceptor subtypes are expressed in white adipocytes.[39] α_2, β_1, β_2 and β_3-ARs belong to the class of seven-transmembrane receptor proteins and they operate through the adenylyl cyclase system. These receptors are coupled to GTP-binding proteins (G-proteins). G-proteins can either stimulate (G$_s$-proteins) or inhibit (G$_i$-proteins) adenylyl cyclase. All β-AR subtypes act through G$_s$-proteins whereas the α_2-AR subtype acts through G$_i$-proteins. The α_1-AR is not linked to the adenylyl cyclase system. Epinephrine and norepinephrine bind to both α- and β-AR. The affinity of epinephrine for the respective receptors ranks in the order $\alpha_2 > \beta_2 > \beta_1 > \beta_3$, that of norepinephrine in the order $\alpha_2 > \beta_1 >= \beta_2 > \beta_3$.[39] After agonist binding to the respective β-AR subtype, adenylyl cyclase is activated and cyclic AMP (cAMP) is synthesized. Because of the increase in cAMP, protein kinase A (PKA) is activated. PKA phosphorylates, i.e., activates hormone-sensitive lipase (HSL). Activated HSL, finally, hydro-lyzes TAGs into free fatty acids and glycerol. PKA not only phosphorylates HSL but also cGMP-inhibited low K_m cAMP-phosphodiesterase (cGI-PDE), glucose transporter, phospho-rylase kinase, glycogen synthase, acyl-CoA carboxylase, and the β_1 and β_2-AR themselves. [92] Phosphorylated cGI-PDE catabolizes cAMP. Therefore, cytosolic cAMP level decreases fol-lowed by a decreased activity of HSL. Phosphorylation of cGI-PDE and the β_1 and β_2-ARs can explain the desensitization/downregulation observed for the β_1 and β_2-ARs after long-term stimulation with respective agonists. [93] Interestingly, the phenomenon of desensitization is ab-sent in β_3-AR. [94,95] Agonist binding to α_2-ARs results in an inhibition of adenylyl cyclase fol-lowed by a decrease in cAMP, inactivation of PKA and, finally, dephosphorylation of HSL. As a consequence, lipolysis is inhibited. A summary of these activating or inhibiting signal cas-cades is given in Figure 8.2.

For detailed studies of the different adrenoreceptor subtypes a variety of specific synthetic ligands have been developed with either agonist or antagonist function. An overview of some of these synthetic ligands is given in Table 8.3.

Numerous in vitro and in vivo studies led to the convincing evidence that regulation of adipose tissue metabolism is not only species-specific but demonstrates also sex- and site-specific differences within a species. During rest or at baseline, this heterogeneity is expressed rather weakly or is absent, at least in men. [96,97] However, stimulation of adipose tissue of normal human subjects with isoproterenol results in an increase in lipolysis which is highest in omental adipose tissue, intermediate in abdominal subcutaneous tissue and lowest in gluteo-femoral adipose tissue. [38] This ranking is primarily a result of different distribution of the adrenoceptor subtypes in the different adipose tissue regions: β_1 and β_2-ARs number is highest in the omen-tal region, intermediate in the abdominal and lowest in the gluteo-femoral region whereas distribution of the α_2-AR ranks in the reversed order. [98-100] An illustration of the site-specific differences in adipose tissue towards β-adrenergic stimulation is given in Figure 8.3.

In omental white adipose tissue of normal-weight subjects, expression of the β_2-AR pre-dominates that of β_1-AR. [101] α_2-AR number correlates positively with fat cell size, i.e., the larger the fat cell the higher the α-AR number. [38,92] In contrast to that, such a relation does not exist for β-AR. [92] Furthermore, α_2-ARs largely outnumber β-ARs in fat cells of certain fat deposits and α_2-ARs exert obviously a tonic inhibition of lipolysis controlled by the SNS. [39,92] In hamsters, for example, epididymal adipocytes are normally more responsive to β-adrenergic stimulation than inguinal adipocytes. [102] However, sympathectomy increased significantly the β-adrenergic lipolytic and α_2 adrenergic antilipolytic responses in inguinal and epididymal adipocytes, respectively, indicating that the sympathetic pathway favors lipolysis in the epid-idymal adipose tissue. [102] In non-obese, healthy men, a sustained sympatho-inhibition by simu-lated microgravity induces an increase in the lipolytic β-adrenergic response in adipose tis-sue. [103] There is some evidence that this hypersensitization is linked to an increase in the postreceptor (cAMP-synthesis) steps of the lipolytic cascade in the adipocyte rather than to changes in β-adrenoceptors. [103]

Adrenoceptors have also an important role in controlling local blood flow in adipose tissue. Vasodilatation is mediated mainly via β_1 and β_2-AR whereas vasoconstriction is caused by activating α_2-AR. [19,104]

Based on the findings of an atypical behavior of brown adipose tissue towards newly developed β-adrenergic drugs, the existence of a third β-AR was postulated. [105] In 1989, a gene was characterized coding for an atypical β receptor which was named β_3-adrenoceptor. [106] Shortly after that, the receptor was also cloned in rat and mice. [107,108] In humans, this receptor is expressed mainly in brown and white adipose tissue followed by stomach, gall bladder, small

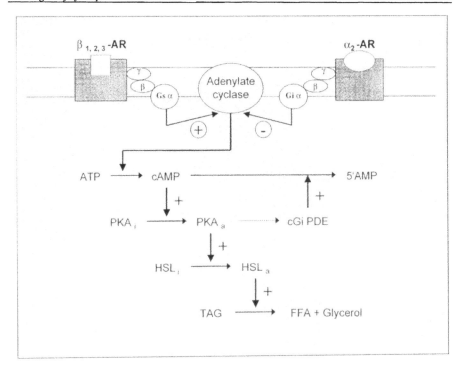

Fig. 8.2 Signaling pathway of the beta-1,-2, and -3 adrenergic receptor pathways. Gs and Gi, stimulating and inhibiting G-proteins, respectively; α, β, γ, subunits of the G-proteins; PKA$_i$ and PKA$_a$; inactive and active protein kinase A, respectively; HSL, hormone-sensitive lipase; TAG, triacylglycerols, FFA, free fatty acids. For explanation of details see text.

intestine, colon, prostate, heart (atrium), and skeletal muscle.[109-111] In rats, the β$_3$-AR is expressed mainly in brown and white adipose tissue but also in the stomach fundus, longitudinal and circular smooth muscle of both ileum and colon and colon submucosa.[112,113] In mice, two splice variants of the β$_3$-AR could be demonstrated differing in their number of amino acids in exon 2: β$_3$a-AR variant (known) with 13 amino acids and β$_3$b-AR variant (new) with 17 amino acids.[114] β$_3$b-AR mRNA is differentially expressed in mouse tissues, with levels relative to β$_3$a-AR mRNA highest in hypothalamus, cortex and white adipose tissue, and lower in ileum smooth muscle and brown adipose tissue.[114] That might be a reason why the β$_3$-AR is differently regulated in brown and white adipose tissue, and in the gastrointestinal tract, at least in the ob/ob mouse.[115] In guinea pigs, the functional expression of the β$_3$-AR is rather weak in both brown and white adipose tissue.[116]

Studies on isolated adipocytes from different adipose sites in humans showed that the β$_3$-AR is functionally expressed mainly in omental adipose tissue.[117] In isolated adipocytes from abdominal and femoral subcutaneous adipose tissue, addition of the unspecific agonist isoproterenol (stimulating β$_1$-, β$_2$-, and β$_3$-AR) combined with alprenolol (blocking β$_1$- and β$_2$-AR) did not result in any stimulation of lipolysis indicating the absence of a functional β$_3$-AR in subcutaneous adipose tissue.[118] However, using CGP 12177, a ligand highly specific for β$_3$-AR, the functional expression of β$_3$-AR was found to be weak but present in isolated adipocytes from subcutaneous adipose tissue.[119] Using the microdialysis technique, it could be

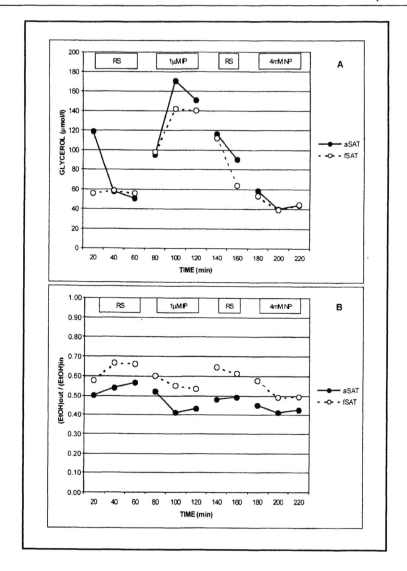

Fig. 8.3 Effect of isoproterenol and nitroprusside on lipolysis (panel A) and blood flow (panel B) in abdominal (aSAT) and gluteo-femoral (fSAT) subcutaneous adipose tissue. Using the microdialysis technique, respective tissues of six nonobese, healthy women were perfused with Ringer's solution (RS) supplemented with 50 mM ethanol (EtOH) at a flow rate of 2 µl/min (microdialysis catheters CMA/60, CMA Microdialysis AB, Solna, Sweden). The effect of a an agent acting on lipolysis and blood flow (isoproterenol, IP) and an agent acting on blood flow only (nitroprusside, NP) was tested. Lipolysis was monitored via measuring dialysate glycerol concentrations, changes in blood flow were monitored using the so-called ethanol-dilution technique. A decrease in the EtOH$_{dial}$ / EtOH$_{perf}$ ratio corresponds to an increase in blood flow and vice versa. Adding of IP induced an increase in dialysate glycerol concentration and this effect was greater in aSAT vs. fSAT (A). Adding of NP induced an decrease in dialysate glycerol concentration because of its vasodilating effect (A, B). IP and NP decreased the EtOH$_{dial}$ / EtOH$_{perf}$ and hence increased blood flow to the same extent. At all experimental steps, there was a tendency of higher blood flow in aSAT vs. fSAT. For more detailed explanations of the microdialysis technique see text.

demonstrated in vivo that all three β-AR subtypes exist concomitantly in human subcutaneous adipose tissue.[120]

Insulin

Insulin inhibits lipolysis stimulated by β-AR agonists. Antilipolysis is achieved mainly by interactions with the signaling cascade of β-ARs at both receptor and postreceptor levels.[92] Insulin binding to its receptor can phosphorylate/activate cGi-PDE via the action of PKB and PI3-kinase.[54] As a consequence, the activities of PKA and HSL are decreased and lipolysis becomes inhibited.[54] Exposure of human fat cells to physiological concentrations of insulin is followed by a rapid and dose-dependent translocation of β-AR from the exterior to the interior of the cell and a subsequent dose-dependent decrease in the lipolytic sensitivity to β-adrenergic agonists, without a change in maximum lipolysis.[121] Insulin down-regulates β-AR expression in 3T3-L1 and 3T3-F442A cultured adipocytes.[122,123] Downregulation is also observed in BAT and WAT of rats with a rise in plasma insulin after transition from the fasted to a fed state.[124] In insulin dependent diabetes mellitus (IDDM), subcutaneous adipose tissue shows an increased β-adrenergic sensitivity indicated by a lipolysis which is markedly enhanced following hypoglycemia, despite a reduced catecholamine secretory response.[125] However, hypoglycemia-induced increase in local adipose tissue blood flow is similar in both normal and diabetic subjects.[125] If interactions of insulin and β-adrenergic drugs are studied, the experimental outcome possibly depends on the sequence the effectors are added. For example, insulin action on glucose uptake and lipolysis is counteracted by β-AR stimulation whereas insulin and β-ARs have synergistic effects on non-oxidative glucose metabolism in human adipose tissue in situ.[126]

Growth Hormone, Thyroid Hormone

Growth hormone (GH) treatment significantly increases catecholamine-induced lipolysis which is blunted after hypophysectomy, and this effect is similar in adipocytes isolated from different fat depots of both female and male rats.[127] In a clinical trial, long-term administration of rhGH to GH-deficient patients resulted in an improved epinephrine-induced lipolytic response in adipocytes isolated from abdominal subcutaneous tissue.[128] In GH-deficient children, chronic systemic growth hormone administration is associated with a redistribution of adipose tissue from an abdominal (android) to a more peripheral (gynoid) distribution.[129] GH therapy was associated with a significant reduction in abdominal fat cell size, a significant reduction in overall basal rates of lipogenesis, and with a variable desensitization of abdominal subcutaneous adipose tissue to the antilipolytic effect of insulin.[129]

Thyroid hormone exerts a direct interaction with β-AR mediated lipolysis in adipose tissue. In humans, hyperthyroidism is associated with an increased catecholamine-induced lipolysis whereas the opposite effect is observed in hypothyroid patients.[130] Both hypothyroid rats and humans show a reduced number of β_1 and β_2-AR. A number of defects at the postreceptor level of β-AR have been reported including uncoupling or potentiation of the inhibitory regulation of the β-adrenergic receptor adenylyl cyclase and increased cGI-PDE activity.[92] The α_2-adrenergic response seems to be unaffected in adipocytes from hypothyroid rats.[131] A hyperthyroid state in the rat potentiates the lipolytic response of white adipocytes to a specific β_3-adrenergic agonist (BRL 37344) and increases the β_3-AR sensitivity without changing β_1-AR number and affinity.[132] On the other hand, hypothyroid male rats show a reduced expression of β_3-AR mRNA in epididymal fat and treatment with T_3 results in an increase in β_3-AR mRNA expression and also in an increased sensitivity of this receptor for the highly specific β_3-AR agonist CL 316,243.[133]

Nutrition, Obesity, Exercise, Aging

Nutrition

Not only the quantity of nutrients but also their quality can influence different characteristics of membrane-bound receptors. Among the different nutrients, the quality of fat taken up with a meal is of special importance. In rats, long-term intake of a diet rich in saturated fat (beef tallow) is followed by a decrease in membrane fluidity in different adipose tissue regions which is accompanied by decreases in both binding of agonist to β-AR and lipolytic response.[134] Feeding of diets rich in n-3 polyunsaturated fatty acids, especially eicosapentaenoic (EPA) and docosahexaenoic acid (DHA), leads to decreasing activities of both lipogenic (LPL) and lipolytic (HSL) enzymes, especially in retroperitoneal fat.[135] In humans, very low caloric diet (VLCD) causes an increase in basal and catecholamine-stimulated lipolysis in subcutaneous adipose tissue and this is the result of an increased β_1-AR sensitivity whereas α_2-AR sensitivity is decreased.[136-138] Variations in α_2-adrenoceptor sensitivity in adipocytes may be predictive of weight loss during VLCD.[138] After massive body weight reduction, crude fat cell membranes from post-gastroplasty patients were hyper-responsive to isoproterenol when compared to those from normal-weight subjects.[139] The response was correlated negatively with cell size and positively with β-AR density and with the ratio of β-AR and stimulatory G-proteins.[139]

Obesity

Development of obesity is not only associated with increases in number and size of fat cells but also with tremendous changes of number and function of the various adrenoceptors. These changes can work in two directions; i) favoring or ii) preventing further increase in fat mass.

Impaired expression of adipocyte β-AR subtypes is a general feature of both genetic and dietary obesity in mice. The degree of obesity is correlated with the extent of loss of β_3-AR and β_1-AR expression in WAT. The distinct endocrine abnormalities associated with these obesity models may be responsible for the degree of impaired adipocyte β-AR expression.[140] In white epididymal adipose tissue of the ob/ob mouse, addition of β-adrenergic agents results in a very low lipolytic activity and this is due mainly to a defective cAMP production which has been associated with deficient levels of some of the isoforms of the G-proteins and also with low expression and functionality of the β_3-AR.[141,142] Adrenalectomy after weaning can restore β_3-AR expression in white adipocytes from ob/ob mice.[143]

In postmenopausal obese women, fat cell size and lipoprotein lipase activity are similar in abdominal and gluteal adipose tissue but the maximal lipolytic responsiveness and sensitivity to isoproterenol is higher in abdominal vs. gluteal adipocytes whereas maximal lipolytic response to a post-AR agent is similar.[144] This is due mainly to a higher β-AR affinity and lower α_2-AR relative to β-AR density in abdominal compared with gluteal adipocytes.[144] β-AR sensitivity is increased in subcutaneous abdominal adipose cells of obese men with a high visceral adipose tissue accumulation compared to those with a low intra-abdominal fat deposition.[145] Lipolytic sensitivity of subcutaneous abdominal adipocytes observed among men with high levels of visceral adipose tissue is also positively correlated with plasma insulin concentrations measured in the fasting state and after an oral glucose load.[145] Recent studies indicate a pathogenic role of visceral β_3-AR in obesity.[146] The rate of FFA and glycerol response to norepinephrine is significantly increased in visceral fat cells from obese when compared to normal subjects and within the obese group the response was higher in men vs. women.[146,147] There are, however, no differences in the lipolytic sensitivity to β_1- or β_2-AR specific agonist between the groups.

Table 8.3. Synthetic ligands for various adrenoceptor subtypes

Receptor type	Agonist ligand	Antagonist ligand
Alpha		
Alpha 1	Phenylephrine Methoxamin Amidephrine	Prazosine (α_1) Phentolamine (α_1, α_2)
Alpha 2	UK 14304 Clonidine	Yohimbine (α_2) Rauwolscine (α_2) Phentolamine (α_1, α_2)
Beta		
Beta, unspecific	Isoproterenol (β_1, β_2, β_3) Metaproterenol (β_1, β_2,β_3)	Bupranolol (β_1, β_2, β_3) Naldolol (β_1, β_2) Propranolol (β_1, β_2)
Beta 1-specific	Dobutamine	Bisoprolol Betaxolol
Beta 2-specific	Fenoterol Procaterol Salbutamol Zinterol	ICI 118 551, IPS 339
Beta 3-specific	BRL 37 344 CGP 12 177 CL 316 243 L-755 507 SR 58 611A Trecadrine	

But, the β_3-AR sensitivity is approximately 50 times enhanced and the coupling efficiency of these receptors is significantly increased whereas α_2-AR sensitivity is significantly decreased in visceral fat cells, at least in obese men.[146,147] Lipolysis induced by agents acting at the adenylyl cyclase and PKA levels is almost two-fold enhanced in men but there is no sex difference in maximum hormone-sensitive lipase activity.[147] These results are confirmed partly in two other studies indicating that upper body obesity is associated not only with metabolic complications such as insulin resistance but also with increased β_3-AR sensitivity.[146-149] Redistribution of β-AR in visceral adipose tissue of obese subject is obviously an efficient mechanism to prevent further increase in visceral fat mass because the β_3-AR is resistant toward desensitization/downregulation by catecholamines.[94,95] The antilipolytic effect of epinephrine and UK-14304 on α_2-AR is greater in abdominal and femoral subcutaneous adipose tissue of obese women when compared to lean ones and femoral fat cell lipolysis in the presence of these agents is

negatively correlated with body fatness indexes.[150] Transgenic mice overexpressing the
β_1-adrenergic receptor in adipose tissue are resistant to obesity because of

1. increased lipolytic rate and
2. induction of the abundant appearance of brown fat cells in subcutaneous white adipose
   tissue.[151]

In familial combined hyperlipidemia (FCHL), the lipolytic sensitivity to α_2-, β_1-, and
β_2-adrenoceptors is normal as is the number and affinity of β_1- and β_2-adrenoceptors
whereas the maximum enzymatic activity of hormone-sensitive lipase is decreased by
approximately 40%.[152]

In 1995, a genetic variation in the β_3-AR was described.[153,154] This variation is characterized by a replacement of tryptophan by arginine (Trp64Arg) in the first intracellular loop of the
receptor.[153,154] The allelic frequency of the mutation was found to be very similar among various populations (about 10%), except for Pima Indians (about 30%).[153-159] However, there are
differences between the populations regarding the functional consequences of the β_3-AR mutation. In one study, Caucasians morbidly obese and heterozygous for the Trp64Arg variant
showed an increased capacity to gain weight when compared to those without the mutation.[153]
In Finns, an association of the receptor variant with features of the insulin resistance syndrome
was found.[155] In three other studies, no substantial evidence was found for a role of the mutation in the pathophysiology of obesity in Caucasians.[156-158] In Pima Indians homozygous for
the mutation, the mean age of onset of non-insulin-dependent diabetes mellitus was significantly lower than in heterozygotes or wildtypes.[154] As a result of a large study including 1122
Japanese men and women, it was concluded that the missense mutation in the β_3-AR may not
be a main determinant of obesity and it may not be deleterious at least in non-obese individuals.[160] However, in obese Japanese either heterozygous or homozygous for the mutation both
visceral fat mass and the tendency to develop insulin resistance were significantly higher when
compared to subjects with the normal variant.[161,162] Interestingly, the mutation was more frequent in subjects with lower serum triglyceride levels.[162] This is in line with a small study of 18
Japanese women demonstrating that the Trp64Arg mutation of the β_3-AR gene deteriorates
lipolysis induced by a β_3-AR agonist in omental adipocytes.[163] However, in subcutaneous adipose tissue of both Pima Indians and Caucasians, no effect of the Trp64Arg β_3-adrenoceptor
variant on in vivo lipolysis could be observed.[164] Studies of the isolated β_3-AR expressed in
Chinese hamster ovary cells showed that the mutant form of the receptor is pharmacologically
and functionally indistinguishable from the wild type β_3-adrenergic receptor.[165] Interestingly,
subjects with the variant, even the heterozygotes, showed a lower resting activity of the autono-
mous nervous system activity than normal subjects.[166]

In the β_2-AR gene, polymorphisms were found in codons 16, 27, and 164.[167] In a small
population of obese women, only the polymorphism in codons 16 and 27 were common. The
Gln27Glu polymorphism was markedly associated with obesity but no significant association
with changes in β_2-AR function was observed.[167,168] In contrast to that, the Arg16Gly polymorphism was not associated with obesity, but subjects with the mutation showed a five-fold
increased agonist sensitivity of the β_2-AR without any change in β_2-AR expression.[167] At least
in Japanese men, the allelic frequency of the Glu27 mutation was found to be about 10%.[168]

Longer-term treatment with the highly specific β_3-adrenoceptor agonist CL 316,243 results in hypertrophy of brown adipocytes in BAT and WAT and a reversal of diet-induced
obesity in rats.[169] The reversal of obesity is accompanied by decrease in enlarged adipocytes, no
loss of a white adipocytes, increase in resting metabolic rate, but no change in food intake.[169]
CL 316,243 can also prevent the development of obesity if given as a supplement to a high-fat
diet by preventing the decline of β_3-AR mRNA in all adipose depots in A/J mice but not in
C57BL/6J mice.[170] Chronic low-dose administration of the β_3-AR agonist SR 58611A led to

a sustained improvement in glucose homeostasis in obese mice, whereas higher dose resulted in increased lipogenesis in obese mice.[171]

Exercise

Exercise generally mobilizes lipid stores to meet the energy requirement of muscle. Short-term submaximal exercise is accompanied by a higher increase in circulating lipid in women vs. men and this appears to be due to a sex difference in the adrenergic regulation of lipid mobilization. In men, exercise activates β-AR as well as α_2-AR, whereas in women only β-AR are activated, at least in abdominal subcutaneous adipose tissue.[172] In men, plasma epinephrine levels are increased 2.5-fold during exercise whereas norepinephrine levels are unchanged which implies activation of antilipolytic α_2-AR.[173] Chronic endurance training results in an improved β_1-AR-mediated lipolytic response whereas the β_2-AR-mediated lipolytic response is unchanged in obese non-diabetic men.[174] Additionally, the antilipolytic effects of α_2-AR and insulin are significantly decreased.[174] Improvement in β_1-AR-mediated lipolytic response is mainly the result of increased number and sensitivity of β_1-AR because no changes could be documented at the post-receptor level.[175] In abdominal subcutaneous adipose tissue, fractional re-esterification of fatty acids within the tissue is 20-30% in the basal state and declines during exercise.[176] Generally, exercise training may help to reduce abdominal obesity and to counteract the aberrant metabolic profile associated with abdominal obesity both directly or as a consequence of body fat loss.[177]

Aging

In both man and rat, aging is associated with a decrease in catecholamine-stimulated lipolysis.[178,179] This decrease is mainly attributed to a decrease in activity of hormone-sensitive lipase.[178,179] In healthy, non-obese subjects, all agents tested (acting at receptor and post-receptor levels) stimulated lipolysis at a 50% lower rate in elderly as compared with young subjects.[178] In epididymal adipocytes from aged male Wistar rats, epinephrine-sensitive lipolysis is decreased and this is possibly due to alterations of the lipolytic pathway distal to the receptor-adenylyl cyclase complex and the generation of cAMP.[179] There is no change in α_2- and β-AR number. However, in epididymal adipocytes from Fischer 344 rats, the expression of both β_3-AR and β_1-AR is decreased age-dependently.[180] This is in line with a reduced effectiveness of the β_3-AR agonist CL 316,243 (osmotic minipumps) regarding stimulation of lipolysis in aged rats.[181] Dehydroepiandrosterone treatment of young adult male rats may help to reduce age-related adiposity directly by increased lipolysis due possibly to a higher response of epididymal adipocytes towards isoproterenol.[182]

Adipose Tissue as a Source of Lactate

Beside its role in lipid metabolism, adipose tissue is also a significant source of lactate production. This has been shown recently in vitro and in vivo and lactate production can account for up to 50 % of glucose metabolism in adipose tissue.[32-35]

Lactate production by rat epididymal adipocytes is mainly under the control of α_1-adrenoceptors.[183] α_1-AR effects are mediated through the classic phosphoinositol-bisphosphate cycle (see Fig. 8.1). In short, after agonist binding to α_1-adrenoceptors, phospholipase C (PLC) is activated which hydrolyzes phosphatidylinositol-4,5-bisphosphate into diacylglycerol (DAG) and inositol-triphosphate (IP3). DAG acts as an second messenger and binds to a protein kinase C (PKC). Binding of DAG to PKC has two effects: increasing affinity of PKC for Ca^{++}-ions and its activity. Before binding of DAG it is necessary that a phospholipid (phosphatidylserin) binds to PKC. Therefore, PKC is also called Ca^{++}- lipid-dependent protein kinase. The PKC-dependent signaling sequence is mediated by a phospholipase-associated G-protein (G_q).

Stimulation of α_1-AR by specific agonists results not only in increased lactate production but also in stimulated glucose uptake by rat epididymal adipocytes.[184] Interestingly, glucose uptake is stimulated by isoproterenol indicating a role of β-AR, too.[184] In lean rats, lactate production is greater in mesenteric than epididymal adipocytes in both fed and fasted state.[185] In obese rats, lactate production differs not significantly between the two regions in the fed state. However, it is higher in mesenteric than epididymal adipocytes in the fasted state.[185] After an overnight fast, lactate release is almost identical from abdominal and subcutaneous adipose tissue in both lean and obese men.[186] Following an oral glucose load, interstitial lactate concentration measured by microdialysis increases significantly in both tissue regions but the increase is significantly higher in lean vs. obese subjects.[186] However, total lactate release from adipose tissue is estimated to be significantly higher in obese vs. lean subjects due to the higher fat mass.[186] Recently, two different aspects of lactate metabolism have been outlined in obesity. First, the association of increased basal lactate levels with increased obesity reflects an increased lactate production from enlarged adipocytes and an increased fat mass. Secondly, the inverse association between acute lactate generation following glucose ingestion and obesity reflects a decreased ability of adipose and/or extra-adipose tissues to convert glucose to lactate due to insulin resistance.[187] It is unclear, if the increased lactate production in adipose tissue of obese subjects is directly related to the development of insulin resistance. Furthermore, it would be interesting to know if chronically increased plasma lactate levels are accompanied by changes in the acid-base balance.

In rats, basal and norepinephrine-stimulated lactate production are significantly increased in adipocytes isolated from old vs. young rats.[188] Norepinephrine- and phenylephrine-sensitivity of lactate production is significantly decreased in adipose tissue of old vs. young rats but there is no difference between the age-groups in maximal lactate production obtained with norepinephrine and phenylephrine indicating that aging is associated with a decrease in α_1-AR number.[188] However, glucose uptake into the adipocyte and its metabolism are stimulated mainly by insulin, and it is unclear how insulin and α_1-AR agonists interact and how changes in that interaction could cause or promote the processes of developing obesity and aging.

Body Fat Distribution and the Role of Sex Steroids

Even at comparable body fat content, men and women differ with respect to body fat distribution. Concerning subcutaneous adipose tissue, men have a larger fat mass in the abdominal area (centripetal, abdominal or android body fat distribution) whereas women tend to have a larger fat mass in the gluteo-femoral area (centrifugal, peripheral or gynoid body fat distribution.)[37] Men display normally a larger visceral fat mass than women.[37] This difference in body fat distribution can be attributed in part to site-specific differences in the action of metabolic regulators such as insulin, catecholamines, and GH (see above). However, a primary role for body fat distribution is played by sex steroids.[189-192] Steroid hormones such as cortisol or sex hormones are rather responsible for long-term adaptation of adipose tissue metabolism forming the permissive basis for acute, short-term effects of the peptide hormones such as insulin and growth hormone, or catecholamine.[192] In vitro, testosterone inhibits LPL even in the presence of cortisol whereas it strongly activates lipid mobilization mainly via enhancing the β-adrenergic response at both receptor and post-receptor levels.[191,192] GH has a potentiating effect. In men, testosterone binds to an androgen receptor and this binding up-regulates the density of androgen receptors. Androgen receptors are present in a higher number in visceral adipose tissue than in subcutaneous adipose, at least in rats, indicating that the effect of testosterone should be more pronounced in visceral than in subcutaneous adipose tissue. In women, a receptor for androgens is also present in adipose tissue but it seems that this receptor is rather occupied by estrogens. Estrogen down-regulates androgen receptors in adipose tissue and this is obviously a mechanism for protecting the tissue from androgen effects. Hyperandrogenic women tend to accumulate visceral fat. Although there is no evidence for a physiological number of receptors for estrogen and progesterone in fat cell culture systems, postmenopausal replacement therapy with estrogen, for example, can prevent accumulation of visceral fat.[191,192] In vivo, administration of testosterone can prevent accumulation of visceral fat mass in obese men or diminish it in hypogonadal men.[191,192] On the other hand, replacement therapy of postmenopausal women with estrogen can prevent or reverse accumulation of visceral fat. Instead of that, triacylglycerols are accumulated in gluteo-femoral subcutaneous adipose tissue.[191,192]

Conclusions

Although white adipose tissue appears as a structurally almost homogeneous organ, there is now a large body of experimental evidence that it is functionally heterogeneous. This heterogeneity depends mainly on the anatomic location of adipose tissue. Generally, metabolic activity is highest in omental adipose tissue, intermediate in the abdominal subcutaneous region and lowest in the gluteo-femoral subcutaneous region. This difference in metabolic activity is mainly due to differences in the equipment of adipocytes of the respective regions with receptors for insulin (activating mainly LPL) or catecholamines (affecting mainly HSL). However, insulin and catecholamines are important for short-term control of adipose tissue metabolism whereas steroid hormones such as glucocorticoids or sex hormones can modulate long-term activity and body fat distribution. There are a lot of genetically (mutations or polymorphisms) or environmental (nutrition, exercise, drugs) factors that can induce changes in structure and function of adipose tissue. Functional changes can include all regulatory levels of adipose tissue metabolism. One of the most critical outcomes of complex functional changes is the accumulation of excess fat mass, i.e., obesity. Once present, obesity itself can induce changes in the control of adipose tissue metabolism leading finally to the development of the so-called metabolic syndrome. Therefore, it can be extremely difficult to distinguish causes from effects in developing obesity. Recent findings indicate that adipose tissue itself can produce a lot of factors important for regulating local metabolism and blood flow. Based on those findings, future studies will help to get a deeper insight not only into the molecular basis of the functional heterogeneity of adipose tissue but also into the role local adipose tissue-derived factors can play in the development of the metabolic syndrome.

References

1. Bray GA. Historical framework for the development of ideas about obesity. In: Bray GA, Bouchard C, James WPT, eds. Handbook of Obesity. New York: Marcel Dekker, 1998:1-29.
2. Vague J. La différenciation sexuelle-facteur déterminant des formes de l'óbésité. Presse Med 1947; 30:339-340.
3. Rodbell M. Metabolism in isolated fat cells. Effect of hormones on glucose metabolism. J Biol Chem 1964; 239:375-380.
4. Wolfe RR, Sidossis LS. Isotopic measurement of substrate flux and cycling. Int J Obes Rel Metab Disord 1993; 17(Suppl 3):S10-S13.
5. Jacobsson B, Holm G, Björntorp P, et al. Influence of cell size on the effects of insulin and noradrenaline on adipose tissue. Diabetologia 1976; 12:69-72.
6. Allen DO, Largis EE, Miller EA et al. Continous monitoring of lipolytic rates in perifused isolated fat cells. J Appl Physiol 1973; 1:125-127.
7. Katocs AS, Elwood EL, Allen DO et al. Perifused fat cells—Effect of lipolytic agents. J Biol Chem 1973; 248:5089-5094.

8. Solomon S, Duckworth WC. Effect of antecedent hormone administrations on lipolysis in the perifused isolated fat cell. J Lab Clin Med 1976; 88:984-994.

9. Turpin BP, Duckworth WC, Solomon SS. Perifusion of isolated rat adipose cells. J Clin Invest 1977; 60:442-448.

10. Boschmann M, Halangk W, Bohnensack R. Interrelation between respiration, substrate supply and redox ratio in perifused, permeabilized rat hepatocytes. Biochim Biophys Acta 1996; 1273:223-230.

11. DiGirolamo M, Thacker SV, Fried SK. Effects of cell density on in vitro glucose metabolism by isolated adipocytes. Am J Physiol 1993; 264:E361-E366.

12. Frayn KN, Coppack SW, Humphreys SM. Subcutaenous adipose tissue metabolism studied by local catheterization. Int J Obes Rel Metab Disord 1993; 17(Suppl 3):S18-S21.

13. Lönnroth P, Smith U. Microdialysis—A novel technique for clinical investigation. J Intern Med 1990; 227:295-300.

14. Arner P, Bolinder J. Microdialysis of adipose tissue. J Intern Med 1991; 230:381-386.

15. Ungerstedt U, Pycock C. Functional correlates of dopamine transmission. Bull Schweiz Akad Med Wiss 1974; 1278:1-13.

16. Bolinder J, Ungerstedt U, Arner P. Long-term continuous glucose monitoring with microdialysis in ambulatory insulin-dependent diabetic patients. Lancet 1993; 342:1080-1085.

17. Horal M, Ungerstedt U, Persson B et al. Metabolic adaptation in IUGR neonates determined with microdialysis—a pilot study. Early Hum Dev 1995; 42:1-14.

18. Hickner R, Rosdahl H, Borg I et al. Ethanol may be used with the microdialysis technique to monitor blood flow changes in skeletal muscle: dialysate glucose concentration is blood flow dependent. Acta Physiol Scand 1991; 143:355-356.

19. Enoksson S, Nordenstrom J, Bolinder J et al. Influence of local blood flow on glycerol levels in human adipose tissue. Int J Obes Relat Metab Disord 1995; 19(5):350-354.

20. Felländer G, Linde B, Bolinder J. Evaluation of the microdialysis ethanol technique for monitoring of subcutaneous adipose tissue blood flow in humans. Int J Obes Relat Metab Disord 1996; 20:220-226.

21. Hellerstein MK, Christiansen M, Kaempfer S et al. Measurement of de novo lipogenesis in humans using stable isotopes. J Clin Invest 1991; 87:1841-1852.

22. Leitch CA, Jones PJ. Measurement of human lipogenesis using deuterium incorporation. J Lipid Res 1993; 34:157-163.

23. Stubbs RJ. Appetite, feeding behavior and energy balance in human subjects. Proc Nutr Soc 1998; 57:341-356.

24. Hudgins LC, Hellerstein M, Seidmann C et al. Human fatty acid synthesis is stimulated by a eucaloric low fat, high carbohydrate diet. J Clin Invest 1996; 97:2081-2091.

25. Cushman SW, Wardzala LJ. Potential mechanism of insulin action on glucose transport in the isolated rat adipose cell. J Biol Chem 1980; 255:4758-4762.

26. Suzuki K, Kono T. Evidence that insulin causes translocation of glucose transport activity to the plasma membrane from an intracellular storage site. Proc Natl Acad Sci USA 1980; 77:2542-2555.

27. Thorens B. Glucose transporters in the regulation of intestinal, renal, and liver glucose fluxes. Am J Physiol 1996; 270:G541-G553.

28. Kahn BB. Lilly lecture 1995. Glucose transport: pivotal step in insulin action. Diabetes 1996; 45:1644-1654.

29. Steinberg D, Vaughan M, Margolis S. Studies of triacylglyceride-biosynthesis in homogenates of adipose tissue. J Biol Chem 1961; 236:1631-1637.

30. Arner P, Liljeqvist L, Östman J. Metabolism of mono and diacylglycerols in subcutaneous adipose tissue of obese and normal weight subjects. Acta Med Scand 1976; 200:187-194.

31. Björntorp P, Sjöström L. Carbohydrate storage in man: speculations and some quantitative considerations. Metabolism 1978; 27(Suppl 2):1853-1865.

32. Mårin P, Rebuffé-Scrive M, Smith U et al. Glucose uptake in human adipose tissue. Metabolism 1987; 36:1154-1160.

33. Jansson PA, Smith U, Lönnroth P. Evidence for lactate production by human adipose tissue in vivo. Diabetologia 1990; 33:253-256.

34. Hagström E, Arner P, Ungerstedt U et al. Subcutaneous adipose tissue: A source of lactate production after glucose ingestion in humans. Am J Physiol 1990; 258:E888-E893.

35. DiGirolamo M, Newby, FD, Lovejoy J. Lactate production in adipose tissue: A regulated function with extra-adipose implications. FASEB J 1992; 6:2405-2412.

36. Frayn KN, Coppack SW, Fielding BA et al. Coordinated regulation of hormone-sensitive lipase and lipoprotein lipase in human adipose tissue in vivo: Implications for the control of fat storage and fat mobilization. Adv Enzyme Regul 1995; 35:163-178.

37. Leibel RL, Edens NK, Fried SK. Physiologic basis for the control of body fat distribution. Annu Rev Nutr 1989; 9:417-443.

38. Arner P. Differences in lipolysis between human subcutaneous and omental adipose tissue. Ann Med 1995; 27:435-438.

39. Lafontan M, Barbe P, Galitzky et al. Adrenergic regulation of adipocyte metabolism. Hum Reprod 1997; 12(Suppl 1):6-20.

40. Eckel RH. Adipose tissue lipoprotein lipase. In: Borensztajn J, ed. Lipoprotein Lipase. Chicago: Evener Publishers, 1987:79-132.

41. Eckel RH. Lipoprotein lipase. A multifunctional enzyme relevant to common metabolic diseases. N Engl J Med 1989; 320:1060-1068.

42. Ong JM, Kern PA. The role of glucose and glycosylation in the regulation of lipoprotein lipase synthesis and secretion in rat adipocytes. J Biol Chem 1989; 264:3177-3182.

43. Kirchgessner TG, Svenson KL, Lusis AJ et al. The sequence of cDNA encoding lipoprotein lipase. A member of a lipase gene family. J Biol Chem 1987; 262:8463-8466.

44. Kern PA, Ong JM, Goers JWF et al. Regulation of lipoprotein lipase immunoreactive mass in isolated human adipocytes. J Clin Invest 1988; 81:398-406.

45. Brun LD, Gagne C, Julien P et al. Familial lipoprotein lipase activity deficiency: study of total body fatness and subcutaneous fat tissue distribution. Metabolism 1989; 38:1005-1009.

46. Weinstock PH, Levak-Frank S, Hudgins LC et al. Lipoprotein lipase controls fatty acid entry into adipose tissue, but fat mass is preserved by endogenous synthesis in mice deficient in adipose tissue lipoprotein lipase. Proc Natl Acad Sci 1997; 94:10261-10266.

47. Tozzo E, Shepherd PR, Gnudi L et al. Transgenic Glut-4 over-expression in fat enhances glucose metabolism: preferential effect on fatty acid synthesis. Am J Physiol 1995; 31:E956-E964.

48. Arner P, Lithell H, Wahrenberg H et al. Expression of lipoprotein lipase in different human subcutaneous adipose tissue regions. J Lipid Res 1991; 32:423-429.

49. Rebuffe-Scrive M, Enk L, Crona N et al. Fat cell metabolism in different regions in women. J Clin Invest 1985; 75:1973-1976.

50. Yost TJ, Eckel RH. Regional similarities in the metabolic regulation of adipose tissue lipoprotein lipase. Metabolism 1992; 41:33-36.

51. Fried SK, Russel CD, Grauso NL et al. Lipoprotein lipase regulation by insulin and glucocorticoid in subcutaneous and omental adipose tissue of obese women and men. J Clin Invest 1993; 92:2191-2198.

52. Kern PA, Marshall S, Eckel RH. Regulation of lipoprotein lipase in primary cultures of isolated human adipocytes. J Clin Invest 1985; 75:199-208.

53. Goldfine ID. The insulin receptor: Molecular biology and transmembrane signaling. Endocrinol Rev 1987; 8:235-255.

54. Smith U. Insulin resistance and signaling in human adipocytes—clinical and molecular aspects. In: Guy-Grand B, Ailhaud G, ed. Progress in Obesity: 8. London, Paris, Sydney: John Libbey & Company, 1999:485-488.

55. Le Marchand-Brustel Y, Tanti JF, Cormont M et al. Glucose entry into adipocyte: A site of hormonal regulation. In: Guy-Grand B, Ailhaud G, ed. Progress in Obesity: 8. London, Paris, Sydney: John Libbey & Company, 1999:77-82.

56. Saltiel AR, Fox JA, Sherline P et al. Insulin-stimulated hydrolysis of a novel glycolipid generates modulators of cAMP phosphodiesterase. Science 1986; 233:967-972.

57. Saltiel AR, Sherline P, Fox JA. Insulin-stimulated diacylglycerol production results from the hydrolysis of a novel phosphatidylinosiol glycan. J Biol Chem 1987; 262:1116-1121.

58. Yost TJ, Jensen DR, Eckel RH. Tissue-specific lipoprotein lipase: Relationships to body composition and body fat distribution in normal weight humans. Obes Res 1993; 1:1-4.

59. Ottosson M, Vikman-Adolsson K, Enerback S et al. The effects of cortisol on the regulation of lipoprotein lipase activity in human adipose tissue. J Clin Endocrinol Metab 1994; 79:820-825.

60. Miller LK, Kral JG, Strain GW et al. Differential binding of dexamethasone to ammonium sulfate precipitates of human adipose tissue cytosols. Steroids 1988; 49:507-522.
61. Rebuffé-Scrive M, Lundholm K, Björntorp P. Glucocorticoid hormone binding to human adipose tissue. Eur J Clin Invest 1985; 15:267-271.
62. Rebuffé-Scrive M, Brönnegård M, Nilsson A et al. Steroid hormone receptors in human adipose tissues. J Clin Endocrinol Metab 1990; 71:1215-1219.
63. Ottosson M, Vikman-Adolfsson K, Enerback S et al. Growth hormone inhibits lipoprotein lipase activity in human adipose tissue. J Clin Endocrinol Metab 1995; 80:936-941.
64. Richelsen B. Effect of growth hormone on adipose tissue and skeletal muscle lipoprotein lipase activity in humans. J Endocrinol Invest 1999; 22(Suppl 5):10-15.
65. Johannsson G, Mårin P, Lonn L et al. Growth hormone treatment of abdominally obese men reduces abdominal fat mass, improves glucose and lipoprotein metabolism, and reduces diastolic blood pressure. J Clin Endocrinol Metab 1997; 82:727-734.
66. Cianflone K, Sniderman A, Walsh M et al. Prufication and characterization of acylation-stimulating protein. J Biol Chem 1989; 264:426-430.
67. Cianflone K. Obesity and the adipocyte. Acylation stimulating protein and the adipocyte. J Endocrinol 1997; 155:203-206.
68. Coppack SW, Yost TJ, Fisher RM et al. Periprandial systemic and regional lipase activity. Am J Physiol 1996; 270:E718-722.
69. Fielding BA, Frayn KN. Lipoprotein lipase and the deposition of dietary fatty acids. Br J Nutr 1998; 60:495-502.
70. Yost TJ, Jensen DR, Haugen BR et al. Effect of dietary macronutrient composition on tissue-specific lipoprotein lipase activity and insulin action in normal-weight subjetcs. Am J Clin Nutr 1998; 68:296-302.
71. Evans K, Clark ML, Frayn KN. Effects of an oral and intravenous fat load on adipose tissue and forearm lipid metabolism. Am J Physiol 1999; 276:E241-E248.
72. Takahashi Y, Ide T. Effect of dietary fats differing in degree of unsaturation on gene expression in rat tissue. Ann Nutr Metab 1999; 43:86-97.
73. Murphy MC, Brooks CN, Rockett JC et al. The quantitation of lipoprotein lipase mRNA in biopsies of human adipose tissue, using the polymerase chain reaction, and the effect of increased consumption of n-3 polyunsaturated fatty acids. Eur J Clin Nutr 1999; 53:441-447.
74. Björntorp P, Smith U. The effect of fat cell size on subcutaneous adipose tissue metabolism. Front Matrix Biol 1976; 2:37
75. Mauriege P, Merette A, Atgie C et al. Regional variation in adipose tissue metabolism of severely obese premenopausal women. J Lipid Res 1995; 36:672-684.
76. LeFebvre AM, Laville M, Vega N et al. Depot-specific differences in adipose tissue gene expression in lean and obese subjects. Diabetes 1998; 47:98-103.
77. Ravussin E, Swinburn BA. Insulin resistance is a result, not a cause of obesity. Socratic debate: the pro side. In: Angel A, Anderson H, Bouchard C et al, ed. Progress in Obesity: 7. London, Paris, Rome, Sydney: John Libbey & Company, 1996:173-178.
78. Schwartz RS, Brunzell JD. Increase of adipose tissue lipoprotein lipase activity with weight loss. J Clin Invest 1981; 67:1425-1430.
79. Kern PA, Ong JM, Saffari B et al. The effects of weight loss on the activity and expression of adipose-tissue lipoprotein lipase in very obese humans. N Engl J Med 1990; 322:1053-1059.
80. Imbeault P, Almeras N, Richard D et al. Effect of a moderate weight loss on adipose tissue lipoprotein lipase activity and expression: existence of sexual variation and regional differences. Int J Obes Relat Metab Disord 1999; 23:957-965.
81. Rebuffe-Scrive M, Basdevant A, Guy-Grand B. Nutritional induction of adipose tissue lipoprotein lipase in obese subjects. Am J Clin Nutr 1983; 37:974-980.
82. Nikkila EA. Role of lipoprotein lipase in metabolic adaptation to exercise and training. In: Borensztajn J, ed. Lipoprotein Lipase. Chicago: Evener Publishers, 1987:187-199.
83. Savard R, Bouchard C. Genetic effects in the response of adipose tissue lipoprotein lipase activity to prolonged exercise. A twin study. Int J Obes 1990; 14:771-777.
84. Seip RL, Angelopoulos TJ, Semenkovich CF. Exercise induces human lipoprotein lipase gene expression in skeletal muscle but not in adipose tissue. Am J Physiol 1995; 268:E229-E236.

85. Simsolo RB, Ong JM, Kern PA. The regulation of adipose tissue and muscle lipoprotein lipase in runners by detraining. J Clin Invest 1993; 92:2124-2130.

86. Mauriege P, Prud'Homme D, Marcotte M et al. Regional differences in adipose tissue metabolism between sedentary and endurance-trained women. Am J Physiol 1997; 273:E497-E506.

87. Lamarche B, Despres JP, Moorjani S et al. Evidence for a role of insulin in the regulation of abdominal adipose tissue lipoprotein lipase response to exercise training in obese women. Int J Obes Relat Metab Disord 1993; 17:255-261.

88. Rowe JW, Minaker KL, Palotta JA et al. Characterization of the insulin resistance of aging. J Clin Invest 1983; 71:1581-1587.

89. Bergman DM, Rogus EM, Busby-Whitehead MJ et al. Predictors of adipose tissue lipoprotein lipase in middle-aged and older men: relationship to leptin and obesity, but not cardiovascular fitness. Metabolism 1999; 48:183-190.

90. Bergo M, Olivecrona G, Olivecrona T. Regulation of adipose tissue lipoprotein lipase in young and old rats. Int J Obes Relat Metab Disord 1997; 21:980-986.

91. Fain JN, García-Sáinz JA. Adrenergic regulation of adipocyte metabolism. J Lipid Res 1983; 24:945-966.

92. Lafontan M, Berlan M. Fat cell adrenergic receptors and the control of white and brown fat cell function. J Lipid Res 1993; 34:1057-1091.

93. Bousquet-Mélou A, Galitzky J, Moreno CM et al. Desensitation of beta-adrenergic responses in adipocytes involves receptor subtypes and cyclic AMP phosphodiesterase. Eur J Pharmacol (Mol Pharmacol) 1995; 289:235-247.

94. Lipworth BJ. Clinical pharmacology of β3-adrenoceptors. Br J Clin Pharmacol 1996; 42:291-300.

95. Farias-Silva E, Grassi-Kassisse DM, Wolf-Nunes V et al. Stress-induced alteration in the lipolytic response to beta-adrenoceptor agonists in rat white adipocytes. J Lipid Res 1999; 40:1719-1727.

96. Jansson PA, Smith U, Lönnroth P. Interstitial glycerol concentrations measured by microdialysis in two subcutaneous regions in humans. Am J Physiol 1990; 258:E918-E922.

97. Jansson PA, Larsson A, Smith U, Lönnroth P. Glycerol production in subcutaneous adipose tissue in lean and obese humans. J Clin Invest 1992; 89:1610-1617.

98. Mauriége P, Galitzky J, Berlan M et al. Heterogeneous distribution of beta- and alpha-2 adrenoceptor binding sites in human fat cells from various fat deposits: functional consequences. Eur J Clin Invest 1987; 17:156-165.

99. Mauriége P, Després JP, Prud'homme D et al. Regional variation in adipose tissue lipolysis in lean and obese men. J Lipid Res 1991; 32:1625-1633.

100. Galitzky J, Lafontan M, Nordenström J, Arner P. Role of vascular alpha-2 adrenoceptors in regulating lipid mobilization from human adipose tissue. J Clin Invest 1993; 91:1997-2003.

101. Deng C, Paoloni-Giacobino A, Kuehne F et al. Respective degree of expression of beta 1-, beta 2- and beta 3-adrenoceptors in human brown and white adipose tissue. Br J Pharmacol 1996; 118:9292-934.

102. Robidoux J, Pirouzi P, Lafond J et al. Site-specific effects of sympathectomy on the adrenergic control of lipolysis in hamster fat cells. Can J Physiol Pharmacol 1995; 73:450-458.

103. Barbe P, Galitzky J, De Glisezinski I et al. Simulated microgravity increases beta-adrenergic lipolysis in human adipose tissue. J Clin Endocrinol Metab 1998; 83:619-625.

104. Barbe P, Millet L, Galitzky J et al. In situ assessment of the role of the beta 1-, beta 2-, and beta 3-adrenoceptors in the control of lipolysis and nutritive blood flow in human subcutaneous adipose tissue. Br J Pharmacol 1996; 117:907-913.

105. Arch JSR, Ainsworth AT, Cawthorne MA et al. Atypical beta-adrenoceptor on brown adipocytes as target for antiobesity drugs. Nature 1984; 309:163-165.

106. Emorine LJ, Marullo S, Briend-Sutren MM et al. Molecular characterization of the human beta 3-adrenergic receptor. Science 1989; 245:1118-1121.

107. Nahmias C, Blin N, Elalouf JM et al. Molecular characterization of the mouse beta 3-adrenergic eceptor: Relationship with the atypical receptor of adipocytes. EMBO J (Eur Mol Biol Organ) 1991; 10:3721-3727.

108. Grannemann JG, Lahners KN, Chaudry A. Molecular cloning and expression of rat beta-3 adrenergic receptor. Mol Pharmacol 1991; 40:895-899.

109. Krief St, Lönnquist F, Raimbault S et al. Tissue distribution of beta 3-adrenergic receptor mRNA in man. J Clin Invest 1993; 91:344-349.

110. Berkowitz DE, Nardone NA, Smiley RM et al. Distribution of beta 3-adrenoceptor mRNA in human tissues. Eur J Pharmacol 1995; 289:223-228.

111. Chamberlain PD, Jennings KH, Paul F et al. The tissue distribution of the human beta 3-adrenoceptor studied using a monoclonal antibody: Direct evidence of the beta 3-adrenoceptor in human adipose tissue, atrium and skeletal muscle. Int J Obes Relat Metab Disord 1999; 23:1057-1065.

112. Muzzin P, Revelli JP, Kuhne F et al. An adipose tissue-specific beta-adrenergic receptor. Molecular cloning and down-regulation in obesity. J Biol Chem 1991; 266:24053-24058.

113. Evans BA, Papaioannou M, Bonazzi VR et al. Expression of beta 3-adrenoceptor mRNA in rat tissues. Br J Pharmacol 1996; 117:210-216.

114. Evans BA, Papaioannou M, Hamilton S et al. Alternative splicing generates two isoforms of the beta 3-adrenoceptor which are differentially expressed in mouse tissues. Br J Pharmacol 1999; 127:1525-1531.

115. Evans BA, Papaioannou M, Anastasopoulos F et al. Differential regulation of beta 3-adrenoceptors in gut and adipose tissue of genetically obese (*ob/ob*) C57BL/6J-mice. Br J Pharmacol 1998; 124:763-771.

116. Atgie C, Tavernier G, D'Allaire F et al. Beta 3-adrenoceptor in guinea pig brown and white adipocytes: low expression and lack of function. Am J Physiol 1996; 271:R1729-1738.

117. Portillo MP, Rocandio AM, Garcia-Calonge MA et al. Lipolytic effects of beta 1-, beta 2- and beta 3-adrenergic agonists in isolated human fat cells from omental and retroperitoneal adipose tissues. Rev Esp Fisiol 1995; 51:193-200.

118. Rosenbaum M, Malbon CC, Hirsch J et al. Lack of beta 3-adrenergic effect on lipolysis in human subcutaneous adipose tissue. J Clin Endocrinol Metab 1993; 77:352-355.

119. Tavernier G, Barbe P, Galitzky J et al. Expression of beta 3-adrenoceptor with low lipolytic action in human subcutaneous white adipocytes. J Lipid Res 1996; 37:87-97.

120. Enocksson S, Shimizu M, Lönnquist F et al. Demonstration of an in vivo functional beta 3-adrenoceptor in man. J Clin Invest 1995; 95:2239-2245.

121. Engfeldt P, Hellmér J, Wahrenberg H et al. Effects of insulin on adrenoceptor binding and the rate of catecholamine-induced lipolysis in isolated human fat cells. J Biol Chem 1988; 263:15553-15560.

123. Olansky L, Pohl SL. Beta-adrenergic desensitation by chronic insulin exposure in 3T3-L1 cultured adipocytes. Metabolism 1984; 33:76-81.

124. Feve B, El Hadri K, Quignard-Boulange A et al. Transcriptional down-regulation by insulin of the beta 3-adrenergic receptor expression in 3T3-F442A adipocytes. A mechanism for repressing the cAMP signaling pathway. Proc Natl Acad Sci USA 1994; 91:5677-5681.

125. Hadri KE, Chadron C, Pairault L et al. Down-regulation of beta 3-adrenergic receptor expression in rat adipose tissue during the fasted/fed transition: Evidence for a role of insulin. Biochem J 1997; 232:359-364.

126. Bolinder J, Sjoberg S, Arner P. Stimulation of adipose tissue lipolysis following insulin-induced hypoglycaemia: evidence of increased beta-adrenoceptor mediated lipolytic response in IDDM. Diabetologia 1996; 39:845-853.

127. Hagström-Toft E, Arner P, Johansson U et al. Effect of insulin on human adipose tissue metabolism in situ. Interactions with beta-adrenoceptors. Diabetologia 1992; 35:664-670.

128. Yang S, Bjorntorp P, Liu X et al. Growth hormone treatment of hypophysectomized rats increases catecholamine-induced lipolysis and the number of beta-adrenergic receptors in adipocytes: No differences in the effects of growth hormone on different fat depots. Obes Res 1996; 4:471-478.

129. Beauville M, Harant I, Crampes F et al. Effect of long-term rhGH administration in GH-deficient adults on fat cell epinephrine response. Am J Physiol 1992; 263:E467-E472.

130. Rosenbaum M, Gertner JM, Leibel RL. Effects of systemic growth hormone (GH) administration on regional adipose tissue distribution and metabolism in GH-deficient children. J Clin Endocrinol Metab 1989; 69:1274-1281.

131. Arner P. Adrenergic receptor function in fat cells. Am J Clin Nutr 1992; 55:S228-S236.

132. Ben Cheikh R, Chomard P, Beltramo JL et al. Influence of prolonged fasting and thyroid hormone on the alpha 2-adrenergic response in isolated epididymal adipocytes of Wistar rats. Eur J Endocrinol 1997; 136:223-229.

133. Germack R, Adli H, Vassy R et al. Triiodthyronine and amiodarone effects on beta-3 adrenoceptor density and lipolytic response to the beta 3-adrenergic agonist BRL 37344 in rat white adipocytes. Fundam Clin Pharmacol 1996; 10:289-297.

134. Fain JN, Coronel EC, Beauchamp MJ et al. Expression of leptin and beta 3-adrenergic receptors in rat adipose tissue in altered thyroid states. Biochem J 1997; 322:145-150.

135. Matsuo T, Sumida H, Suzuki M. Beef tallow diet decreases beta-adrenrgic receptor binding and lipolytic activities in different adipose tissues of rat. Metabolism 1995; 44:1271-1277.

136. Raclot T, Groscolas R, Langin D et al. Site-specific regulation of gene expression by n-3 polyunsaturated fatty acids in rat white adipose tissue. J Lipid Res 1997; 38:1963-1972.

137. Hellström L, Reynisdottir S, Langin D et al. Regulation of lipolysis in fat cells of obese women during long-term hypocaloric diet. Int J Obes Relat Metab Disord 1996; 20:745-752.

138. Barbe P, Stich V, Galitzky J et al. In vivo increase in beta-adrenergic lipolytic response in subcutaneous adipose tissue of obese subjects submitted to a hypocaloric diet. J Clin Endocrinol Metab 1997; 82:63-69.

139. Hellström L, Rössner S, Hagström-Toft E et al. Lipolytic catecholamine resistance linked to alpha 2-adrenoceptor sensitivity—A metabolic predictor of weight loss in obese subjects. Int J Obes Relat Metab Disord 1997; 21:314-320.

140. Kaartinen JM, LaNoue KF, Martin LF et al. Beta-adrenergic responsiveness of adenylate cyclase in human adipocyte plasma membranes in obesity and after massive weight reduction. Metabolism 1995; 44:1288-1292.

141. Collins S, Daniel KW, Rohlfs EM. Depressed expression of adipocyte beta-adrenergic receptors is a common feature of congenital and diet-induced obesity in rodents. Int J Obes Relat Metab Disord 1999; 23:669-677.

142. Begin-Heick N. Beta-adrenergic receptors and G-proteins in the *ob/ob* mouse. Int J Obes relat Metab Disord 1996; 20 (Suppl 3):S32-35.

143. Begin-Heick N. Of mice and women: The beta 3-adrenergic receptor, leptin and obesity. Biochem Cell Biol 1996; 74:615-622.

144. Gettys TW, Watson PM, Seger L et al. Adrenalectomy after weaning restores beta 3-adrenergic receptor expression in white adipocytes from C57BL/6J-ob/ob mice. Endocrinol 1997; 138:2697-26704.

145. Lönnquist F, Thorne A, Large V et al. Sex differences in visceral fat lipolysis and metabolic complications of obesity. Arterioscler Thromb Vasc Biol 1997; 17:1472-1480.

146. Bergman DM, Nicklas BJ, Rogus EM et al. Regional differences in adrenoceptor binding and fat cell lipolysis in obese, postmenopausal women. Metabolism 1998; 47:467-473.

147. Mauriege P, Brochu M, Prud'homme D et al. Is visceral adiposity a significant correlate of subcutaneous adipose cell lipolysis in men? J Clin Endocrinol Metab 1999; 84:736-742.

148. Lönnquist F, Thome A, Nilsell K et al. A pathogenic role of visceral fat beta 3-adrenoceptors in obesity. J Clin Invest 1995; 95:1109-1116.

149. Hoffstedt J, Wahrenberg H, Thorne A et al. The metabolic syndrome is related to beta 3-adrenoceptor sensitivity in visceral adipose tissue. Diabetologia 1996; 39:838-844.

150. Hoffstedt J, Arner P, Hellers G et al. Variation in adrenergic regulation of lipolysis between omental and subcutaneous adipocytes from obese and non-obese men. J Lipid Res 1997; 38:795-804.

151. Reynisdottir S, Eriksson M, Angelin B et al. Impaired activation of adipocyte lipolysis in familial combined hyperlipidemia. J Clin Invest 1995; 95:2161-2169.

152. Mauriege P, Prud'homme D, Lemieux S et al. Regional differences in adipose tissue lipolysis from lean and obese women: Existence of postreceptor alterations. Am J Physiol 1995; 269:E341-E350.

153. Soloveva V, Graves RA, Resenick MM et al. Transgenic mice overexpressing the beta 1-adrenergic receptor in adipose tissue are resistant to obesity. Mol Endocrinol 1997; 11:27-38.

154. Clement K, Vaisse C, Manning BS et al. Genetic variation in the beta 3-adrenergic receptor and an increased capacity to gain weight in patients with morbid obesity. N Engl J Med 1995; 333:352-354.

155. Walston J, Silver K, Bogardus C et al. Time of onset of non-insulin-dependent diabetes mellitus and genetic variation in the beta 3-adrenergic receptor gene. N Engl J Med 1995; 333:343-347.

156. Widen E, Lehto M, Kanninen T et al. Association of a polymorphism in the beta-3 adrenergic receptor gene with features of the insulin resistance syndrome in Finns. N Engl J Med 1995; 333:348-351.

157. Li LS, Lönnquist F, Luthman H et al. Phenotypic characterization of the Trp64Arg polymorphism in the beta 3-adrenergic receptor gene in normal weight and obese subjects. Diabetologia 1996; 39:857-860.

158. Arii K, Suehiro T, Yamamoto M et al. Trp64Arg mutation of beta 3-adrenergic receptor and insulin sensitivity in subjects with glucose intolerance. Intern Med 1997; 36:603-606.

159. Tchernof A, Starlin RD, Walston JD et al. Obesity-related phenotypes and the beta 3-adrenoceptor gene variant in the postmenopausal women. Diabetes 1999; 48:1425-1428.

160. Ongphiphadhanakul B, Rajatanavin R, Chanprasertyothin S et al. Relation of the beta 3-adrenergic receptor gene mutation to total body fat but not percent body fat and insulin levels in Thais. Metabolism 1999; 48:564-567.

161. Yuan X, Yamada K, Koyama K et al. Beta-3 adrenergic receptor gene polymorphism is not a major genetic determinant of obesity and diabetes in Japanese general population. Diabetes Res Clin Pract 1997; 37:1-7.

162. Sakane N, Yoshida T, Umekawa T et al. Beta 3-adrenergic receptor polymorphism: a genetic marker for visceral fat obesity and the insulin resistance syndrome. Diabetologia 1997; 40:200-204.

163. Kim-Motoyama H, Yasuda K, Yamaguchi T et al. A mutation of the beta 3-adrenergic receptor is associated with visceral obesity but decreased serum triglyceride. Diabetologia 1997; 40:469-472.

164. Umekawa T, Yoshida T, Sakane N et al. Trp64Arg mutation of beta 3-adrenoceptor gene deteriorates lipolysis induced by beta 3-adrenoceptor agonist in human omental adipocytes. Diabetes 1999; 48:117-120.

165. Snitker S, Odeleye OE, Hellmér J et al. No effect of the Trp64Arg beta 3-adrenoceptor variant on in vivo lipolysis in subcutaneous adipose tissue. Diabetologia 1997; 40:838-842.

166. Candelore MR, Deng L, Tota LM et al. Pharmacological characterization of a recently described human beta 3-adrenergic receptor mutant. Endocrinol 1996; 137:2638-2641.

167. Shihara N, Yasuda K, Moritani T et al. The association between Trp64Arg polymorphism of the beta 3-adrenergic receptor and autonomic nervous system activity. J Clin Endocrinol Metab 1999; 84:1623-1627.

168. Large V, Hellström L, Reynisdottir S et al. Human beta 2-adrenoceptor gene polymorphisms are highly frequent in obesity and associated with altered adipocyte beta 2-adrenoceptor function. J Clin Invest 1997; 100:3005-3013.

169. Mori Y, Kim-Motoyama H, Ito Y et al. The Gln27Glu beta 2-adrenergic receptor variant is associated with obesity due to subcutaneous fat accumulation in Japanese men. Biochem Biophys Res Comm 1999; 258:138-140.

170. Ghorbani M, Claus TH, Himms-Hagen J. Hypertrophy of brown adipocytes in brown and white adipose tissues and reversal of diet-induced obesity in rats treated with a beta 3-adrenoceptor agonist. Biochem Pharmacol 1997; 54:121-131.

171. Collins S, Daniel KW, Petro AE et al. Strain-specific response to beta 3-adrenergic receptor agonist treatment of diet-induced obesity in mice. Endocrinol 1997; 138:405-413.

172. Williams CA, Shih MF, Taberner PV. Sustained improvement in glucose homeostasis in lean and obese mice following chronic administration of the beta-3 agonist SR 58611A. Br J Pharmacol 1999; 128:1586-1592.

173. Hellström L, Blaak E, Hagström-Toft E. Gender differences in adrenergic regulation of lipid mobilization during exercise. Int J Sports Med 1996; 17:439-447.

174. Stich V, De Glisezinski I, Crampes F et al. Activation of antilipolytic alpha 2-adrenergic receptors by epinephrine during exercise in human adipose tissue. Am J Physiol 1999; 277:R1076-R1083.

175. De Glisezinski I, Crampes F, Harant I et al. Endurance training changes in lipolytic responsiveness of obese adipose tissue. Am J Physiol 1998; 275:E951-E956.

176. Stich V, De Glisezinski I, Galitzky J et al. Endurance training increases the beta-adrenergic lipolytic response in subcutaneous adipose tissue in obese subjects. Int J Obes Relat Metab Disord 1999; 23:374-381.

177. Hodgetts V, Coppack SW, Frayn KN et al. Factors controlling fat mobilization from human subcutaneous adipose tissue during exercise. J Appl Physiol 1991; 71:445-451.

178. Buemann B, Tremblay A. Effects of exercise training on abdominal obesity and related metabolic complications. Sports Med 1996; 21:191-212.

179. Dax EM, Partilla JS, Gregerman RI. Mechanism of the age-related decrease of epinephrine-stimulaed lipolysis in isolated rat adipocytes: β-adrenergic receptor binding, adenylate cyclase activity, and cyclic AMP accumulation. J Lipid Res 1981; 22:934-943.

180. Lönnquist F, Nyberg B, Wahrenberg H et al. Catecholamine-induced lipolysis in adipose tissue of the elderly. J Clin Invest 1990; 85:1614-1621.

181. Gettys TW, Rohlfs EM, Prpic V et al. Age-dependent changes in beta-adrenergic receptor subtypes and adenyly cyclase activation in adipocytes from Fischer 344 rats. Endocrinology 1995; 136:2022-2032.

182. Kumar MV, Moore RL, Scarpace PJ. Beta 3-adrenergic regulation of leptin, food intake, and adiposity is impaired with age. Pflüger's Arch 1999; 438:681-688.

183. Tagliaferro AR, Ronan AM, Payne J et al. Increased lipolysis to beta-adrenergic stimuation after dehydroepiandrosterone treatment in rats. Am J Physiol 1995; 268:R1374-R1380.

184. Faintrenie G, Géloën A. Alpha 1-adrenergic regulation of lactate production by white adipocytes. J Pharmacol Exp Ther 1996; 277:235-238.

185. Faintrenie G, Géloën A. Alpha 1-adrenergic stimulation of glucose uptake in rat white adipocytes. J Pharmacol Exp Ther 1998; 286: 607-610.

186. King JL, DiGirolamo M. Lactate production from glucose and response to insulin in perifused adipocytes from mesenteric and epididymal regions of lean and obese rats. Obes Res 1998; 6:69-75.

187. Jansson PA, Larsson A, Smith U et al. Lactate release from the subcutaneous tissue in lean and obese men. J Clin Invest 1994; 93:240-246.

188. Lovejoy J, Mellen B, DiGirolamo M. Lactate generation following glucose ingestion: relation to obesity, carbohydrate tolerance and insulin sensitivity. Int J Obes 1990: 14:843-855.

189. Faintrenie G, Géloën A. Effect of aging on norepinephrine and phenylephrine stimulated lactate production by white adipocytes. Obes Res 1997; 5:100-104.

190. Björntorp P. Adipose tissue distribution and function. Int J Obes 1991; 15:67-81.

191. Björntorp P. Visceral obesity: A 'civilizaion syndrome'. Obes Res 1993; 1:206-222.

192. Björntorp P. The regulation of adipose tissue distribution in humans. Int J Obes Relat Metab Disord 1996; 20:291-302.

193. Björntorp P. Hormonal control of regional fat distribution. Human Reprod 1997; 12(Suppl 1):21-25.

CHAPTER 9

White Adipose Tissue as a Secretory and Endocrine Organ:

Leptin and Other Secreted Proteins

Paul Trayhurn, Nigel Hoggard, D. Vernon Rayner

We are currently undergoing a revolution in our perspectives on the physiological role of white adipose tissue (WAT). Traditionally, white fat has been viewed primarily as a long-term energy storage organ. The deposition of triacylglycerol with twice the energy density of carbohydrate and with little associated water (in the region of 15%) provides a concentrated and highly efficient energy reserve. The central role of WAT in the storage of lipid has resulted in lipogenesis and lipolysis being viewed as the major metabolic processes associated with the organ, and much is known about the nature and regulation of these pathways. A second role widely attributed to WAT, at least for the subcutaneous deposits, is that of thermal insulation and in the case of the blubber of aquatic mammals such as whales the tissue may even have a mechanical role. Brown fat, the other form of adipose tissue, has the generation of heat, both thermoregulatory and in relation to the regulation of energy balance, as its primary function (see Chapters 4 and 5 by Klaus).

The overall picture of the central functions of WAT has changed radically in recent years, with white fat now being viewed as a major endocrine organ, secreting both a critical hormone and a number of protein factors which play either a local or systemic role. By virtue of the secretion of these diverse factors it is now evident that WAT is actively involved in a range of physiological processes, including the regulation of energy balance and the control of blood pressure and of vascular homeostasis. In this chapter we present an overview of the endocrine and secretory role of WAT. The protein secreted by white fat which has been principally responsible for altering our views on the functional role of the tissue is the cytokine-like hormone leptin and this factor is therefore the principal focus of the chapter.

Fatty Acids and Steroid Secretions

Before considering leptin and other proteins it should be emphasized that fatty acids are quantitatively the most important secretion from WAT (Fig. 9.1) and these result from the stimulation of lipolysis. A number of factors which influence lipolysis in WAT have long been recognized, particularly from studies on isolated adipocytes.[1,2] These include insulin, glucagon, ACTH, growth hormone, epinephrine (adrenaline) and norepinephrine (noradrenaline).[1] White fat provides a store of cholesterol and it is also implicated in the metabolism of steroid hormones.[3] The tissue does not synthesize steroid hormones as such, but it does express

Adipose Tissues, edited by Susanne Klaus. ©2001 Eurekah.com.

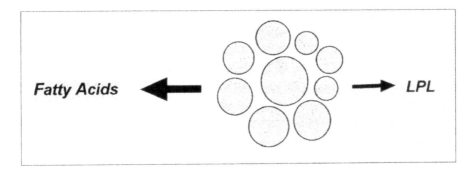

Fig. 9.1.   The "classical' view of the secretory functions of white adipose tissue. LPL, lipoprotein lipase.

enzymes which are involved in the metabolism of both glucocorticoids and of sex hormones which are subsequently released. Estrone is converted to estradiol and androstenedione to testosterone, while androgens can be aromatized to estrogens.[3]

There is a general presumption that physiologically the sympathetic nervous system (SNS) is the key regulator of the breakdown of triacylglycerols.[1] This is based on several lines of evidence: (i) denervation leads to an increase in adipose tissue mass; (ii) electrical stimulation of the sympathetic nerves to WAT results in fatty acid release; (iii) the pharmacological abolition of sympathetic activity inhibits the mobilization of lipid.[1] In addition, more recent studies measuring norepinephrine turnover have demonstrated that there is a marked sympathetic activation in white fat in specific situations in which there is net lipolysis, namely cold exposure and fasting.[4,5] In fasted animals this sympathetic activation is selective to WAT since norepinephrine turnover falls in other tissues such as brown fat and the heart.[6,7] Overall, there is strong evidence that the SNS plays a dominant role in the control of fatty acid release from white fat.

Central to the argument for a pivotal role for catecholamines from the sympathetic system in the control of lipolysis is the relatively extensive presence of adrenoceptors, both α and β, on the plasma membrane of white adipocytes.[8,9] β_1, β_2 and β_3-adrenoreceptors are present, and these provide the means by which lipolysis can be stimulated by norepinephrine or epinephrine. In rodents, the β_3-adrenoceptor is viewed as the principle receptor-subtype through which the stimulation of lipolysis occurs.[9,10] In humans, on the other hand, the β_2-adrenoceptor is much more important. Although direct evidence for the presence of the β_3-subtype in human adipose tissue has recently been reported,[11] there is in practice uncertainty on the extent to which the β_3-adrenoceptor is implicated in lipolysis in humans.

Leptin

Leptin (from the Greek leptos, thin or small), which is also termed OB protein, was discovered in 1994 by Friedman and colleagues[12] with the identification of the mutant gene which underlies the development of the profound obesity of the genetically obese (*ob/ob*) mouse. The analysis of the obese mutant, which is the most extensively studied animal model in obesity research, has provided a route for the elucidation of a previously unknown regulatory protein and its associated physiological system.

The *ob* gene encodes a protein of 18,000 Dalton containing a signal sequence which is cleaved to produce the mature cytokine-like hormone of 16,000 Daltons.[12] The initial conception, based on the lipostatic theory of the regulation of energy balance,[13] was that leptin is secreted from WAT in proportion to the amount of the tissue with the hormone being viewed as a satiety factor acting on receptors in the hypothalamus. The picture that has emerged over the past 6 years is considerably more complex, however; indeed, it is not even clear that satiety can be regarded as 'the central function' of leptin.

Leptin Production in Adipose Tissue

On its discovery the *ob* gene was initially shown to be strongly expressed in WAT[12] and this formed the basis for the presumption that the encoded protein provides the molecular basis for lipostasis. It is, however, now evident that leptin is produced in several sites additional to WAT. Synthesis occurs in brown adipose tissue,[14-18] the stomach,[19] placenta,[20-22] mammary gland,[23.24] ovarian follicles[25] and certain fetal organs such as the heart and bone/collagen.[20.26] Furthermore, expression can be induced in skeletal muscle[27] and there is even some indication that leptin may be synthesized in the brain.[28] The functional significance of leptin production in organs other than white fat is not certain, particularly as there is unlikely to be a substantive contribution to the total circulating pool of the hormone. A local rather than systemic effect is probable in some of the more confined sites of leptin synthesis. This is apparent in the case of the stomach where the leptin produced may be linked to the release of cholecystokinin.[19] A potential function in relation to some localized sites of production, such as the placenta, may be angiogenesis given that leptin has now been identified as an angiogenic factor (see below).[29.30]

Despite the various sites of expression of the *ob* gene, there are strong reasons for believing that WAT is quantitatively the principal site of leptin synthesis and the major determinant of the circulating level of the hormone. This is evident from the correlation between plasma leptin and body fatness in both humans and experimental animals.[31-34] It is also implicit in the observation that transgenic mice with little adipose tissue have very low circulating leptin levels.[35.36]

WAT is, of course, a heterogeneous organ which is distributed in different parts of the body, and this is an unusual situation with respect to the production of a hormone. The *ob* gene is expressed in each of the major adipose tissue sites, but there are substantive differences in the level of *ob* mRNA between depots.[37-39] There are also differences between species and according to developmental stage. In adult rodents, the level of *ob* mRNA is highest in the gonadal and perirenal adipose tissue and lowest in the subcutaneous sites.[39] In humans, however, the subcutaneous depot exhibits higher levels of *ob* mRNA than omental fat.[37.40.41] Rodent studies indicate that there are substantial developmental changes; in suckling rats, in contrast to the mature animal, the level of *ob* mRNA is much higher in the subcutaneous adipose tissue than in the internal fat.[42] On the assumption that relative levels of mRNA reflect relative rates of leptin production, it would seem that in the suckling rodent subcutaneous adipose tissue is the main site of production of the hormone but that after weaning the internal depots are much more important. The functional significance of these species and age-related differences is not easy to characterize.

Regulation of Leptin Production

The emerging picture of the factors which influence the expression of the *ob* gene and the production of leptin relates principally to adipocytes and the major WAT depots. The main determinant of the circulating level of leptin is the amount of white fat. Thus leptin levels are higher in obese animals than in lean, regardless of the form of obesity—with the exception of the *ob/ob* mouse which does not produce the functional hormone by virtue of the mutation in the *ob* gene.[34.43] As noted earlier, in humans there is a strong correlation between circulating leptin and body mass index (and other indices of fatness). However, acute regulation of leptin synthesis is superimposed on the endogenous production associated with the amount of WAT.

A number of factors which can affect expression of the *ob* gene and leptin production have been identified, both by in vivo and by cell culture studies (Fig. 9.2). The first major influence to be demonstrated was that of nutritional status, with fasting in rodents leading to a marked fall in the level of *ob* mRNA and the amount of circulating leptin—changes which are rapidly reversed on refeeding.[34,39,44,45] These acute responses to fasting and refeeding do not reflect alterations in adipose tissue mass since the size of the major fat depots alters little during the duration of most such experiments. A fall in leptin level and production on fasting is also apparent in humans, and this has been shown to be relatively rapid in studies in which leptin has been measured in the direct venous drainage from WAT.[46,47] Changes in leptin levels also occur with chronic under- and over-nutrition.[48]

Acute exposure to cold leads to a suppression of *ob* gene expression and a decline in circulating leptin level, responses which are reversed by exposure to the warm.[34,49,50] Cold adaptation appears to result in the maintenance of low leptin levels and this may be linked to the hyperphagia that is required to maintain energy balance in the cold-adapted animal.[50] Initially, exercise training was considered to lead to a fall in *ob* gene expression and circulating leptin.[51] However, the effects of exercise seem complex; in many cases any changes in leptin may simply reflect alterations in fat mass.[52,53] Falls in leptin in severe exercise probably reflect the effects of major changes in sympathetic activity (see below).

An intriguing proposal is that there is direct nutrient sensing in both adipose tissue and muscle leading to inducible expression of the *ob* gene through the hexosamine biosynthetic pathway.[27] At a hormonal level, insulin and glucocorticoids have been widely shown to stimulate leptin production.[44,54-57] There are regulatory effects of sex hormones with testosterone inhibiting production and being linked to the lower circulating leptin level in men as compared with women.[58,59] Estrogens, on the other hand, stimulate leptin production.[60] Stimulation is also reported by lipopolysaccharide administration and by the cytokines interleukin-1 and TNF-α.[61-64] Other agents that affect leptin production include thiazolidinediones, which activate peroxisome proliferator-activator receptor-γ.[65,66] It is unclear whether there is a significant effect of either growth hormone or thyroid hormones; however, the variability in the responses obtained suggests that these hormones do not have a major effect on expression of the *ob* gene.[43]

Some of the most potent effects on *ob* gene expression and leptin production are observed with catecholamines. One of the earliest findings was that acute administration of norepinephrine or the β-adrenoceptor agonist, isoprenaline, inhibits *ob* gene expression in WAT leading to a reduction in circulating leptin levels.[15,43,49] Epinephrine also induces a rapid and marked fall in circulating leptin.[67] Studies on human subjects are consistent with those on experimental animals, isoprenaline inducing an acute suppression of plasma leptin.[68] The suppressive effects of catecholamines obtained in whole-body studies are paralleled by results in vitro on adipocytes or adipocyte cell culture systems,[69-71] indicating that there is a direct interaction between catecholamines and WAT in the control of leptin production.

The inhibition of leptin synthesis by catecholamines in rodents appears to occur primarily through β_3-adrenoceptors, since both in vivo and in vitro studies have shown that selective β_3-agonists (e.g., BRL 35153A, CL316243) have a powerful suppressive effect on *ob* gene expression with a rapid reduction in circulating leptin.[10,15,70,72,73] This has led to the view that the SNS constitutes a negative feedback loop to white fat, regulating leptin synthesis by inhibiting *ob* gene transcription through the β_3-adrenoceptor.[10,49,72] While such a view may generally be correct in terms of the role of the sympathetic system, the nature of the β-adrenoceptor subtype involved may vary according to species. In particular, in humans the significance of β_3-adrenoceptors in the regulation of adipose tissue metabolism is uncertain, and β_1/β_2-adrenoceptors may be much more important., particularly the β_2.[9] Nevertheless, there is evidence that the β_3-adrenoceptor may play some role in the control of lipolysis in human adipocytes.[74]

At the level of the adipocyte, it is apparent that the larger the cell the greater the level of *ob* mRNA and presumably the amount of leptin that is synthesized. The mechanisms that lead to higher rates of production within an expanding cell have not been identified. The possibilities include a volume stretch effect or a specific effect of fatty acids. However, since the lipolytic sensitivity to catecholamines falls in obesity, the production of leptin may increase in response to decreased sympathetic stimulation/responsiveness as fat cells enlarge.

A Central Regulatory Role for the Sympathetic Nervous System in Leptin Production?

Although as we have described there are a number of factors which can acutely influence leptin production, an important challenge is the identification of those which are physiologically important—and whether any particular factor plays a pivotal role. We have argued previously that norepinephrine released from the SNS is the dominant physiological regulator of leptin production with the sympathetic system playing a tonic role.[43,75] There is a parallel with lipolysis in that while several hormones can influence the process, in vivo the SNS is considered the key regulator, as noted earlier. Apart from any direct effects on leptin production of norepinephrine secreted from the sympathetic nerve endings within white fat, there is the possibility of an indirect interaction via the sympathetic innervation of the adrenals through the release of epinephrine.

Studies aimed at assessing the physiological significance of the SNS in the control of leptin production have utilized the blockade of catecholamine synthesis with α-methyl-*p*-tyrosine, an inhibitor of tyrosine hydroxylase (the rate limiting enzyme in catecholamine synthesis). Administration of this compound leads to a rapid increase (up to 8-fold within 6-10 h) in plasma leptin in mice.[75] Thus blockade of norepinephrine production leads to marked hyperleptinemia and this appears to be due to increased synthesis since the level of *ob* mRNA in epididymal white fat is increased by treatment with α-methyl-*p*-tyrosine. This is consistent with the proposition that sympathetic tone may be the central regulator of leptin production in WAT through modulation of the transcription of the *ob* gene. In contrast, in human studies no effect of α-methyl-*p*-tyrosine on circulating leptin has been observed.[76] However, the apparent difference between the human and the rodent studies could reflect the difficulty in delivering a dose of the compound to humans sufficient to ensure that catecholamine synthesis is fully blocked.

Further evidence for a central role physiologically for the sympathetic system in the regulation of leptin production comes from studies on the effects of β-adrenoceptor antagonists. The administration of a combination of the general β-adrenoceptor antagonist, propranolol, and the selective β_3-antagonist, SR 59230A, inhibits the fall in circulating leptin level occurring with cold exposure and on fasting (Rayner et al, unpublished).[77] This suggests that the reduction in leptin production with both these conditions is mediated primarily through sympathetic activation. Nevertheless, the regulatory role of the SNS appears complex, as recent studies employing 6-hydroxydopamine have suggested.[78] Indeed, the effects of α-methyl-*p*-tyrosine may involve a central as well as a peripheral inhibition of norepinephrine synthesis and this may complicate the interpretation of studies utilizing the compound.

Sympathetic Innervation of White Adipose Tissue

It is widely recognized that brown adipose tissue is densely innervated by sympathetic nerve endings and that these are adjacent to the adipocytes themselves as well as to blood vessels. The sympathetic system plays a central regulatory role in the generation of heat in

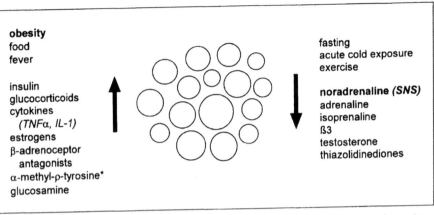

obesity
food
fever

insulin
glucocorticoids
cytokines
  (TNFα, IL-1)
estrogens
β-adrenoceptor
  antagonists
α-methyl-ρ-tyrosine*
glucosamine

fasting
acute cold exposure
exercise

noradrenaline *(SNS)*
adrenaline
isoprenaline
ß3
testosterone
thiazolidinediones

Fig. 9.2. A schematic view of the major factors which influence leptin production in white adipose tissue. IL-1, interleukin-1; SNS, sympathetic nervous system; TNF-α, tumour necrosis factor α.

brown fat, particularly in the stimulation of lipolysis and the activation of the mitochondrial proton conductance pathway (see Chapter 4 by Klaus). The extent to which WAT is innervated has, in contrast, been much less clear. The issue has recently been discussed in a review by Bartness[79] and elsewhere in this book (see Chapter 6 by Bartness et al). There is now compelling evidence for a direct sympathetic innervation of WAT, this including the adipocytes themselves as well as the vasculature. Evidence for a sympathetic innervation of WAT comes primarily from histological and fluorescence immunohistochemical studies.[79] Direct sympathetic nerve endings adjacent to adipocytes have been clearly demonstrated using fluorescence immunocytochemistry for catecholamines.[80] Despite these observations, it is evident that the innervation of WAT is much less extensive than that of brown fat.

There is also evidence that the SNS is involved in the control of fat cell number since unilateral denervation of inguinal WAT leads to an increase in the number of adipocytes relative to the contralateral fat pad in which the innervation remains intact.[81] Thus the sympathetic system appears to play a trophic role in the tissue.[79, 81] Studies employing a retrovirus technique have been used to map the source of the sympathetic innervation to WAT.[82] This technique depends upon the virus being taken up by nerve endings following injection into the tissue and then being translocated back to the roots of the sympathetic innervation. Such studies have demonstrated that the sympathetic innervation of white fat has its origins in regions of the hypothalamus, including the paraventricular nucleus and the dorsomedial hypothalamus, as well as in the suprachiasmatic nucleus.[82]

The interaction between the sympathetic system and leptin is two-way, the hormone stimulating sympathetic activity in WAT and other tissues through its hypothalamic receptors,[83,84] with afferent signals from leptin sensors in white fat exerting a reflex effect.[85] Thus, leptin may be seen to regulate its own synthesis by modulating sympathetic activity in WAT, high leptin levels leading to a stimulation of the SNS which in turn lowers production of the hormone.

Pulsatile Release of Leptin

Circulating leptin levels show a distinct diurnal variation in both humans and rodents. The levels rise during the afternoon and reach a peak in the early hours of the morning, then declining towards dawn.[86] In humans, evidence has been presented that the leptin rhythm is entrained to the meal pattern.[87] However, the circadian rhythm persists during continuous enteral nutrition, albeit at a reduced amplitude[88] and is related to the day/night activity pattern and to changes in body temperature, possibly through the sympathetic system. The circadian rhythm for leptin in rodents may have a different explanation as the zenith occurs at a time of high activity and intake.

Leptin is released in a pulsatile manner. In humans, estimates of the ultraradian frequency vary from 3-32 per 24 h depending on the experimental conditions and the conditions of analysis with the number of pulses detected increasing with the frequency of sampling.[86, 88-91] Sampling at least every 7 min for 4 h detects approximately 30 pulses per day, with a pulse therefore lasting on average 48 min.[91] The documentation of ultraradian and circadian pulsatility raises two important and related questions:

1. how is pulsatility from a given adipose tissue depot regulated, and
2. how is the pulsatile release from several adipose tissue depots coordinated?

It is, of course, unlikely that pulsatility could simply be reflective of the removal of leptin from the blood, so the most likely explanation for the phenomenon lies in the control of secretion – either by a stimulation or through an inhibition. One possibility is that pulsatility is due to the periodic firing of the SNS and given the emphasis that we have given to the regulatory role of the sympathetic system in leptin production there are attractions to such an idea; certainly it provides the necessary facility for coordination.[52]

Leptin Receptor and Neuroendocrine Interactions

The leptin receptor, which was cloned one year after the discovery of leptin itself,[92] exists in a series of splice variants as a result of alternative splicing.[93] One or more of the receptor isoforms—Ob-Ra, Ob-Rb etc—are found in most tissues.[94,95] The Ob-Ra isoform is considered to play a transport role, especially with respect to the passage of leptin across the blood-brain barrier.[96, 97] One splice variant, Ob-Rb, is a long form with an intracellular signaling domain which interacts with the JAK-STAT system.[93,98] Ob-Rb is regarded as the key signaling form of the leptin receptor, although signaling may also occur through other receptor variants via the MAP kinase pathway.[99]

The Ob-Rb form is found at high levels in regions of the hypothalamus such as the arcuate nucleus and paraventricular nucleus.[100,101] Thus those parts of the brain long-associated with the control of energy balance are a major target for leptin. It is, however, increasingly apparent that Ob-Rb is also distributed widely in other parts of the brain, such as the hindbrain, thalamus and cortex, as well as in peripheral tissues.[100,102,103] The diverse peripheral tissues in which Ob-Rb has now been identified include the kidneys, adrenal medulla, ß-cells of the pancreas, the placenta and adipose tissue.[20,94,104-106] The leptin receptor can mediate leptin internalization via coated pits, leading to receptor down-regulation, and degradation can occur in lysosomes.[107,108]

The wide distribution of the leptin receptor, both peripherally and centrally, indicates that there are a range of target tissues for leptin. Ob-Ra and Ob-Rb isoforms have been identified in WAT itself, implying that the hormone can have either an autocrine or paracrine role within the tissue.[109,110] Indeed lipolysis may be stimulated by leptin in white adipocytes and may attenuate the responses to insulin.[111,112] An important local role for leptin through its receptors within WAT may relate to the proposed function as an angiogenic factor;[29,30] extensive remodeling of the vasculature is required as adipose tissue expands and contracts with the development and reversal of obesity or during any change which leads to major alterations in fat mass.

Leptin interacts with several neuroendocrine systems in the hypothalamus[103] (Fig 9.3). The initial focus was on neuropeptide Y, reflecting the potent stimulatory effect of this brain peptide on food intake; neuropeptide Y is down-regulated by leptin.[113,114] Other neuroendocrine targets now recognized for leptin are glucagon-like peptide-1,[115] the pro-opiomelanocortin system,[116] corticotrophin releasing hormone[114,117] and CART (cocaine- and amphetamine-regulated transcript).[118] These neuropeptides, which have an inhibitory effect on food intake, are each up-regulated by leptin.[115-118] Evidence for an interaction between leptin and different neuropeptide systems comes both from the colocalization of the leptin receptor with specific peptides within neurons and from functional studies. The current perspective is that leptin affects multiple neuroendocrine systems involved in the control of food intake in an integrated manner, there being an inhibition of intake through a combination of the up-regulation of satiety neuropeptides and a down-regulation of stimulatory influences on intake (see ref. 119).

Functions of Leptin

Leptin is a powerful satiety factor,[120-122] operating centrally as described in the previous section, and the initial perspective was that the suppression of food intake is the principal function of the hormone. This view has now changed substantially in that there are a wide range of actions which leptin has been reported to elicit. The hormone also affects energy balance through the stimulation—or prevention of a reduction—in energy expenditure.[120-124] However, the link with energy expenditure may have been exaggerated through studies on mice which use torpor as an adaptive response to conserve energy.[52]

The functions attributable to leptin in practice go far beyond an energy balance paradigm (Fig. 9.4). The hormone acts as a major signal to the reproductive system, the administration of leptin restoring the infertility of female *ob/ob* mice and accelerating the time of the first estrus in normal mice.[125-127] Leptin has also been implicated as a factor in angiogenesis; leptin receptors are present on human endothelial cells and the hormone has been shown to induce angiogenesis both in vitro and in vivo.[29,30] Further major functions with which leptin is involved are hematopoiesis[128] and as a signal to the immune system.[129]

Because of the powerful suppressive effect of leptin on food intake in *ob/ob* mice, there has been much interest in dissecting the possible influences of the hormone on satiety. Rodents, of course, eat discrete meals and these are taken predominantly during the nocturnal active phase. It is known that the hyperphagia of genetically obese *ob/ob* mice and *fa/fa* rats is due mainly to increased meal size.[130,131] The decrease in food intake after administration of leptin to normal rats is due to a decrease in meal size, an effect which is seen with single or multiple central i.c.v. infusions or after daily subcutaneous treatment for several days; it starts after a short lag phase and is long lasting.[132-135] In humans, a number of studies have shown that plasma leptin concentrations do not change acutely at the period around meal times.[52,87,136] Thus, it seems unlikely that leptin can convey the message of satiation to the brain towards the end of a meal, i.e., it is not a short-term satiety signal.

In addition to major physiological processes, leptin has been reported to influence a diverse spectrum of metabolic functions. These range from brain development to platelet aggregation, bone formation, insulin gene transcription, the inhibition of insulin secretion by pancreatic β-cells, the stimulation of sugar transport and the induction of metallothionein gene expression.[104,137-142] Indeed, leptin would seem to be pervasive in its actions and the range of processes which the hormone is reputed to influence is continuing to expand very rapidly. It is likely, however, that some of the many reported responses to leptin represent epiphenomena and some may well reflect secondary or indirect effects; certainly they require independent verification.

In any discussion on the function of leptin the question of binding proteins has to be considered. A high proportion of plasma leptin is bound to specific and/or nonspecific proteins in the circulation and greater levels of circulating free leptin have been reported in obese

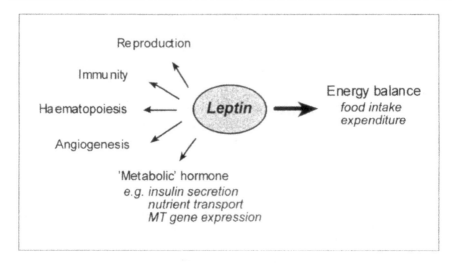

Fig. 9.3. Diagrammatic representation of the major physiological functions currently ascribed to leptin. MT, metallothionein.

individuals.[143] In both rodents and humans, leptin is mainly cleared by the kidneys where it is degraded rather than being excreted unchanged in the urine.[110,144,145]

Is There a Unifying View of the Function of Leptin?

In view of the plethora of functions described for leptin, it is pertinent to ask whether there is a unifying concept for the actions of this hormone. The first serious attempt to provide an integrating hypothesis for the role of leptin was the proposal that it is principally a starvation signal. Ahima et al[146] showed that the administration of recombinant leptin to fasted mice reverses, or substantially attenuates, a number of the hormonal and neuroendocrine changes which occur on fasting. These include alterations in the thyroid, adrenal and gonadal axes in males and the starvation-induced delay in ovulation in females.[146] In addition, the fasting-induced falls in energy expenditure are prevented in mice by leptin.[124] Subsequent studies have expanded the concept of leptin as a starvation signal, most notably in the demonstration that the immunosuppressive effects of food deprivation can be reversed in mice by treatment with leptin.[129] Since fasting leads to a selective increase in sympathetic activity in WAT,[5] with a fall in leptin synthesis, the SNS may be the conduit through which food deprivation results in leptin-dependent neuroendocrine and immune cell adaptations.

A recent proposal suggests that a core role of leptin is to channel fatty acids into adipose tissue and limit triacylglycerol deposition in other tissues.[147] The function and viability of nonadipocytes can be compromised when the triacylglycerol content extends beyond the normal physiological range. Evidence for this challenging concept includes the effects of triacylglycerol deposition on pancreatic β-cells in Zucker *fa/fa* rats, which have a mutation in the leptin receptor.[147] Lipid is deposited in the pancreatic β-cells and other nonadipose cells of these animals in parallel with the deposition of lipid in adipose tissue, and initially insulin production rises until a point where lipoapoptosis occurs and frank diabetes ensues. Such a process could explain the profound hyperinsulinemia of transgenic lipodystrophic mice, and the reversal of insulin resistance and diabetes on the administration of leptin.[35,36,148]

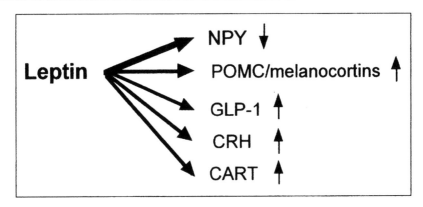

Fig. 9.4. A view of the major neuroendocrine interactions of leptin. CART, cocaine- and amphetamine-regulated transcript; CRH, corticotrophin releasing hormone; GLP-1, glucagon-like peptide-1; POMC, pro-opiomelanocortin.

Transgenic lipodystrophic mice in practice provide a further perspective on the criticality of WAT in its endocrine role. One of the characteristics of the animals is low levels of leptin mRNA in the residual adipose tissue together with greatly reduced circulating leptin levels. Since the administration of leptin results in a normalization of plasma glucose and the circulating insulin also falls towards normal, adipose tissue through the provision of leptin would seem to be essential for normal glucose homeostasis and insulin sensitivity.[35]

Human Leptin and Leptin Receptor Mutations

Several polymorphisms in the leptin and leptin receptor genes have been described but there is as yet no major association with obesity.[149-153] However, three reports have now identified mutations in the coding region of these genes in conjunction with severe obesity.[154-156] In the first report, obesity was described in two young children with a mutation in codon 133 of the mRNA for leptin.[154] In the others, which involved older subjects, mutations in the leptin or leptin receptor genes were associated with hypogonadism as well as with marked obesity.[155,156] Thus there are human equivalents of the *ob/ob* and the *db/db* mouse; indeed one of the reports on human subjects has identified a mutation in the same codon (105) as that causing a premature stop codon in the *ob/ob* mouse.[93,155,157,158]

The significance of these findings is not the presence of specific mutations per se, but the demonstration that the leptin system is as important physiologically for the regulation of body fat and energy balance in man as it is in rodents. The studies also indicate that leptin from adipose tissue is a critical signal for the maturation of the reproductive system in humans, again paralleling the situation in experimental animals. However, humans with mutations in the leptin gene or its receptor do not exhibit all the defects of their rodent counterparts. Thus there is neither hypothermia nor evidence of a low energy expenditure, and no indication of a defective hypothalamic-pituitary-adrenal, -thyroid, or growth hormone axis, each of which are observed in the obese mutant rodents.[154,159] Despite these species differences it should be noted that in the Turkish pedigree family with mutant leptin,[155] 7 out of 11 obese children died after infections, perhaps demonstrating the importance of leptin in the response to immune challenges.

A rapidly growing range of clinical conditions outside obesity have been examined in which evidence for a link with leptin has been sought. Cushing's disease and chronic renal disease, for example, are associated with high leptin levels while congenital lipodystrophy and anorexia are characterized by low leptin (see ref. 52). Some of the large number of clinical studies are based on a clear hypothesis while others reflect a nonspecific approach; in some cases (e.g., anorexia nervosa) changes in leptin seem to essentially reflect alterations in body fat rather a defined clinical pathology.

Leptin Sensitivity and Resistance

In principle, for leptin to function as a key signal of the state of the adipose stores in energy balance regulation its effects depend both on the amount of the hormone and the sensitivity of the end organ receptor systems. With the recognition that obesity is not characterized by reduced levels of leptin, but on the contrary the levels are generally elevated, the concept of leptin resistance as a key factor in the biological effects of the hormone has become embedded.[31,32,160] Indeed, human obesity has been widely considered as resulting from leptin resistance. This concept has, however, been challenged as representing an oversimplification.[161] Nevertheless, some modulation in the response to leptin at the receptor level and downstream from the receptor would be expected in common with other hormones. Thus it is likely that changes in sensitivity may have a role in transmitting the effects of leptin. In the extreme situation of leptin receptor mutations, which have been observed in a small number of human subjects and which characterize the *db/db* mouse and the *fa/fa* rat, resistance to the hormone is clearly central to the development of the obesity and reproductive disorders.[93,156-158]

One of the complexities in determining whether resistance, or substantive changes in sensitivity, are an important component in the transmission of the leptin signal is the wide range of biological effects that are now observed. Resistance therefore has to be defined in terms of the specific effect in question. This can be illustrated by the transgenic lipodystrophic mice referred to earlier. These animals have low leptin levels consequent upon the paucity of WAT, but in contrast to *ob/ob* mice (from which the functional hormone is completely absent) food intake is similar to that of wild-type mice.[35] This suggests that only a limited amount of leptin is required for the effects of the hormone on food intake and energy balance to be realized. However, hyperglycemia, hyperinsulinemia and insulin resistance are present in the transgenic animals and for the restoration of these parameters to normal the administration of exogenous leptin so that plasma levels are similar to wild-type mice is required – at least this is the case for one of the two models that have been developed.[35]

Human trials where leptin has been administered to obese subjects have shown little effect on body weight,[162] and this is consistent with the view that only a low amount of leptin is required for the energy balance actions of the hormone. The sole report demonstrating a substantial effect of leptin on body weight in humans relates to the treatment of one of the first identified children with a leptin mutation.[159]

Adipose Tissue Secreted Proteins

Although leptin is a major, indeed critical, factor secreted from white fat, it is far from being the only protein released from the tissue. A range of other protein factors are secreted, as summarized in Fig. 9.5. They include the cytokines TNF-α and interleukin-6, adipsin, adiponectin, angiotensinogen, plasminogen activator inhibitor-1, cholesteryl ester transfer protein and retinol binding protein. The enzyme lipoprotein lipase, which is responsible for the delivery of fatty acids from circulating triacylglycerols, has of course long been recognized as a secreted product of adipocytes.

The physiological role of each of the various protein secretions from WAT is not established, but it is recognized that they could in principle play either an autocrine, paracrine or endocrine role – or a combination thereof. To date, because obesity has been and remains a key focus in studies on adipose tissue, greater emphasis has been placed on the pathological significance of changes in the production of different factors in the face of an expanded adipose mass. In this section we outline a number of the key protein factors secreted from WAT; for other recent reviews see refs. 3, 163. The challenge in the longer term is to provide a rationale in functional terms as to why particular factors are secreted from WAT, especially when they are but one element in a process involving a number of different proteins (as in vascular hemostasis).

Methodological Considerations

WAT is a heterogeneous organ, not only in terms of differences between individual sites but also with respect to the cells which are present within a given depot. Indeed, the tissue consists of several cell types in addition to mature white adipocytes, the stromal-vascular fraction which includes fibroblasts and macrophages accounting for at least half of the total cells in the organ.[164] Thus if adipose tissue is found to express a particular gene, it is necessary to establish whether the expression occurs within the adipocytes themselves. The most common way to do this is to separate adipocytes from the stromal-vascular component by collagenase digestion and then to probe for the mRNA of interest in the two fractions. Alternatively, techniques such as in situ hybridization and immunohistochemistry for the direct localization of the mRNA or protein, respectively, within a cell type can be employed.

On a further methodological point, both in vivo and in vitro studies can be readily performed with adipose tissue in investigating potential secretory proteins, although each approach clearly has its own specific objectives. For in vitro studies mature adipocytes may be harvested and incubated for short periods of time. Alternatively, either primary cell culture can be employed by inducing fibroblastic preadipocytes to differentiate into adipocytes, or murine clonal cell lines (e.g., 3T3-L1 or F442A) can be utilized. Primary cell culture is particularly valuable in the direct investigation of human adipocytes. In vivo approaches such as microdialysis and the collection of the venous drainage from WAT provide a valuable means by which the production of secreted proteins from the tissue can be determined. This has proved particularly fruitful in human studies.

Proteins of Lipid and Lipoprotein Metabolism

Several proteins which play an important role in lipid and lipoprotein metabolism are released from white adipocytes. The enzyme lipoprotein lipase is secreted and this was in effect the first identified protein secretory product of adipocytes. Lipoprotein lipase is responsible for the breakdown of circulating triacylglycerols, in the form of chylomicrons and very low density lipoproteins, to fatty acids. Following transport into the adipocyte the fatty acids are subsequently re-esterified to triacylglycerol. A number of factors which regulate the expression of the lipoprotein lipase gene and the level and activity of the protein have been identified, with insulin playing an important role. Several reviews on lipoprotein lipase in adipose tissue—its function and regulation—have been presented, to which the reader is referred.[165,166] Other secreted proteins from WAT that are directly involved in lipid and lipoprotein metabolism include cholesteryl ester transfer protein and apolipoprotein E, the former playing an important role in the accumulation of cholesterol ester by adipose tissue (see refs. 3,167,168). Indeed, WAT is a substantial site of cholesteryl ester transfer protein synthesis.[168]

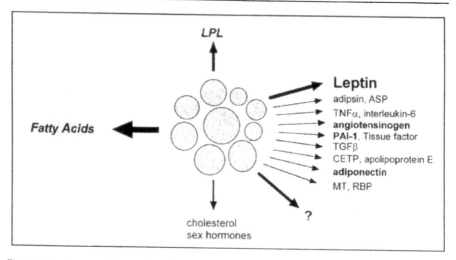

Fig. 9.5. Proteins secreted from white adipose tissue. ASP, acylation stimulating protein; CETP, cholesteryl ester transfer protein; LPL, lipoprotein lipase; MT, metallothionein; PAI-1, plasminogen activator inhibitor-1; RBP, plasma retinol binding protein; TGF-β, transforming growth factor β; TNF-α, tumor necrosis factor α.

Angiotensinogen

Among the various protein factors additional to leptin secreted by white fat, angiotensinogen has attracted much attention. White adipocytes are regarded as an important source of angiotensinogen and may be second only to the liver in this regard.[169] Angiotensinogen is the substrate for renin in the renin-angiotensin system, which plays a central role in blood pressure regulation both by affecting renal function and by modulating vascular tone. The activation product of angiotensinogen, angiotensin II, stimulates the production and release of prostacyclin which acts as a potent signal in the differentiation of preadipocytes to adipocytes.[170] Expression of the angiotensinogen gene in WAT is inhibited by fasting and by catecholamines,[171] and this has parallels with leptin and the *ob* gene. There are similar parallels to leptin in the effects of glucocorticoids, which stimulate angiotensinogen gene expression and the production of the secreted protein.[172]

Insulin, principally on the basis of studies on cultured adipose cells, appears to inhibit angiotensinogen production, both at the level of gene expression and of protein secretion.[173] Correspondingly, there is also an association with insulin resistance in that in this situation the level of angiotensinogen mRNA is increased.[173] There is an elevation in circulating angiotensinogen in obesity and this is thought to reflect the rise in adipose tissue mass. Thus hypertension in obese subjects may result from increased secretion of angiotensinogen from the expanded adipose tissue depots. Recent studies indicate that adipose tissue expresses both the angiotensin converting enzyme and type 1 angiotensin receptor genes, as well as angiotensinogen.[174-176] This suggests the presence of a local renin-angiotensin system in adipose tissue, with potentially important physiological and pathological implications.

Plasminogen Activator Inhibitor-1 and Tissue Factor

WAT secretes at least two proteins involved in the fibrinolytic system and vascular hemostasis, namely tissue factor and plasminogen activator inhibitor-1 (PAI-1), the latter being the focus of considerable recent interest. Tissue factor is the key cellular initiator of the coagulation cascade and it acts as a cell-surface receptor for the activation of factor VII. The gene encoding tissue factor is expressed in white fat and expression is higher in *ob/ob* mice than in normal animals.[177] The administration of transforming growth factor β (TGF-β), which is itself expressed in adipose tissue, increases the level of tissue factor mRNA in white fat.[177]

There is an increased incidence of cardiovascular disease in obesity and this has been linked to the rise in the circulating level of PAI-1.[178] Indeed, there is a correlation between plasma PAI-1 levels and body mass index,[179] with visceral fat being a particular predictor of circulating PAI-1 protein in the obese.[180,181] WAT is now regarded as a significant site of the production of PAI-1, and the tissue may well be the source of the elevated levels in obesity.[182-184] As its name implies, PAI-1 inhibits the activation of plasminogen and is the primary physiological inhibitor of this proenzyme, plasminogen being the precursor of plasmin which breaks down fibrin.

The PAI-1 gene has been shown to be expressed and PAI-1 protein released in WAT, both human and rodent.[182,185,186] A number of factors are recognized to stimulate either PAI-1 gene expression or the production of the protein by adipose tissue. TGF-β, TNF-α and insulin are each stimulatory, with some synergistic effects being recorded.[187,188] On the other hand, there is also evidence that catecholamines may suppress production with β₃-adrenoceptor agonists being inhibitory in rodents.[189,190] This suggests that the sympathetic system may play a role in the regulation of PAI-1 production similar to that proposed for lipolysis and the synthesis of leptin.

Cytokines and Growth Factors

Several important 'classical' cytokines and growth factors are synthesized in WAT, particularly TNF-α, interleukin-6 and TGF-β. TNF-α production is increased in obesity and the cytokine has been implicated in the development of insulin resistance in the adipocyte in the obese by altering insulin signaling through an autocrine or paracrine action.[191] The cytokine may also spill out from adipose tissue to contribute to the circulating pool. In addition to the production of TNF-α itself, soluble receptors for the cytokine are synthesized in white fat.[192-194] Given the multiplicity of effects now attributed to TNF-α and interleukin-6, including a role in the regulation of the production of other adipose tissue-derived factors,[193,195] these cytokines may be particularly important in the integration of the function of the tissue. Interleukin-6 has recently been identified as being produced in WAT, both the mRNA and the protein being demonstrated in the tissue, with increased levels of production in obesity.[3,194]

The TGF-β gene is expressed in white fat and the level of both the mRNA and the protein are increased in genetically obese rodents (*ob/ob* and *db/db*) compared with their lean counterparts.[187] TGF-β is released from adipocytes and TNFα stimulates gene expression and the production of the protein.[187] An interaction between different protein secretions from WAT seems to occur with TNF-α stimulating the synthesis of TGF-β which in turn leads to an increase in the production of PAI-1, as noted above.

Adipsin and Acyl Stimulation Protein

After lipoprotein lipase, adipsin was the first major protein secreted from white fat to be identified.[196] At the time there was considerable excitement on its discovery as a factor which was expressed in a differentiation-dependent manner in adipocyte cell lines.[196] Expression of the adipsin gene was initially found to be greatly decreased in obesity in animal models, with reduced levels of the circulating protein, leading to the view that it was a key lipostatic signal in the regulation of energy balance.[197,198] However, adipsin, which is a serine protease, is not reduced in human obesity and it is a part of the alternative complement pathway (complement factor D).

Adipsin is not the only protein of the alternative complement system synthesized by WAT. Recent interest has centered on acylation stimulating protein, or C3ades-Arg, which is derived from the C3 complex through the action of adipsin, factor B and a carboxypeptidase. Several roles have been proposed for acylation stimulating protein in lipid metabolism.[199] Studies on transgenic mice lacking acylation stimulating protein through a deficiency of C3, support the hypothesis that acylation stimulating protein is important in the postprandial clearance of triacylglycerols.[200,201] The protein stimulates the uptake and esterification of fatty acids into white adipocytes and has further metabolic effects on the cells.[199] Although these effects are local to adipose tissue, with acylation stimulating protein playing an autocrine or paracrine role, the factor may also act in an endocrine manner on other tissues.

Retinol Binding Protein

Retinol is stored in WAT and the gene encoding plasma retinol binding protein is strongly expressed in adipocytes.[202-204] Indeed, it is suggested that retinol binding protein is one of the most abundant transcripts in both rodent and human adipose tissue;[202,205] the cellular retinol binding protein gene is also expressed in WAT, though not at the same high levels.[203,204] Cell culture studies indicate that the plasma retinol binding protein is secreted from adipocytes where it presumably contributes in part to the total circulating pool.[203,204] However, quantitatively the liver and kidney have been regarded as the main sites of retinol binding protein production.[206] Recent studies in our group suggest that in obesity, at least in the case of the *ob/ob* mouse, expression of the retinol binding protein gene is not increased (Keeley, unpublished results). Overall, it is evident that adipocytes are dynamically involved in the storage and metabolism of retinoids (see ref. 207).

Adiponectin

Adiponectin is one of the most recently discovered secretory proteins from WAT. It was identified from a cDNA, apM1 (adipose most abundant gene transcript 1), reflecting a gene which is abundantly and specifically expressed in adipose tissue. Adiponectin has homology with collagen VIII and collagen X and with complement factor C1q.[208] It is suggested that adiponectin may modulate endothelial adhesion molecules and there are proposals that the protein is involved in the link between atherosclerosis and obesity.[209] In contrast to many of the proteins secreted by adipose tissue, expression of the adiponectin gene and the circulating levels of the protein fall in obesity.[210] Much is likely to emerge on this intriguing adipose-derived factor in the near future.

Metallothionein—A Novel Secretory Product?

Metallothionein (MT), a low molecular weight (6,000 Dalton) stress response and metal-binding protein, has long been recognized to be produced in several tissues, including liver and kidney.[211] It has recently been shown that the MT gene is also expressed in brown fat where it is suggested that the protein may play an antioxidant role.[212] Our recent studies have demonstrated that both the MT-1 and MT-2 genes are additionally expressed in WAT, the mRNA being present in several fat depots with no major site differences (Trayhurn and Beattie, unpublished). The MT genes are expressed in the adipocytes themselves rather than the stromal-vascular cells.

Subsequent studies have indicated that the level of MT-1 mRNA in WAT is unaltered by fasting or by the administration of norepinephrine. However, injection of a β_3-agonist induced a significant increase in MT-1 mRNA level, suggesting that expression of the MT gene can be subject to adrenergic activation, given a sufficiently potent stimulus. In in vitro studies, the differentiation of fibroblastic preadipocytes to adipocytes in primary culture was associated with the release of MT protein into the medium. This release occurred before the secretion of leptin and was not a reflection of the general leakage of cell contents. These observations raise the intriguing possibility that MT could be a secretory product of adipocytes, notwithstanding the absence of a signal sequence in the protein.

Within adipocytes, MT may protect fatty acids from oxidative damage, given the general postulate that the protein has an antioxidant function. If MT is secreted from adipocytes, one possibility is that it plays a signaling role and there is some initial evidence for such a proposition.[213]

Coda

It is clear that white adipocytes are endocrine and secretory cells of considerable complexity, despite the apparent simplicity implied by the very large proportion of the intracellular content contributed exclusively by triacylglycerols. The discovery of leptin has been pivotal in terms of the changed perspective on the biological functions of white fat. Unsurprisingly, most recent studies on the signaling and secretory role of the tissue have been directed towards leptin and the changes that occur in obesity. It is, however, important that the full range of protein signals and secretions from white fat are explored together with their significance in different physiological and pathological conditions. It is also critical that an integrative view of the homeostatic function of WAT is developed in view of the diverse nature of the factors that it releases. Integration requires an appreciation of the heterogeneity of white fat in terms of the different depots.

It seems highly probable that other proteins secreted from WAT will be discovered in the near future; certainly there is no reason to suppose that all such secretions have already been identified. Both gene-based approaches and proteomics are currently being harnessed in the search for novel secreted factors. Proteomics in particular may prove especially valuable with the explosion of interest in the application of this methodology in the rapidly emerging era of 'functional genomics.' Obesity will undoubtedly remain a key driver in research on adipose tissue, given the continuing escalation of the incidence of the disorder in both developed and developing countries.

Acknowledgments

We are grateful to our colleagues in the Molecular Physiology Group at the Rowett Research Institute for their support and to the Scottish Executive Rural Affairs Department for funding.

References

1. Hales CN, Luzio JP, Siddle K. Hormonal control of adipose tissue lipolysis. Biochem Soc Trans 1978; 43:97-135.
2. Vernon RG, Clegg RA. The metabolism of white adipose tissue in vivo and in vitro. In: Cryer A, Van RLR, eds. New Perspectives in Adipose Tissue: Structure, Function and Development. London: Butterworths, 1985:65-86.
3. Mohamed-Ali V, Pinkney JH, Coppack SW. Adipose tissue as an endocrine and paracrine organ. Int J Obesity 1998; 22:1145-1158.
4. Garofalo MAR, Kettelhut IC, Roselino JES et al. Effect of acute cold exposure on norpinephrine turnover rates in rat white adipose tissue. J Autonom Nerv Sys 1996; 60:206-208.
5. Migliorini RH, Garofalo MAR, Kettelhut IC. Increased sympathetic activity in rat white adipose tissue during prolonged fasting. Am J Physiol 1997; 272:R656-R661.
6. Young JB, Landsberg L. Suppression of sympathetic nervous system during fasting. Science 1976; 196:1473-1475.
7. Landsberg L, Young JB. The role of the sympathoadrenal system in modulating energy expenditure. Clinics Endocrinol Metab 1984; 13:475-499.
8. Lafontan M, Berlan M. Fat cell adrenergic receptors and the control of white and brown fat cell function. J Lipid Res 1993; 34:1057-1091.
9. Sennitt MV, Kaumann AJ, Molenaar P et al. The contribution of classical ($\beta(1/2-)$) and atypical β-adrenoceptors to the stimulation of human white adipocyte lipolysis and right atrial appendage contraction by novel $\beta(3)$-adrenoceptor agonists of differing selectives. J Pharmacol Exp Therap 1998; 287:1135.
10. Giacobino JP. Role of the $\beta(3)$-adrenoceptor in the control of leptin expression. Horm Metabol Res 1996; 28:633-637.
11. Chamberlain PD, Jennings KH, Paul F et al. The tissue distribution of the human β3-adrenoceptor studied using a monoclonal antibody: Direct evidence of the b-adrenoceptor in human adipose tissue, atrium and skeletal muscle. Int J Obesity 1999; 23:1057-1065.
12. Zhang YY, Proenca R, Maffei M et al. Positional cloning of the mouse obese gene and its human homolog. Nature 1994; 372:425-432.
13. Kennedy GC. The role of depot fat in the hypothalamic control of food intake in the rat. Proc Roy Soc Lond B Biol Sci 1953; 140:578-592.
14. Deng CJ, Moinat M, Curtis L et al. Effects of β-adrenoceptor subtype stimulation on obese gene messenger ribonucleic acid and on leptin secretion in mouse brown adipocytes differentiated in culture. Endocrinology 1997; 138:548-552.
15. Moinat M, Deng CJ, Muzzin P et al. Modulation of obese gene-expression in rat brown and white adipose tissues. FEBS Lett 1995; 373:131-134.
16. Dessolin S, Schalling M, Champigny O et al. Leptin gene is expressed in rat brown adipose tissue at birth. FASEB J 1997; 11:382-387.
17. Cinti S, Frederich RC, Zingaretti MC et al. Immunohistochemical localization of leptin and uncoupling protein in white and brown adipose tissue. Endocrinology 1997; 138:797-804.
18. Klingenspor M, Dickopp A, Heldmaier G et al. Short photoperiod reduces leptin gene expression in white and brown adipose tissue of djungarian hamsters. FEBS Lett 1996; 399:290-294.
19. Bado A, Levasseur S, Attoub S et al. The stomach is a source of leptin. Nature 1998; 394:790-793.
20. Hoggard N, Hunter L, Duncan JS et al. Leptin and leptin receptor mRNA and protein expression in the murine fetus and placenta. Proc Natl Acad Sci USA 1997; 94:11073-11078.
21. Masuzaki H, Ogawa Y, Sagawa N et al. Nonadipose tissue production of leptin: Leptin as a novel placenta- derived hormone in humans. Nature Med 1997; 3:1029-1033.
22. Hassink SG, de Lancey E, Sheslow DV et al. Placental leptin: An important new growth factor in intrauterine and neonatal development? Pediatrics 1997; 100:E11-E16.
23. Smith-Kirwin SM, O'Connor DM, Johnston J et al. Leptin expression in human mammary epithelial cells and breast milk. J Clin Endocrinol Metab 1998; 83:1810-1813.
24. Aoki N, Kawamura M, Matsuda T. Lactation-dependent down regulation of leptin production in mouse mammary gland. Biochim Biophys Acta 1999; 1427:298-306.
25. Cioffi JA, Van Blerkom J, Antczak M et al. The expression of leptin and its receptors in preovulatory human follicles. Mol Human Reprod 1997; 3:467-472.

26. Hoggard N, Hunter L, Trayhurn P et al. Leptin and reproduction. Proc Nutr Soc 1998; 57:421-427.

27. Wang JL, Liu R, Hawkins M et al. A nutrient-sensing pathway regulates leptin gene expression in muscle and fat. Nature 1998; 393:684-688.

28. Esler M, Vaz M, Collier G et al. Leptin in human plasma is derived in part from the brain, and cleared by the kidneys. Lancet 1998; 351:879.

29. Bouloumie A, Drexler HCA, Lafontan M et al. Leptin, the product of *ob* gene, promotes angiogenesis. Circulation Res 1998; 83:1059-1066.

30. Sierra-Honigmann MR, Nath AK, Murakami C et al. Biological action of leptin as an angiogenic factor. Science 1998; 281:1683-1686.

31. Considine RV, Sinha MK, Heiman ML et al. Serum immunoreactive leptin concentrations in normal-weight and obese humans. New Engl J Med 1996; 334:292-295.

32. Ostlund RE, Yang JW, Klein S et al. Relation between plasma leptin concentration and body fat, gender, diet, age, and metabolic covariates. J Clin Endocrinol Metab 1996; 81:3909-3913.

33. Maffei M, Halaas J, Ravussin E et al. Leptin levels in human and rodent—Measurement of plasma leptin and ob rna in obese and weight-reduced subjects. Nature Med 1995; 1:1155-1161.

34. Hardie LJ, Rayner DV, Holmes S et al. Circulating leptin levels are modulated by fasting, cold exposure and insulin administration in lean but not Zucker *(fa/fa)* rats as measured by ELISA. Biochem Biophys Res Commun 1996; 223:660-665.

35. Shimomura I, Hammer RE, Ikemoto S et al. Leptin reverses insulin resistance and diabetes mellitus in mice with congenital lipodystrophy. Nature 1999; 401:73-76.

36. Moitra J, Mason MM, Olive M et al. Life without white fat: A transgenic mouse. Genes Devel 1998; 12:3168-3181.

37. Masuzaki H, Ogawa Y, Isse N et al. Human obese gene-expression—Adipocyte-specific expression and regional differences in the adipose-tissue. Diabetes 1995; 44:855-858.

38. Masuzaki H, Ogawa Y, Shigemoto M et al. Adipose tissue-specific expression of the obese *(ob)* gene in rats and its marked augmentation in genetically obese-hyperglycemic Wistar fatty rats. Proc Japan Acad Series B Phys Biol Sci 1995; 71:148-152.

39. Trayhurn P, Thomas MEA, Duncan JS et al. Effects of fasting and refeeding on ob gene-expression in white adipose-tissue of lean and obese *(ob/ob)* mice. FEBS Lett 1995; 368:488-490.

40. Hube F, Lietz U, Igel M et al. Difference in leptin mRNA levels between omental and subcutaneous abdominal adipose tissue from obese humans. Horm Metab Res 1996; 28:690-693.

41. Montague CT, Prins JB, Sanders L et al. Depot- and sex-specific differences in human leptin mRNA expression: Implications for the control of regional fat distribution. Diabetes 1997; 46:342-347.

42. Rayner DV, Dalgliesh GD, Duncan JS et al. Postnatal development of the *ob* gene system: Elevated leptin levels in suckling *fa/fa* rats. Am J Physiol 1997; 273:R446-R450.

43. Trayhurn P, Duncan JS, Hoggard N et al. Regulation of leptin production: A dominant role for the sympathetic nervous system? Proc Nutr Soc 1998; 57:413-419.

44. Becker DJ, Ongemba LN, Brichard V et al. Diet-induced and diabetes-induced changes of *ob* gene-expression in rat adipose-tissue. FEBS Lett 1995; 371:324-328.

45. Macdougald OA, Hwang CS, Fan HY et al. Regulated expression of the obese gene-product (leptin) in white adipose-tissue and 3T3-L1 adipocytes. Proc Natl Acad Sci USA 1995; 92:9034-9037.

46. Boden G, Chen X, Mozzoli M et al. Effect of fasting on serum leptin in normal human subjects. J Clin Endocrinol Metab 1996; 81:3419-3423.

47. Coppack SW, Pinkney JH, Mohamed-Ali V. Leptin production in human adipose tissue. Proc Nutr Soc 1998; 57:461-470.

48. Seeley RJ, Schwartz MW. Neuroendocrine regulation of food intake. Acta Paediat 1999; 88:58-61.

49. Trayhurn P, Duncan JS, Rayner DV. Acute cold-induced suppression of *ob* (obese) gene-expression in white adipose-tissue of mice—Mediation by the sympathetic system. Biochem J 1995; 311:729-733.

50. Bing C, Pickavance L, Wang Q et al. Role of hypothalamic neuropeptide Y neurons in the defective thermogenic response to acute cold exposure in fatty Zucker rats. Neurosci 1997; 80:277-284.

51. Zheng D, Wooter MH, Zhou Q et al. The effect of exercise on *ob* gene expression. Biochem Biophys Res Commun 1996; 225:747-750.

52. Himms-Hagen J. Physiological roles of the leptin endocrine system: Differences between mice and humans. Critical Rev Clin Lab Sci 1999; 36:575-655.

53. Perusse L, Collier G, Gagnon J et al. Acute and chronic effects of exercise on leptin levels in humans. J Appl Physiol 1997; 83:5-10.

54. Saladin R, De Vos P, Guerre-Millo M et al. Transient increase in obese gene-expression after food-intake or insulin administration. Nature 1995; 377:527-529.

55. De Vos P, Saladin R, Auwerx J et al. Induction of *ob* gene-expression by corticosteroids is accompanied by body-weight loss and reduced food-intake. J Biol Chem 1995; 270:15958-15961.

56. Leroy P, Dessolin S, Villageois P et al. Expression of *ob* gene in adipose-cells—Regulation by insulin. J Biol Chem 1996; 271:2365-2368.

57. Wabitsch M, Jensen PB, Blum WF et al. Insulin and cortisol promote leptin production in cultured human fat cells. Diabetes 1996; 45:1435-1438.

58. Wabitsch M, Blum WF, Muche R et al. Contribution of androgens to the gender difference in leptin production in obese children and adolescents. J Clin Invest 1997; 100:808-813.

59. Elbers JMH, Asscheman H, Seidell JC et al. Reversal of the sex difference in serum leptin levels upon cross-sex hormone administration in transsexuals. J Clin Endocrinol Metab 1997; 82:3267-3270.

60. Shimizu H, Shimomura Y, Nakanishi Y et al. Estrogen increases in vivo leptin production in rats and human subjects. J Endocrinol 1997; 154:285-292.

61. Grunfeld C, Zhao C, Fuller J et al. Endotoxin and cytokines induce expression of leptin, the ob gene product, in hamsters—A role for leptin in the anorexia of infection. J Clin Invest 1996; 97:2152-2157.

62. Zumbach MS, Boehme MWJ, Wahl P et al. Tumor necrosis factor increases serum leptin levels in humans. J Clin Endocrinol Metab 1997; 82:4080-4082.

63. Sarraf P, Frederich RC, Turner EM et al. Multiple cytokines and acute inflammation raise mouse leptin levels: Potential role in inflammatory anorexia. J Exp Med 1997; 185:171-175.

64. Kirchgessner TG, Uysal K, Wiesbrock SM et al. Tumor necrosis factor-alpha contributes to obesity-related hyperleptinemia by regulating leptin release from adipocytes. J Clin Invest 1997; 100:2777-2782.

65. Kallen CB, Lazar MA. Antidiabetic thiazolidinediones inhibit leptin (*ob*) gene expression in 3T3-L1 adipocytes. Proc Natl Acad Sci USA 1996; 93:5793-5796.

66. De Vos P, Lefebvre AM, Miller SG et al. Thiazolidinediones repress *ob* gene expression in rodents via activation of peroxisome proliferator-activated receptor gamma. J Clin Invest 1996; 98:1004-1009.

67. Mostyn A, Rayner DV, Trayhurn P. Effects of adrenaline, noradrenaline and isoprenaline on circulating leptin levels in mice. Proc Nutr Soc 1999; 58:69A.

68. Pinkney JH, Coppack SW, Mohamed-Ali V. Effect of isoprenaline on plasma leptin and lipolysis in humans. Clin Endocrinology 1998; 48:407-411.

69. Mitchell SE, Rees WD, Hardie LJ et al. *Ob* gene expression and secretion of leptin following differentiation of rat preadipocytes to adipocytes in primary culture. Biochem Biophys Res Commun 1997; 230:360-364.

70. Gettys TW, Harkness PJ, Watson PM. The $\beta(3)$-adrenergic receptor inhibits insulin-stimulated leptin secretion from isolated rat adipocytes. Endocrinology 1996; 137:4054-4057.

71. Hardie LJ, Guilhot N, Trayhurn P. Regulation of leptin production in cultured mature white adipocytes. Horm Metab Res 1996; 28:685-689.

72. Trayhurn P, Duncan JS, Rayner DV et al. Rapid inhibition of *ob* gene expression and circulating leptin levels in lean mice by the β3-adrenoceptor agonists BRL 35135A and ZD2079. Biochem Biophys Res Commun 1996; 228:605-610.

73. Mantzoros CS, Qu DQ, Frederich RC et al. Activation of β3-adrenergic receptors suppresses leptin expression and mediates a leptin-independent inhibition of food intake in mice. Diabetes 1996; 45:909-914.

74. Enocksson S, Shimizu M, Lönnqvist F et al. Demonstration of an in vivo functional β3-adrenoceptor in man. J Clin Invest 1995; 95:2239-2245.

75. Rayner DV, Simon E, Duncan JS et al. Hyperleptinaemia in mice induced by administration of the tyrosine hydroxylase inhibitor α-methyl-*p*-tyrosine. FEBS Lett 1998; 429:395-398.

76. Zimmermann RC, Krahn L, Rahmanie N et al. Prolonged inhibition of presynaptic catecholamine synthesis does not alter leptin secretion in normal-weight men women. Human Reprod 1998; 13:822-825.
77. Evans BA, Agar L, Summers RJ. The role of the sympathetic nervous system in the regulation of leptin synthesis in C57Bl/6 mice. FEBS Lett 1999; 444:149-154.
78. Sivitz WI, Fink BD, Morgan DA et al. Sympathetic inhibition, leptin, and uncoupling protein subtype expression in normal fasting rats. Am J Physiol 1999; 277:E668-E677.
79. Bartness TJ, Bamshad M. Innervation of mammalian white adipose tissue: Implications for the regulation of total body fat. Am J Physiol 1998; 275:R1399-R1411.
80. Rebuffé-Scrive M. Neuroregulation of adipose tissue: Molecular and hormonal mechanisms. Int J Obesity 1991; 15:83-86.
81. Youngström TG, Bartness TJ. White adipose tissue sympathetic nervous system denervation increases fat pad mass and fat cell number. Am J Physiol 1998; 275:R1488-R1493.
82. Bamshad M, Aoki VT, Adkison MG et al. Central nervous system origins of the sympathetic nervous system outflow to white adipose tissue. Am J Physiol 1998; 275:R291-R299.
83. Haynes WG, Sivitz WI, Morgan DA et al. Sympathetic and cardiorenal actions of leptin. Hypertension 1997; 30:619-623.
84. Haynes WG, Morgan DA, Walsh SA et al. Receptor-mediated regional sympathetic nerve activation by leptin. J Clin Invest 1997; 100:270-278.
85. Niijima A. Afferent signals from leptin sensors in the white adipose tissue of the epididymis, and their reflex effect in the rat. J Autonom Nerv Sys 1998; 73:19-25.
86. Sinha MK, Ohannesian JP, Heiman ML et al. Nocturnal rise of leptin in lean, obese, and noninsulin-dependent diabetes-mellitus subjects. J Clin Invest 1996; 97:1344-1347.
87. Schoeller DA, Cella LK, Sinha MK et al. Entrainment of the diurnal rhythm of plasma leptin to meal timing. J Clin Invest 1997; 100:1882-1887.
88. Simon C, Gronfier C, Schlienger JL et al. Circadian and ultradian variations of leptin in normal man under continuous enteral nutrition: Relationship to sleep and body temperature. J Clin Endocrinol Metab 1998; 83:1893-1899.
89. Saad MF, Riad-Gabriel MG, Khan A et al. Diurnal and ultradian rhythmicity of plasma leptin: Effects of gender and adiposity. J Clin Endocrinol Metab 1998; 83:453-459.
90. Sinha MK, Sturis J, Ohannesian J et al. Nocturnal rise and pulsatile secretion of leptin in humans. Diabetes 1996; 45:386-386.
91. Licinio J, Mantzoros C, Negrao AB et al. Human leptin levels are pulsatile and inversely related to pituitary-adrenal function. Nature Med 1997:575-579.
92. Tartaglia LA, Dembski M, Weng X et al. Identification and expression cloning of a leptin receptor, Ob-R. Cell 1995; 83:1263-1271.
93. Lee GH, Proenca R, Montez JM et al. Abnormal splicing of the leptin receptor in diabetic mice. Nature 1996; 379:632-635.
94. Hoggard N, Mercer JG, Rayner DV et al. Localization of leptin receptor mRNA splice variants in murine peripheral tissues by RT-PCR and in situ hybridization. Biochem Biophys Res Commun 1997; 232:383-387.
95. Ghilardi N, Ziegler S, Wiestner A et al. Defective STAT signaling by the leptin receptor in diabetic mice. Proc Natl Acad Sci USA 1996; 93:6231-6235.
96. Golden PL, Maccagnan TJ, Pardridge WM. Human blood-brain barrier leptin receptor—Binding and endocytosis in isolated human brain microvessels. J Clin Invest 1997; 99:14-18.
97. Bjorbaek C, Elmquist JK, Michl P et al. Expression of leptin receptor isoforms in rat brain microvessels. Endocrinology 1998; 139:3485-3491.
98. Baumann H, Morella KK, White DW et al. The full-length leptin receptor has signaling capabilities of interleukin 6-type cytokine receptors. Proc Natl Acad Sci USA 1996; 93:8374-8378.
99. Bjorbaek C, Uotani S, da Silva B et al. Divergent signaling capacities of the long and short isoforms of the leptin receptor. J Biol Chem 1997; 272:32686-32695.
100. Mercer JG, Hoggard N, Williams LM et al. Localization of leptin receptor mRNA and the long form splice variant (Ob-Rb) in mouse hypothalamus and adjacent brain regions by in situ hybridization. FEBS Lett 1996; 387:113-116.

101. Håkansson ML, Brown H, Ghilardi N et al. Leptin receptor immunoreactivity in chemically defined target neurons of the hypothalamus. J Neurosci 1998; 18:559-572.

102. Mercer JG, Moar KM, Hoggard N. Localization of leptin receptor (Ob-R) messenger ribonucleic acid in the rodent hindbrain. Endocrinology 1998; 139:29-34.

103. Woods SC, Seeley RJ, Porte D et al. Signals that regulate food intake and energy homeostasis. Science 1998; 280:1378-1383.

104. Emilsson V, Liu YL, Cawthorne MA et al. Expression of the functional leptin receptor mRNA in pancreatic islets and direct inhibitory action of leptin on insulin secretion. Diabetes 1997; 46:313-316.

105. Trayhurn P, Hoggard N, Mercer JG et al. Leptin: Fundamental aspects. Int J Obesity 1999; 23:22-28.

106. Kieffer TJ, Keller RS, Leech CA et al. Leptin suppression of insulin secretion by the activation of ATP-sensitive K+ channels in pancreatic β-cells. Diabetes 1997; 46:1087-1093.

107. Barr VA, Lane K, Taylor SI. Subcellular localization and internalization of the four human leptin receptor isoforms. J Biol Chem 1999; 274:21416-21424.

108. Uotani S, Bjorbaek C, Tornoe J et al. Functional properties of leptin receptor isoforms internalization and degradation of leptin and ligand-induced receptor downregulation. Diabetes 1999; 48:279-286.

109. Kielar D, Clark JSC, Ciechanowicz A et al. Leptin receptor isoforms expressed in human adipose tissue. Metabolism 1998; 47:844-847.

110. Meyer C, Robson D, Rackovsky N et al. Role of the kidney in human leptin metabolism. Am J Physiol 1997; 273:E903-E907.

111. Frühbeck G, Aguado M, Gomez-Ambrosi J et al. Lipolytic effect of in vivo leptin administration on adipocytes of lean and *ob/ob* mice, but not *db/db* mice. Biochem Biophys Res Commun 1998; 250:99-102.

112. Siegrist-Kaiser CA, Pauli V, Juge-Aubry CE et al. Direct effects of leptin on brown and white adipose tissue. J Clin Invest 1997; 100:2858-2864.

113. Stephens TW, Basinski M, Bristow PK et al. The role of neuropeptide-Y in the antiobesity action of the obese gene-product. Nature 1995; 377:530-532.

114. Schwartz MW, Seeley RJ, Campfield LA et al. Identification of targets of leptin action in rat hypothalamus. J Clin Invest 1996; 98:1101-1106.

115. Goldstone AP, Mercer JG, Gunn I et al. Leptin interacts with glucagon-like peptide-1 neurons to reduce food intake and body weight in rodents. FEBS Lett 1997; 415:134-138.

116. Schwartz MW, Seeley RJ, Woods SC et al. Leptin increases hypothalamic pro-opiomelanocortin mRNA expression in the rostral arcuate nucleus. Diabetes 1997; 46:2119-2123.

117. Uehara Y, Shimizu H, Ohtani K et al. Hypothalamic corticotropin-releasing hormone is a mediator of the anorexigenic effect of leptin. Diabetes 1998; 47:890-893.

118. Kristensen P, Judge ME, Thim L et al. Hypothalamic CART is a new anorectic peptide regulated by leptin. Nature 1998; 393:72-76.

119. Schwartz MW, Baskin DG, Kaiyala KJ et al. Model for the regulation of energy balance and adiposity by the central nervous system. Am J Clin Nutr 1999; 69:584-596.

120. Campfield LA, Smith FJ, Guisez Y et al. Recombinant mouse OB protein—Evidence for a peripheral signal linking adiposity and central neural networks. Science 1995; 269:546-549.

121. Halaas JL, Gajiwala KS, Maffei M et al. Weight-reducing effects of the plasma-protein encoded by the obese gene. Science 1995; 269:543-546.

122. Pelleymounter MA, Cullen MJ, Baker MB et al. Effects of the obese gene-product on body-weight regulation in *ob/ob* mice. Science 1995; 269:540-543.

123. Stehling O, Döring H, Ertl J et al. Leptin reduces juvenile fat stores by altering the circadian cycle of energy expenditure. Am J Physiol 1996; 271:R1770-R1774.

124. Döring H, Schwarzer K, Nuesslein-Hildesheim B et al. Leptin selectively increases energy expenditure of food-restricted lean mice. Int J Obesity 1998; 22:83-88.

125. Chehab FE, Lim ME, Lu RH. Correction of the sterility defect in homozygous obese female mice by treatment with the human recombinant leptin. Nature Genet 1996; 12:318-320.

126. Chehab FF, Mounzih K, Lu RH et al. Early onset of reproductive function in normal female mice treated with leptin. Science 1997; 275:88-90.

127. Barash IA, Cheung CC, Weigle DS et al. Leptin is a metabolic signal to the reproductive system. Endocrinology 1996; 137:3144-3147.
128. Gainsford T, Willson TA, Metcalf D et al. Leptin can induce proliferation, differentiation, and functional activation of hemopoietic cells. Proc Natl Acad Sci USA 1996; 93:14564-14568.
129. Lord GM, Matarese G, Howard LK et al. Leptin modulates the T-cell immune response and reverses starvation-induced immunosuppression. Nature 1998; 394:897-901.
130. Alingh-Prins A, de Jong-Nagelsmit A, Keijser J et al. Daily rhythms of feeding in the genetically obese and lean Zucker rats. Physiol Behav 1986; 38:423-6.
131. Strohmayer AJ, Smith GP. The meal pattern of genetically obese (*ob/ob*) mice. Appetite 1987; 8:111-23.
132. Kahler A, Geary N, Eckel LA et al. Chronic administration of OB protein decreases food intake by selectively reducing meal size in male rats. Am J Physiol 1998; 275:R180-R185.
133. Hulsey MG, Lu HX, Wang TL et al. Intracerebroventricular (icv) administration of mouse leptin in rats: Behavioral specificity and effects on meal patterns. Physiol Behav 1998; 65:445-455.
134. Flynn MC, Scott TR, Pritchard TC et al. Mode of action of OB protein (leptin) on feeding. Am J Physiol 1998; 275:R174-R179.
135. Eckel LA, Langhans W, Kahler A et al. Chronic administration of OB protein decreases food intake by selectively reducing meal size in female rats. Am J Physiol 1998; 275:R186-R193.
136. Pratley RE, Nicolson M, Bogardus C et al. Plasma leptin responses to fasting in Pima indians. Am J Physiol 1997; 273:E644-E649.
137. Lostao MP, Urdaneta E, Martinez-Anso E et al. Presence of leptin receptors in rat small intestine and leptin effect on sugar absorption. FEBS Lett 1998; 423:302-306.
138. Eyckerman S, Waelput W, Verhee A et al. Analysis of TYR to PHE and *fa/fa* leptin receptor mutations in the PC12 cell line. Eur Cytokine Network 1999; 10:549-556.
139. Steppan CM, Swick AG. A role for leptin in brain development. Biochem Biophys Res Commun 1999; 256:600-602.
140. Seufert J, Kieffer TJ, Habener JF. Leptin inhibits insulin gene transcription and reverses hyperinsulinemia in leptin-deficient *ob/ob* mice. Proc Natl Acad Sci USA 1999; 96:674-679.
141. Nakata M, Yada T, Soejima N et al. Leptin promotes aggregation of human platelets via the long form of its receptor. Diabetes 1999; 48:426-429.
142. Ducy P, Amling M, Takeda S et al. Leptin inhibits bone formation through a hypothalamic relay: A central control of bone mass. Cell 2000; 100:197-207.
143. Houseknecht KL, Mantzoros CS, Kuliawat R et al. Evidence for leptin binding to proteins in serum of rodents and humans: Modulation with obesity. Diabetes 1996; 45:1638-1643.
144. Cumin F, Baum HP, Levens N. Mechanism of leptin removal from the circulation by the kidney. J Endocrinol 1997; 155:577-585.
145. Zeng JB, Patterson BW, Klein S et al. Whole body leptin kinetics and renal metabolism in vivo. Am J Physiol 1997; 273:E1102-E1106.
146. Ahima RS, Prabakaran D, Mantzoros C et al. Role of leptin in the neuroendocrine response to fasting. Nature 1996; 382:250-252.
147. Unger RH, Zhou YT, Orci L. Regulation of fatty acid homeostasis in cells: Novel role of leptin. Proc Nat Acad Sci USA 1999; 96:2327-2332.
148. Shimomura I, Hammer RE, Richardson JA et al. Insulin resistance and diabetes mellitus in transgenic mice expressing nuclear SREBP-1c in adipose tissue: Model for congenital generalized lipodystrophy. Genes Develop 1998; 12:3182-3194.
149. Chagnon YC, Chung WK, Pérusse L et al. Linkages and associations between the leptin receptor (lepr) gene and human body composition in the Québec family study. Int J Obesity 1999; 23:278-286.
150. Comings DE, Gade R, MacMurray JP et al. Genetic variants of the human obesity (*ob*) gene: Association with body mass index in young women, psychiatric symptoms, and interaction with the dopamine D-2 receptor (drd2) gene. Mol Psychiatry 1996; 1:325-335.
151. Considine RV, Considine EL, Williams CJ et al. The hypothalamic leptin receptor in humans: Identification of incidental sequence polymorphisms and absence of the *db/db* mouse and *fa/fa* rat mutations. Diabetes 1996; 45:992-994.

152. Echwald SM, Sorensen TD, Sorensen TIA et al. Amino acid variants in the human leptin receptor: Lack of association to juvenile onset obesity. Biochem Biophys Res Commun 1997; 233:248-252.

153. Francke S, Clement K, Dina C et al. Genetic studies of the leptin receptor gene in morbidly obese french caucasian families. Human Genet 1997; 100:491-496.

154. Montague CT, Farooqi IS, Whitehead JP et al. Congenital leptin deficiency is associated with severe early- onset obesity in humans. Nature 1997; 387:903-908.

155. Strobel A, Issad T, Camoin L et al. A leptin missense mutation associated with hypogonadism and morbid obesity. Nature Genet 1998; 18:213-215.

156. Clement K, Vaisse C, Lahlou N et al. A mutation in the human leptin receptor gene causes obesity and pituitary dysfunction. Nature 1998; 392:398-401.

157. Chen H, Charlat O, Tartaglia LA et al. Evidence that the diabetes gene encodes the leptin receptor—Identification of a mutation in the leptin receptor gene in *db/db* mice. Cell 1996; 84:491-495.

158. Chua SC, Chung WK, Wupeng XS et al. Phenotypes of mouse diabetes and rat fatty due to mutations in the *ob* (leptin) receptor. Science 1996; 271:994-996.

159. Farooqi IS, Jebb SA, Langmack G et al. Effects of recombinant leptin therapy in a child with congenital leptin deficiency. New Engl J Med 1999; 341:879-884.

160. Caro JF, Kolaczynski JW, Nyce MR et al. Decreased cerebrospinal-fluid/serum leptin ratio in obesity: A possible mechanism for leptin resistance. Lancet 1996; 348:159-161.

161. Arch JRS, Stock MJ, Trayhurn P. Leptin resistance in obese humans: Does it exist and what does it mean? Int J Obesity 1998; 22:1159-1163.

162. Heymsfield SB, Greenberg AS, Fujioka K et al. Recombinant leptin for weight loss in obese and lean adults—A randomized, controlled, dose-escalation trial. J Am Med Assoc 1999; 282:1568-1575.

163. Ailhaud G. The adipocyte: A secretory and endocrine cell. M S-Med Sci 1998; 14:858-864.

164. Hausman GJ. The comparative anatomy of adipose tissue. In: Cryer A, Van RLR, eds. New Perspectives in Adipose Tissue: Structure, Function and Development. London: Butterworths; 1985:1-21.

165. Enerback S, Gimble JM. Lipoprotein lipase gene expression—Physiological regulators at the transcriptional and post-transcriptional level. Biochim Biophys Acta 1993; 1169:107-125.

166. Fielding BA, Frayn KN. Lipoprotein lipase and the disposition of dietary fatty acids. Br J Nutr 1998; 80:495-502.

167. Radeau T, Robb M, McDonnell M et al. Preferential expression of cholesteryl ester transfer protein mRNA by stromal-vascular cells of human adipose tissue. Biochim Biophys Acta 1998; 1392:245-253.

168. Radeau T, Robb M, Lau P et al. Relationship of adipose tissue cholesteryl ester transfer protein (CETP) mRNA to plasma concentrations of CETP in man. Atherosclerosis 1998; 139:369-376.

169. Safonova I, Aubert J, Négrel R et al. Regulation by fatty acids of angiotensinogen gene expression in preadipose cells. Biochem J 1997; 322:235-239.

170. Zorad S, Fickova M, Zelezna B et al. The role of angiotensin II and its receptors in regulation of adipose tissue metabolism and cellularity. Gen Physiol Biophys 1995; 14:383-391.

171. Jones BH, Standridge MK, Taylor JW et al. Angiotensinogen gene expression in adipose tissue: Analysis of obese models and hormonal and nutritional control. Am J Physiol 1997; 273:R236-R242.

172. Aubert J, Darimont C, Safonova I et al. Regulation by glucocorticoids of angiotensinogen gene expression and secretion in adipose cells. Biochem J 1997; 328:701-706.

173. Aubert J, Safonova I, Négrel R et al. Insulin down-regulates angiotensinogen gene expression and angiotensinogen secretion in cultured adipose cells. Biochem Biophys Res Commun 1998; 250:77-82.

174. Harp JB, DiGirolamo M. Components of the renin-angiotensin system in adipose-tissue—Changes with maturation and adipose mass enlargement. J Gerontol Ser A-Biol Sci Med Sci 1995; 50:B270-B276.

175. Engeli S, Gorzelniak K, Kreutz R et al. Co-expression of renin-angiotensin system genes in human adipose tissue. J Hypertension 1999; 17:555-560.

176. Karlsson C, Lindell K, Ottosson M et al. Human adipose tissue expresses angiotensinogen and enzymes required for its conversion to angiotensin II. J Clin Endocrinol Metab 1998; 83:3925-3929.

177. Samad F, Pandey M, Loskutoff DJ. Tissue factor gene expression in the adipose tissues of obese mice. Proc Natl Acad Sci USA 1998; 95:7591-7596.

178. Juhan-Vague I, Alessi MC. PAI-1, obesity, insulin resistance and risk of cardiovascular events. Thromb Haemostasis 1997; 78:656-660.
179. Alessi MC, Peiretti F, Morange P et al. Production of plasminogen activator inhibitor 1 by human adipose tissue—Possible link between visceral fat accumulation and vascular disease. Diabetes 1997; 46:860-867.
180. Shimomura I, Funahashi T, Takahashi M et al. Enhanced expression of PAI-1 in visceral fat: Possible contributor to vascular disease in obesity. Nature Med 1996; 2:800-803.
181. Janand-Delenne B, Chagnaud C, Raccah D et al. Visceral fat as a main determinant of plasminogen activator inhibitor 1 level in women. Int J Obesity 1998; 22:312-317.
182. Lundgren CH, Brown SL, Nordt TK et al. Elaboration of type-1 plasminogen activator inhibitor from adipocytes—A potential pathogenetic link between obesity and cardiovascular disease. Circulation 1996; 93:106-110.
183. Samad F, Yamamoto K, Loskutoff DJ. Distribution and regulation of plasminogen activator inhibitor-1 in murine adipose tissue in vivo—Induction by tumor necrosis factor- a and lipopolysaccharide. J Clin Invest 1996; 97:37-46.
184. Samad F, Loskutoff DJ. The fat mouse: A powerful genetic model to study elevated plasminogen activator inhibitor 1 in obesity/NIDDM. Thromb Haemostasis 1997; 78:652-655.
185. Eriksson P, Reynisdottir S, Lönnqvist F et al. Adipose tissue secretion of plasminogen activator inhibitor-1 in non obese and obese individuals. Diabetologia 1998; 41:65-71.
186. Cigolini M, Tonoli M, Borgato L et al. Expression of plasminogen activator inhibitor-1 in human adipose tissue: A role for TNF-α? Atherosclerosis 1999; 143:81-90.
187. Samad F, Yamamoto K, Pandey M et al. Elevated expression of transforming growth factor-β in adipose tissue from obese mice. Mol Med 1997; 3:37-48.
188. Sakamoto T, Woodcock-Mitchell J, Marutsuka K et al. TNF-α and insulin, alone and synergistically, induce plasminogen activator inhibitor-1 expression in adipocytes. Am J Physiol 1999; 276:C1391-C1397.
189. Halleux CM, Tran S, Declerck PJ et al. Camp and catecholamines inhibit PAI-1 gene expression and production in human adipose tissue. Int J Obesity 1999; 23(Suppl 5):S25.
190. Gottschiling-Zeller H, Birgel M, Hauner H. Effect of β-adrenoceptor-agonists on plasminogen activator inhibitor-1 secretion from human adipocytes in suspension culture. Int J Obesity 1999; 23, Suppl 5:S25.
191. Hotamisligil GS, Shargill NS, Spiegelman BM. Adipose expression of tumor necrosis factor-α— Direct role in obesity-linked insulin resistance. Science 1993; 259:87-91.
192. Uysal KT, Wiesbrock SM, Hotamisligil GS. Functional analysis of tumor necrosis factor (TNF) receptors in TNF-α-mediated insulin resistance in genetic obesity. Endocrinology 1998; 139:4832-4838.
193. Sethi JK, Hotamisligil GS. The role of TNF-α in adipocyte metabolism. Semin Cell Dev Biol 1999; 10:19-29.
194. Mohamed-Ali V, Goodrick S, Bulmer K et al. Production of soluble tumor necrosis factor receptors by human subcutaneous adipose tissue in vivo. Am J Physiol 1999; 277:E971-E975.
195. Samad F, Uysal KT, Wiesbrock SM et al. Tumor necrosis factor alpha is a key component in the obesity-linked elevation of plasminogen activator inhibitor 1. Proc Natl Acad Sci USA 1999; 96:6902-6907.
196. Cook KS, Groves DL, Min HY et al. A developmentally regulated mRNA from 3T3 adipocytes encodes a novel serine protease homologue. Proc Natl Acad Sci USA 1985; 82:6480-6484.
197. Flier JS, Cook KS, Usher P et al. Severely impaired adipsin expression in genetic and acquired obesity. Science 1987; 237; 405-408.
198. Cook KS, Min HY, Johnson D et al. Adipsin: A circulating serine protease homolog secreted by adipose tissue and sciatic nerve. Science 1987; 237:402-.
199. Cianflone K, Maslowska M, Sniderman AD. Acylation stimulating protein (ASP), an adipocyte autocrine: New directions. Semin Cell Dev Biol 1999; 10:31-41.
200. Murray I, Sniderman AD, Havel PJ et al. Acylation stimulating protein (ASP) deficiency alters postprandial and adipose tissue metabolism in male mice. J Biol Chem 1999; 274:36219-36225.
201. Murray I, Sniderman AD, Cianflone K. Mice lacking acylation stimulating protein (ASP) have delayed postprandial triglyceride clearance. J Lipid Res 1999; 40:1671-1676.

202. Makover A, Soprano DR, Wyatt ML et al. Localization of retinol-binding protein mRNA in the rat kidney and in perinephric fat tissue. J Lipid Res 1989; 30:171-180.
203. Zovich DC, Orologa A, Okuno M et al. Differentiation-dependent expression of retinoid-binding proteins in BFC-1 beta adipocytes. J Biol Chem 1992; 267:13884-138890.
204. Tsutsumi C, Okuno M, Tannous L et al. Retinoids and retinoid-binding protein expression in rat adipocytes. J Biol Chem 1992; 267:1805-1810.
205. Montague CT, Prins JB, Sanders L et al. Depot-related gene expression in human subcutaneous and omental adipocytes. Diabetes 1998; 47:1384-1391.
206. Blomhoff R, Green MH, Berg T et al. Transport and storage of vitamin A. Science 1990; 250:399-404.
207. Wolf G. Uptake of retinoids by adipose tissue. Nutr Rev 1994; 52:356-358.
208. Maeda K, Okubo K, Shimomura I et al. Analysis of an expression profile of genes in the human adipose tissue. Gene 1997; 190:227-235.
209. Ouchi N, Kihara S, Arita Y et al. Novel modulator for endothelial adhesion molecules— Adipocyte-derived plasma protein adiponectin. Circulation 1999; 100:2473-2476.
210. Arita Y, Kihara S, Ouchi N et al. Paradoxical decrease of an adipose-specific protein, adiponectin, in obesity. Biochem Biophys Res Commun 1999; 257:79-83.
211. Bremner I, Beattie JH. Metallothionein and the trace minerals. Ann Rev Nutr 1990; 10:63-83.
212. Beattie JH, Black DJ, Wood AM et al. Cold-induced expression of the metallothionein-1 gene in brown adipose tissue of rats. Am J Physiol 1996; 270:R971-R977.
213. El Refaey H, Ebadi M, Kuszynski CA et al. Identification of metallothionein receptors in human astrocytes. Neurosci Lett 1997; 231:131-134.

CHAPTER 10

Adipose Tissue Pathology in Human Obesity

Hans Hauner, Thomas Skurk

Human obesity is characterized by an excess of adipose tissue mass that has potential adverse health consequences and may finally result in a reduced quality of life and life expectancy.[1] In normal-weight subjects the adipose tissue organ constitutes between 10 and 20% of total body mass in males and between 15 and 25% in females. Although many techniques are available to determine body fat mass all of them have specific limitations. Therefore, for clinical practice and epidemiological studies simple anthropometric measures are widely used such as the body mass index (BMI = body mass (in kg) divided by body height (in m) squared; i.e., kg/m^2) which was found to be a good surrogate marker of body fat mass.

According to WHO recommendations obesity is defined as a BMI $\geq$ 30 kg/m^2 independent of gender. A BMI between 25 and 29.9 kg/m^2 indicates a preobese state and can already induce characteristic health problems. In the industrialized world between 10 and 25% of the adult population have a BMI $\geq$ 30 kg/m^2 and, in addition, up to 40% are preobese or overweight according to this criteria. Thus, obesity represents an epidemic that affects most populations and its prevalence is increasing worldwide.[2]

Among the many complications that are associated with obesity, disturbances of the metabolic and cardiovascular system are of particular relevance. Excess body fat promotes the development of type 2 diabetes mellitus and dyslipidemia, favors other cardiovascular risk factors such as hypertension and impaired fibrinolysis and may therefore result in an increased incidence of cardiovascular complications.[1] To date, the precise causal relationship between obesity and these disturbances is only poorly understood, but there is growing evidence that alterations in fat cell function may directly or indirectly contribute to the development of these unfavorable health hazards.

The cellular architecture of adipose tissue is unique: fully developed adipocytes with the characteristic signet-ring morphology represent the largest constituent of adipose tissue. This cell type makes up approximately two thirds of all cells present in this organ and due to its exceptional volume more than 90% of total adipose tissue mass. Interestingly, there is a high variation in the size of these cells independent of total body weight. On average, obese subjects have significantly larger adipocytes than lean individuals. The mean fat cell diameter in obese subjects is in the range of 100 to 120 μm, whereas in normal-weight subjects fat cells reach a mean diameter of approximately 70 to 80 μm. As fat cells have a spherical shape, their volume can be easily calculated from the diameter: it can rise from average values of 300 to 500 pl in lean to more than 1000 pl in obese subjects. The nonadipocyte cell fraction consists mainly of specific adipocyte precursor cells which are able to undergo differentiation in vitro and—probably,

Adipose Tissues, edited by Susanne Klaus. ©2001 Eurekah.com.

but still unproven—in vivo under appropriate conditions. The other cellular components of adipose tissue are still poorly defined. For example, this fraction also includes endothelial cells and leukocytes. In addition, it is known from many studies that the cellular composition of adipose tissue can vary substantially according to age, anatomical location and nutritional state.

It is important to note that the proportion of body fat changes throughout life. Studies of body composition at different ages indicate a typical pattern in both genders. From birth to the end of the first year of life body fat mass increases from approximately 10% to 30% of total mass. Then, there is a continuous decrease in body fat until early puberty when a rapid rise in fat mass is seen in girls but not in boys. While the changes of body fat mass during childhood and adolescence appear to be mainly under genetic and hormonal control, the increase in body fat during adulthood is largely a result of the modern lifestyle, i.e., due to the lack of exercise and due to overnutrition, and can vary substantially. This age-dependent increase in body fat in adult life is not observed in populations that follow a traditional lifestyle. However, even in western societies the genetic background will finally determine who will become overweight or obese under an unhealthy environment of a high-fat, energy-dense diet and of low physical activity due to comfortable technical advances. Two recent studies have clearly illustrated how differently individuals respond to a defined overnutrition thereby confirming the important role of genes.[3,4] In this sense, obesity is the consequence of a complex interaction between genes and lifestyle factors.

Adipose Tissue Cellularity in Obese Humans

Previous studies of adipose tissue cellularity in humans have suggested that the first response to chronic overnutrition is an increase in fat cell size, also termed hypertrophic adipose tissue growth. This is the main morphological feature that can be detected in adult people with a moderate form of obesity. In severe forms of obesity there appears to be an additional hyperplastic growth which may begin if the existing fat cells reach a critical cell volume.[5] Then, dormant adipocyte precursor cells are recruited by as yet unknown signals which are possibly generated by enlarged fat cells.[6] However, this attractive hypothesis of a paracrine regulation of fat cell differentiation still awaits proof by future studies, although interesting data on a possible contribution of macrophage colony-stimulating factor (MCSF) as a mediator of adipose tissue growth has been presented recently.[7] A simple scheme of the cellular changes occurring during the transition from normal weight to extreme obesity and back to a post-obese state is illustrated in Figure 10.1. A somewhat different picture is seen in childhood-onset obesity. In obese children, there appears to be a parallel increase in fat cell number and size from the early beginning of a positive energy balance.[8] Studies in nonobese and obese subjects demonstrated that fat cell hypertrophy is predominant in adulthood-onset obesity, whereas childhood-onset obesity is mainly characterized by fat cell hyperplasia. This hypercellularity of adipose tissue appears to occur in two distinct periods early in life: within the first few years and from age nine to 13 years (Table 10.1).[9]

After weight loss due to therapeutic intervention or due to other causes a shrinkage of the fat cell size can be found, whereas a decrease in fat cell number does obviously not occur, at least under short-term observation. There is only one long-term follow-up study related to this topic which showed that prolonged and marked weight reduction by gastric surgery in extremely obese subjects is also associated with a reduction in the number of adipocytes.[10] An interesting question is whether and if so how depleted adipocytes contribute to the frequent, almost inevitable weight relapse after dieting. It has been demonstrated that fat cells shrunk by weight loss measures secrete less leptin, express more lipoprotein lipase, and become more insulin-sensitive. All these changes may contribute to a rapid weight regain and repletion of the existing adipocytes.

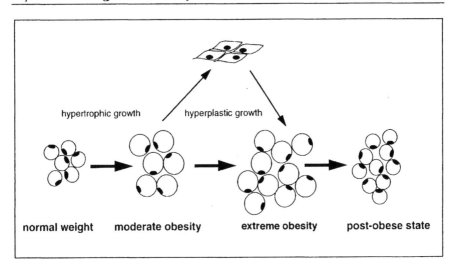

Fig. 10.1. Adipose tissue cellularity in obesity.

Adipose Tissue Development in Human Obesity

There are only sparse data concerning adipose tissue growth in lean and obese humans. Using serum-containing culture conditions, Roncari and coworkers were the first to systematically address this question. They reported an exaggerated replication and differentiation of cultured adipocyte precursor cells from massively obese subjects as compared to lean controls and postulated from these results that the former may suffer from a specific intrinsic defect that induces hyperplasia and promotes weight gain.[11] In a subsequent study from Sweden using a similar culture system, however, no difference was reported in both the replication and differentiation rate of cultured stromal-vascular cells isolated from adipose tissue of obese and lean subjects clearly arguing against a genetic predisposition at the cellular level.[12] A similar study from our laboratory also failed to detect a substantial difference in the differentiation rate between moderately obese and lean adults thereby supporting the observation from Sweden although these findings do not fully rule out that extremely obese subjects have a different growth pattern.[13]

Using chemically defined, hormone-supplemented culture conditions which allow a much higher in vitro differentiation of adipocyte precursor cells, we were also unable to reveal a significantly higher proliferation rate in cultures from obese versus lean subjects. Here, the obese group comprised only individuals with a BMI $\geq$ 40 kg/m^2 who underwent vertical gastric banding. The differentiation capacity appeared to be somewhat lower in cultures from obese subjects (unpublished data). An explanation for this latter phenomenon may be that in the obese state and in an environment that promotes obesity there is a high pressure on adipose precursor cells from obese subjects to undergo differentiation due to higher circulating concentrations of adipogenic factors such as insulin as well as of substrates such as triglycerides.[14] In addition, there may be a more adipogenic local environment, e.g., due to the release of proliferation and differentiation promoting signals from enlarged fat cells. These changes may lead to a shift in the balance between adipogenic and antiadipogenic factors and, subsequently, to a higher conversion of preadipose into adipose cells. In this context, it is interesting to stress that also more detailed animal studies failed to uncover consistent differences in the proliferation

Table 10.1. Adipose tissue cellularity in different forms of human obesity

Onset/severity of obesity	Hyperplasia	Hypertrophy
Childhood	++	+
Adulthood		
- moderate form	(+)	+
- severe form	+ - ++	++

and differentiation capacity of stromal cells from obese vs. lean rodents.[6] For example, in young Zucker rats differentiation and lipid accumulation were similar in preadipocytes from lean and obese animals. However, in older animals obese-derived preadipocytes differentiated poorly which may provoke the speculation that the adipocyte precursor pool may become exhausted in the course of obesity.[15] Taken together, the general view is that both proliferation and differentiation of adipocyte precursor cells is not defective in human obesity. The described changes of adipose tissue cellularity are very likely the consequence of an excess intake of exogenous energy substrates and factors that promote lipid storage and preadipocyte differentiation.

Adipose Tissue Function in Human Obesity

Adipose tissue is the central organ for energy storage in the form of triglycerides. Therefore, the main metabolic function of adipocytes is uptake and storage of fatty acids on the one hand and rapid mobilization of these energy substrates upon demand on the other. These processes seem to be mutually exclusive and are regulated in a complex manner. It is obvious that defects at the level of triglyceride storage and hydrolysis may contribute to the expansion of the adipose tissue mass.

Lipid Accumulation

Little information is available on alterations in lipid storage in human obesity. Studies using isotope labeling techniques are sparse and could not unequivocally report a higher glucose and fatty acid uptake and incorporation into triglycerides in obese versus lean humans.[16] In contrast, much more data have been published on the expression and activity of lipoprotein lipase (LPL), the key enzyme for triglyceride uptake in adipose tissue. LPL activity was found to be closely related to BMI and adipocyte size. However, if LPL activity is expressed per gram adipose tissue no difference is apparent between obese and lean subjects.[16] Thus, it is the greater adipose tissue mass that is responsible for the increased capacity to store lipids. This statement is in agreement with the results of a report that did not show increased levels of LPL mRNA in the obese state.[17] It is well known that LPL enzyme activity is subject to posttranslational regulation by a variety of factors, among them insulin. Studies in obese subjects have shown that the effect of insulin on LPL activity is reduced which may be considered as another manifestation of the insulin resistance syndrome. In addition, this deficiency of LPL responsiveness to insulin and high-carbohydrate meals may represent an adaptive mechanism to limit triglyceride deposition.[18] Weight loss is known to be associated with a rapid and pronounced increase in LPL activity which may be a mechanism to rapidly replace the loss of adipose tissue.[18]

Acylation stimulating protein (ASP) identical to C3adesArg is a product of the interaction of three proteins of the alternative complement pathway, i.e., C3, factor B and adipsin which are all synthesized by adipocytes.[19] Today, ASP is known to be a major determinant of triglyceride synthesis in adipose tissue and, therefore, may be critically involved in fatty acid entry and release. Studies in humans clearly indicate that ASP plasma levels are 3-fold higher in obese as compared to lean individuals.[20] Although the biological role of ASP is far from being fully understood, there is growing evidence that this protein increases and supports triglyceride synthesis in the adipocyte which includes glucose transport and reesterification of fatty acids and, on the other side, decreases the activity of hormone-sensitive lipase. The magnitude of the effect of ASP seems to depend on receptor number and affinity which are higher in subcutaneous versus omental adipocytes and greater in fat cells from females than males. As ASP is produced in a differentiation-dependent fashion by fat cells it appears that ASP is part of a positive feedback system that allows a highly efficient triglyceride storage. Additional insight into the physiological role of ASP comes from a ASP-deficient mouse model. Lack of ASP was found to cause a marked delay in postprandial triglyceride clearance. ASP (-/-) mice were also characterized by a decrease in leptin and in specific adipose tissue depots versus wildtype mice on both low fat and high fat diets indicating that ASP could play an important role in the pathogenesis of obesity, but is unlikely to be a primary cause of obesity.[21]

Lipolysis

Metabolic studies using isotope techniques have shown that free fatty acid concentrations as well as the overall rate of lipolysis are increased in obese subjects.[22] However, when related to the total fat mass, obese subjects appear to have a normal or in some studies even decreased rate of lipolysis. Thus, enlargement of body fat mass rather than changes at the cellular level may be the cause of an overall increased rate of lipolysis in obesity.[16] It is, however, noteworthy that enlarged fat cells have an increased basal rate of lipolysis. When the increase in cell volume is taken into account, the difference in lipolysis between obese and lean humans will disappear.[16] Nevertheless, due to their expanded body fat mass obese individuals have significantly higher free fatty acid concentrations and a higher fatty acid turnover, particularly if they have an abdominal type of fat distribution.[23]

Fatty acid release from adipocytes occurs mainly under the control of catecholamines. Several studies reported a blunted catecholamine-induced lipolysis in obese subjects.[22] Similar findings were obtained in in vitro studies using isolated human adipocytes. It was described that this catecholamine resistance may be due to a reduced expression and function of β_2-adrenoceptors as well as due to an increased function of α_2-adrenoceptors.[16] Furthermore, an additional defect in the ability of cyclic AMP to activate lipolysis was found in a recent study in obese subjects with insulin resistance.[24] Published data from both in vivo and in vitro studies concerning the antilipolytic role of insulin in human obesity is rather conflicting.[22,16] A key component of the lipolytic pathway is the expression and activity of hormone-sensitive lipase (HSL). In a recent study in massively obese subjects, a defect in HSL expression was observed that was linked to an impaired lipolytic capacity in subcutaneous adipocytes.[25]

Microdialysis is a very elegant technique to perform direct measurements of adipose tissue metabolism under more physiological conditions compared to isolated fat cells. By implanting a semipermeable catheter it is possible to introduce substances into adipose tissue and, at the same time, to collect samples of the interstitial fluid for the analysis of molecules of interest (see also Chapter 8).[26] Such studies demonstrated that glycerol production as a measure of lipolysis

Table 10.2. Characteristics of insulin resistance in obesity

- reduced insulin binding
- reduced insulin-stimulated glucose transport
- reduced non-oxidative glucose disposal
- increased hepatic glucose production
- increased basal lipolysis
- reduced antilipolytic action of insulin
- reduced LPL expression and activity
- reduced activation of the sympathetic nervous system

from the subcutaneous tissue is increased in obesity.[28] Such studies also clearly established a predominant adrenergic regulation of lipolysis mediated via β_1-, β_2-, and β_3-adrenoceptors.[27] In addition, lactate release from glucose degradation is higher in obese than lean individuals. Both glycerol and lactate enter the gluconeogenetic pathway in the liver and may favor an increased hepatic glucose output.

Insulin Action

Obesity is commonly associated with insulin resistance, although the development of this phenomenon depends on additional determinants such as fat distribution pattern, age, family history of hypertension and type 2 diabetes among others.[29] As insulin influences the utilization of all three macronutrients, i.e., glucose, protein and lipid, insulin resistance not only affects carbohydrate metabolism but also lipid and protein metabolism. It is well characterized by a variety of clinical studies that obesity is associated with a decrease in insulin sensitivity particularly with regard to glucose disposal, that is usually compensated by an oversecretion of insulin from the pancreas.[30] The characteristic features of insulin resistance in obesity as summarized in Table 10.2 include reduced glucose transport, reduced nonoxidative glucose disposal, increased hepatic glucose production, increased basal lipolysis, and reduced LPL expression and activity among others.

Studies at the fat cell levels have unraveled a number of defects in obese subjects, as summarized in Table 10.3. Again, most of these alterations depend on fat cell size. Previous studies demonstrated a reduced insulin binding to its receptor on the fat cell membrane as well as a reduced insulin receptor kinase activity. Insulin-stimulated glucose transport in adipocytes from obese humans is blunted, at least partially due to a reduced expression of the insulin-regulated GLUT4 transporter isoform. It turned out in more recent studies that IRS-1 (insulin receptor substrate-1) expression and phosphoinositol-3-kinase activity are also reduced in fat and muscle cells from morbidly obese subjects. All these findings are secondary phenomena as most of them are reversible after weight reduction.[31] If and to which extent other components of the insulin signaling cascade are affected by obesity is subject of current investigations.

Cytokines

It was originally reported by Hotamisligil and coworkers that fat cells are a source of tumor necrosis factor-alpha (TNF-α) and that adipocytes from obese rodents express higher levels of the cytokine than adipocytes from lean control animals.[32] Of similar interest was the observation that the overexpression of TNF-α was linked to a bundle of metabolic disturbances characteristic of the insulin resistance syndrome which were partially reversible after neutralization of the cytokine by a IgG fusion protein containing soluble TNF receptor domains. These findings among others led to the attractive and still intensely discussed hypothesis that TNF-α may act as the mediator of obesity-linked insulin resistance.[33] Subsequent studies in man demonstrated only moderately higher adipose tissue TNF-α mRNA levels in obese vs. lean subjects.[34,35,36] In addition, adipose expression of the two TNF receptor subtypes and serum concentrations of both soluble TNF receptors were higher in obese vs. lean subjects indicating that the complete TNF-α system was up-regulated in the obese state.[37,38] However, in contrast to the various rodent models of obesity, serum levels of TNF-α were not elevated in obese vs. lean humans.[38]

TNF-α is now known to have profound catabolic effects on adipose tissue development and function. This cytokine prevents the differentiation of human adipocyte precursor cells and causes delipidation of mature fat cells.[39] Furthermore, TNF-α was found to induce lipolysis, to suppress GLUT4 and LPL expression and to interfere with the insulin signaling cascade at the level of IRS-1 and phosphoinositol-3 kinase, thereby blocking the insulin-induced stimulation of glucose transport in human adipocytes.[40,41] Using a microperfusion technique in subcutaneous abdominal adipose tissue, Orban et al found a positive correlation between interstitial TNF-α concentrations and lipolysis which may suggest that the cytokine participates in a local auto/paracrine feedback system potentiating lipolysis and reducing the antilipolytic action of insulin.[42] Although the action of locally produced TNF-α appears to be restricted to adipose tissue, this system appears to be of physiological importance as overexpression of the cytokine in enlarged adipocytes may result in a feedback inhibition of adipose tissue growth at the expense of insulin resistance and of subsequent metabolic disturbances.[37]

Interleukin-6 (IL-6) is another cytokine expressed by adipose cells that has only recently gained attention. A clinical study had indicated that serum IL-6 concentrations are positively correlated with the body mass index.[43] A subsequent in vitro study revealed that adipose tissue contributes to IL-6 production.[44] By measuring arterio-venous differences in the concentrations of IL-6 over a subcutaneous adipose tissue bed in the postabsorptive state, Mohamed-Ali et al demonstrated a significant release of IL-6 from adipose tissue into the circulation which may be indicative of a systemic action of the cytokine.[45] However, the consequences of the overproduction of IL-6 by enlarged fat cells are currently unknown. There is some evidence that give rise to the speculation that IL-6 is involved in the wasting of adipose tissue during cachexia. Studies from our laboratory established that human fat cells express the full IL-6 receptor system and that IL-6 is able to increase basal and isoproterenol-stimulated lipolysis while glucose uptake was not affected (Päth et al, manuscript submitted).

At present, there are many efforts to detect other cytokines at the fat cell level. It is not unlikely that fat cells are able to produce and release other cytokines which may function as components of the local cross-talk in adipose tissue. One already mentioned example is macrophage colony-stimulating factor.[7]

Leptin

Leptin is a recently discovered 16,000 Dalton protein that is almost exclusively produced by white and brown adipocytes.[46] This cytokine-like protein is rapidly secreted by fat cells, circulates in blood in a bound and unbound fraction and acts primarily as a satiety factor in the hypothalamus after crossing the blood-brain-barrier. Serum concentrations of leptin closely reflect the size of the body fat depots but are also subject to short-term regulation by hormones and metabolites.[47] In view of the elevated serum levels of leptin obesity can be considered as a

Table 10.3. Alterations in the expression and function of specific genes in human obesity

Lipid accumulation ↓	lipoprotein lipase ↓
Insulin action ↓	GLUT4 ↓, IRS-1 ↓
Lipolysis ↑	hormone-sensitive lipase ↑
Cytokines ≠	TNF-α ↑ IL-6 ↑
Fibrinolysis ↓	PAI-1 ↑

result of resistance to leptin although the evidence for its support is sparse.[48] The cellular and molecular basis of leptin resistance in both obese rodents and humans is currently not known.

An interesting finding in this context is the wide biological variation of serum leptin concentrations at a given BMI. It was speculated that at least a subgroup of obese individuals may have a relative leptin deficiency, possibly on a specific genetic background, and may be particularly susceptible to diet-induced obesity.[47] As a logical consequence, these individuals could benefit from leptin administration. However, the effect of subcutaneous administration of leptin on body weight appears to be only modest.[49] It was also reported that the entry of leptin into the cerebrospinal fluid may be reduced in some obese subjects, particularly when the capacity of the presumed transport system is exceeded. Mutations of the human leptin and leptin receptor gene are apparently extremely rare but in the few cases reported lack of biologically active leptin or functional receptors was associated with early development of extreme obesity characterized by excessive hyperphagia. This is convincing proof that leptin is a central regulator of energy homeostasis in humans.

During the last years, several peripheral actions of leptin were reported including an inhibitory effect on insulin secretion.[50] But it is also noteworthy that leptin and insulin were repeatedly found to be positively correlated with BMI. Other studies also suggested that leptin may be involved in the pathophysiology of insulin resistance.[51] Irrespective of some controversies, there is growing evidence that leptin is implicated in the regulation of glucose disposal by insulin in the liver as well as in other tissues. There are some reports that leptin may interfere at early steps of the insulin signaling system. However, the contributing mechanisms are poorly understood and there is some uncertainty that this interaction is physiologically significant.[47]

Plasminogen Activator Inhibitor-1 (PAI-1)

Plasminogen activator inhibitor-1 (PAI-1) is the main regulator of the endogenous fibrinolytic pathway and elevated concentrations of PAI-1 are now considered to be a feature of the metabolic syndrome (see Table 10.3).[52] Markedly elevated plasma PAI-1 activity is a common finding in obese subjects and was also reported in a prospective study to be a risk factor for thrombembolic complications.[53] Recent studies suggested that substantial amounts of PAI-1 are released by human adipocytes and that adipocytes from obese subjects have both a greater PAI-1 gene expression and a higher protein secretion than fat cells from lean individuals.[54,55,57]

Interestingly, omental adipocytes secrete significantly more PAI-1 than subcutaneous adipocytes even after controlling for differences in cell size. Animal studies by Loskutoff et al originally revealed that TGF-β1 and TNF-α are potent stimulators of PAI-1 production.[56] Recent data from our laboratory suggest that both TNF-α and TGF-β1 are also important regulators of adipose PAI-1 expression in human adipose tissue.[57] It is tempting to speculate that the local overexpression of TNF-α in adipose tissue from obese subjects contributes to the up-regulation of PAI-1 antigen production. However, the regional expression pattern cannot be fully explained by differences in TNF-α expression, as relative TNF mRNA levels are higher in subcutaneous than in omental adipose tissue.[37] In another study, TGF-β1 mRNA was detectable in fat cell cultures by using a semi-quantitative RT-PCR technique indicating that TGF-β1 also plays a major role in PAI-1-regulation in human adipose tissue (Birgel et al, in press). However, it is currently unknown whether the increased PAI-1 production in adipocytes from obese subjects is also mediated by increased local TGF-β1 expression. It should be noted that the elevated production and secretion of insulin may also be involved in the high rate of PAI-1 production in adipocytes.[58] Thus, elevated PAI-1 production appears to be a consequence of multiple alterations in fat cell function in the obese state.

Renin-Angiotensin System in Human Adipose Tissue

It was only recently reported that human adipose tissue expresses a local renin-angiotensin system that includes all components such as angiotensinogen, angiotensin converting enzyme (ACE), renin as well as the type 1 angiotensin receptor.[59,60] Angiotensin II, the cleavage product of angiotensinogen, is well known to be an important regulator of blood pressure and of salt and water balance. Thus, the question that may immediately arise is whether adipose expression of the renin-angiotensin system is associated with the development of hypertension in obesity. Indirect evidence for this hypothesis was provided by the observation that circulating angiotensinogen levels are related to BMI on the one hand and to blood pressure on the other.[61] However, such association studies do not allow to assess a causal relationship between these parameters. Surprisingly, in animal models of obesity, angiotensinogen mRNA levels were decreased by 50 to 80% in adipose tissues of obese vs. lean rodents.[62] However, this finding does not exclude the possibility that steps distal to angiotensinogen expression may lead to an increased supply of angiotensin II, the biologically active degradation product. To date, there are no quantitative data on the expression of these genes in obese subjects as compared to lean individuals. Thus, it remains a scientific challenge to elucidate the role of the renin-angiotensin system in human adipose tissue, particularly with regard to the pathophysiology of obesity-associated hypertension. A completely different aspect of the role of angiotensin II is that this local hormone could function as an adipogenic factor which may increase the expression of lipogenic enzymes such as fatty acid synthetase and thereby the cellular accumulation of triglycerides.[62] However, up to now there is not sufficient evidence to support this hypothesis.

The Peroxisome Proliferator-Activated Receptor-γ (PPAR-γ)

The nuclear receptor called peroxisome proliferator-activated receptor γ (PPARγ) is a member of a family of transcription factors that plays a central role in the control of genes that are involved in glucose and lipid metabolism.[63] PPARγ is predominantly expressed in adipose tissue of both rodents and humans. Therefore, one of the first questions that may arise is whether there is an altered expression of the gene in the obese state. Such studies revealed that mRNA levels are only moderately increased in human obesity and type 2 diabetes mellitus but are highly regulated over a four-fold range by nutritional factors.[64,65] Interestingly, an obesity-linked increased expression of PPARγ in the omental adipose tissue was observed without a significant change of PPARγ expression in the subcutaneous adipose depot.[66] The implications of this depot-specific alteration in the pathophysiology of obesity and its complications remain to be determined. Fascinating new data may emerge from recent investigations of mutations in the PPARγ gene. Mutations with "gain of function" were found to be associated with severe obesity. However, there are only rare cases which may explain only a small proportion of obesity.[68,67] Very recently, two "loss-of-function" mutations of the PPARγ gene were identified that were associated with insulin resistance and type 2 diabetes indicating that PPARγ is a central regulator of normal insulin sensitivity.[69] The latter and other studies on the molecular basis of insulin resistance have focused on a possible role of PPARγ. A strong argument for this hypothesis is that thiazolidinediones, a new class of antidiabetic drugs, are potent activators of this receptor. Administration of these compounds increase insulin sensitivity and lower blood glucose in diabetic patients. However, the currently available information concerning the role of PPARγ in obesity-associated insulin resistance is still puzzling. Therefore, new approaches are required to better define the implications of a dysfunction of PPARγ in the metabolic syndrome.

Changes of Fat Cell Function by Fasting and After Weight Loss

Fasting represents a state of acute and marked shortage of nutrient (calorie) supply and subsequent mobilization of endogenous energy stores. In contrast, stable weight reduction is characterized by a new balance between energy intake and consumption at a lower level. Due to this substantial difference alterations in fat cell gene expression and function between these two states of energy homeostasis are not necessarily the same. Irrespective of this aspect the bulk of studies has established that most if not all metabolic disturbances associated with obesity are reversible after weight reduction. The extent of change usually depends on the amount of weight lost.

Enlarged fat cells exhibit the phenomenon of insulin resistance in terms of glucose transport which may be due to both reduced sensitivity and reduced responsiveness to the hormone. Both disturbances are improved by weight reduction. It was also reported that adipocytes from obese subjects produce less GLUT4 which is completely restored after weight reduction.[70,71] A similar impaired insulin action on glucose transport activity was reported in muscle from obese humans. This defect was almost fully reversible after substantial weight loss by gastric surgery.[72] However, there are additional mechanisms that may contribute to the improved insulin sensitivity in obese subjects following weight loss such as a reduction in the activity of protein-tyrosine phosphatases.[73] However, it should be noted again that despite this reduction in insulin-dependent glucose transport at the single cell level total glucose uptake in adipose tissue is elevated in obesity due to the enlarged adipose tissue mass. At the same time, fasting induces an up-regulation of lipoprotein lipase expression and activity, a mechanism that serves to compensate for the loss of energy and could contribute to the high risk of weight relapse after a dietary treatment regime.[18]

Concerning lipolysis, the blunted response of adipose tissue to catecholamines is reversed by weight reduction. Similarly, most steps of the lipolytic pathway including the antilipolytic action of insulin are largely normalized after weight loss.[16] For example, in a study of severely obese subjects, weight loss was associated with a decrease in the rate of appearance of glycerol and palmitate. Weight loss also caused a decrease in adipocyte hormone-sensitive lipase by 38% without any consistent changes in the concentrations of GTP-dependent regulatory proteins.[74] But there is also conflicting data from other studies. Hellstrom and coworkers reported that the maximal antilipolytic effect of insulin and the ability of insulin to stimulate lipogenesis

Table 10.4. Complications of human obesity and possibly involved fat cell products

Type 2 diabetes mellitus	TNF-α, TGF-β1, IL-6
Hypertension	angiotensinogen
Thromembolic complications	PAI-1

is almost completely blunted in obese women during a long-term very low calorie diet, whereas basal lipolysis by fat cells was two-fold increased and the lipolytic response to catecholamines remained unchanged.[75]

One of the mediators of the metabolic changes associated with weight reduction may be TNF-α. It was recently shown that adipose TNF expression decreases significantly with weight loss.[34] As TNF is known to inhibit LPL activity and as there was an inverse relationship between TNF expression and LPL activity, this finding may be an explanation for the increase in LPL activity during fasting. However, in a recent study in obese women undergoing a 3-week very low calorie diet TNF-α mRNA levels measured by quantitative RT-PCR in subcutaneous adipose tissue samples increased by 78 %.[77] As TNF-α promotes lipolysis it could also be argued the other way around in the sense that TNF-α supports fatty acid mobilization.

Acute fasting is well known to induce a dramatic decrease in leptin expression and serum concentrations by up to 60%.[48,76] When a new stable weight is obtained, leptin levels tend to rise but remain closely related to body fat mass.[48] However, it was also reported that the individual leptin response to fasting and weight loss varies considerably for as yet unknown reasons.

Little is known about the regulation of PAI-1 during fasting or after weight loss. Although fasting is associated with a decrease in plasma PAI-1 concentrations, a surprising up-regulation of PAI-1-expression in subcutaneous adipose tissue was recently found in obese subjects undergoing a 3-week very low calorie diet.[78] Whether this up-regulation holds true under weight-stable conditions after weight loss is currently unknown. No data are available on the effect of fasting or stable weight loss on the expression of the renin-angiotensin system in human adipose tissue, although weight loss is accompanied by a parallel decrease in blood pressure. However, in young Sprague-Dawley rats, angiotensinogen mRNA was dramatically reduced after fasted conditions.[79]

As PPARγ plays a key role in adipogenesis and adipocyte gene expression, the interesting question may arise how dietary factors affect the expression of this transcription factor. In a study in seven obese subjects who were fed a low calorie diet until 10% weight loss was achieved mean expression of PPARγ mRNA fell by 25% after the reduction in body weight, but then increased to pretreatment levels after 4 weeks of weight maintenance.[65] In another study in obese women on a very low calorie diet, a rather modest decrease in adipose tissue PPARγ mRNA expression by 13% was observed indicating that dietary factors have little influence on the expression of this transcription factor.[77] In contrast, the two recently discovered uncoupling protein homologues, UCP2 and UCP3, are at least partially regulated by the level of energy intake. Both UCP2 and UCP3 mRNAs were found to increase 2- to 2.5-fold under a hypocaloric diet of approximately 1000 kcal/day suggesting a role for these proteins in the metabolic adaptation to energy restriction.[80] However, these findings were only partially confirmed in a recent study on the effect of stable weight reduction on UCP2 and UCP3 expression in muscle and adipose tissue.[81]

Taken together, there is overwhelming evidence today that fasting and weight loss are associated with marked changes in fat cell function and gene expression directed back to normal baseline levels. This seems to be the case for practically all unfavorable alterations which may promote health complications on the long-term. For this reason, weight loss is the first-line choice of treatment in obese subjects with one or more clinical disorders such as diabetes, dyslipidemia and hypertension.

Conclusions

This review addressed a variety of characteristic changes in adipose tissue growth and function associated with human obesity. It should be stressed that the changes described are only partially due to an altered metabolism of the enlarged fat cells but are also the result of an excess adipose tissue mass per se without changed function at the fat cell level. Most of these changes, if not all, are believed to be a consequence of obesity, as they were found to be normalized by weight loss. These new insights into adipose tissue metabolism may be very valuable as they offer a better understanding of the development of the typical complications linked with obesity (Table 10.4). It is obvious from the published literature that local cytokine overproduction and action, increased supply of free fatty acids by enlarged fat depots, and possibly overproduction of leptin may contribute to the development of insulin resistance and type 2 diabetes mellitus. Local expression of angiotensinogen and other components of the renin-angiotensin system could be involved in the pathophysiology of hypertension, one of the most frequent complications of obesity, although much work remains to be done to get a clear picture. Last but not least, increased production of PAI-1 and possibly of other components of the fibrinolytic system could be responsible for the increased risk of thrombembolic complications reported in obese individuals and undoubtedly deserves more attention.

Our new understanding of the physiology of adipose tissue metabolism may have other important clinical implications. Despite many efforts our current arsenal to treat obesity safely and successfully, also on a long-term basis, is rather limited. It is, therefore, desirable to focus our attention and to allocate our clinical resources to those obese who are at high risk of developing complications. A better knowledge of the pathological alterations taking place in adipose tissue during human obesity may also help to develop more precise strategies to reduce the risk of complications. A better targeting of prevention and intervention actions at the underlying pathophysiological processes could improve health even in the absence of significant weight reduction.

References

1. Pi-Sunyer FX. Medical hazards of obesity. Ann Intern Med 1993;119:655-660.
2. World Health Organization. Obesity—Preventing and managing the global epidemic. Report of a WHO consultation on obesity. WHO Geneva 1998.
3. Bouchard C, Tremblay A, Despres JP, Nadeau A, Lupien PJ, Theriault G, Dussault J, Moorjani S, Pineault S, Fournier G. The response of long-term overfeeding in identical twins. N Engl J Med 1990;322:1477-1482.
4. Levine AS, Eberhardt NL, Jensen MD. Role of nonexercise activity thermogenesis in resistance to fat gain in humans. Science 1999;283:212-214
5. Hirsch J, Batchelor B. Adipose tissue cellularity in human obesity. Clin Endocrinol Metab 1976; 5:299-311.
6. Ailhaud G, Hauner H. Development of white adipose tissue. In: Bray GA, Bouchard C, James WPT, eds. Handbook of Obesity. 1st ed. New York: Marcel Dekker, 1998:359-378.

7. Levine JA, Jensen MD, Eberhardt NL, O'Brien T. Adipocyte macrophage colony-stimulating factor is a mediator of adipose tissue growth. J Clin Invest 1998; 101:1557-1564.

8. Knittle JL, Timmers K, Ginsberg-Fellner F et al. The growth of adipose tissue in children and adolescents. Cross-sectional and longitudinal studies of adipose cell number and size. J Clin Invest 1979; 63:239-246.

9. Salans LB, Cushman SW, Weismann RE. Studies of human adipose tissue. Adipose cell size and number in nonobese and obese patients. J Clin Invest 1973; 52:929-941.

10. Näslund I, Hallgren P, Sjöström L. Fat cell weight and number before and after gastric surgery for morbid obesity. Int J Obes 1988; 12:191-197

11. Roncari DAK, Lau DCW, Kindler S. Exaggerated replication in culture of adipocyte Precursors from massively obese persons. Metabolism 1981; 30:425-427.

12. Pettersson P, Van RLR, Karlsson M et al. Adipocyte precursor cells in obese and nonobese humans. Metabolism 1985; 34:808-812.

13. Hauner H, Wabitsch M, Pfeiffer EF. Differentiation of adipocyte precursor cells from obese and nonobese adult women and from different adipose tissue sites. Horm Metab Res 1988; 19(Suppl):35-39.

14. Hauner H, Wabitsch M, Zwiauer K et al. Adipogenic activity in sera from obese children before and after weight reduction. Am J Clin Nutr 1989; 50:63-67.

15. Gregoire FM, Johnson PR, Greenwood MRC. Comparison of the adipoconversion of preadipocytes derived from lean and obese Zucker rats in serum-free cultures. Int J Obes 1995;19:664-670.

16. Arner P, Eckel RH. Adipose tissue as a storage organ. In: Bray GA, Bouchard C, James WPT, eds. Handbook of Obesity. 1st ed. New York: Marcel Dekker, 1998:379-395.

17. Ong JM, Kern PA. Effect of feeding and obesity on lipoprotein lipase activity, immunoreactive protein, and messenger RNA levels in human adipose tissue. J Clin Invest 1989; 84:305-311.

18. Eckel RH. Lipoprotein lipase. A multifunctional enzyme relevant to common metabolic disorders. N Engl J Med 1989; 320:1060-1068.

19. Maslowska M, Cianflone K, Rosenbloom M. Acylation stimulating protein (ASP): role in adipose tissue. In: Guy-Grand B, Ailhaud G, eds. Progress in Obesity Research: 8. London: John Libbey, 1999:65-70.

20. Sniderman AD, Cianflone K, Eckel RH. Levels of acylation stimulating protein in obese women before and after moderate weight loss. Int J Obes 1991; 15:327-332.

21. Murray I, Sniderman AD, Havel PJ et al. Acylation stimulating protein (ASP) deficiency alters postprandial and adipose tissue metabolism in male mice. J Biol Chem 1999; 274:36219-36225.

22. Coppack SW, Jensen MD, Miles JM. In vivo regulation of lipolysis in humans. J Lipid Res 1994; 35:177-193.

23. Jensen MD, Haymond MW, Rizza RA et al. Influence of body fat distribution on free fatty acid metabolism in obesity. J Clin Invest 1989; 83:1168-1173.

24. Reynisdottir S, Ellerfeldt K, Wahrenberg H et al. Multiple lipolysis defects in the insulin (metabolic) syndrome. J Clin Invest 1994; 93:1590-1599.

25. Large V, Reynisdottir S, Langin D et al. Decreased expression and function of adipocyte hormone-sensitive lipase in subcutaneous fat cells of obese subjects. J Lipid Res 1999; 40:2059-2066.

26. Arner P, Kriegholm E, Engfeldt P et al. Adrenergic regulation of lipolysis in situ at rest and during exercise. J Clin Invest 1990; 85:893-898.

27. Enocksson S, Shimizu M, Lönnqvist F et al. Demonstration of an in vivo functional β_3-adrenoceptor in man. J Clin Invest 1995; 95:2239-2245.

28. Jansson PA, Larsson A, Smith U et al. Glycerol production in subcutaneous adipose tissue in lean and obese humans. J Clin Invest 1992; 89:1610-1617.

29. Kissebah AH, Krakower GR. Regional adiposity and morbidity. Physiol Rev 1994; 74:761-811.

30. Ferrannini E, Natali A, Bell P et al. Insulin resistance and hypersecretion in obesity. J Clin Invest 1997; 100:1166-1173.

31. Virkamäki A, Ueki K, Kahn CR. Protein-protein interaction in insulin signaling and the molecular mechanisms of insulin resistance. J Clin Invest 1999; 103:931-943.

32. Hotamisligil GS, Shargill NS, Spiegelman BM. Adipose expression of tumor necrosis factor-alpha: direct role in obesity-linked insulin resistance. Science 1993; 259:87-91.

33. Hotamisligil GS, Spiegelman BM. Tumor necrosis factor α: A key component of the obesity-diabetes link. Diabetes 1994; 43:1271-1278

34. Kern PA, Saghizadeh M, Ong JM et al. The expression of tumor necrosis factor in human adipose tissue, regulation by obesity, weight loss and relationship to lipoprotein lipase. J Clin Invest 1995; 95:2111-2119.

35. Hotamisligil GS, Arner P, Caro JF et al. Increased adipose tissue expression of tumor necrosis factor-alpha in human obesity and insulin resistance. J Clin Invest 1995; 95:2409-2415.95

36. Hube F, Birgel M, Lee Y-M et al. Expression pattern of tumor necrosis factor receptors in subcutaneous and omental human adipose tissue: Role of obesity and noninsulin-dependent diabetes mellitus. Eur J Clin Invest 1999; 29:672-678.

37. Hube F, Hauner H. The role of TNF-α in human adipose tissue: Prevention of weight gain at the expense of insulin resistance. Horm Metab Res 1999; 31:626-631.

38. Hauner H, Bender M, Haastert B et al. Plasma concentration of TNF-α and its soluble receptors in obese subjects. Int J Obes 1998; 22:1239-1243.

39. Petruschke T, Hauner H. Tumor necrosis factor-α prevents the differentiation of human adipocyte precursor cells and causes delipidation of newly developed fat cells. J Clin Endocrinol Metab 1993; 76:742-747.

40. Hauner H, Petruschke T, Russ M et al. Effects of tumor necrosis factor alpha (TNF-α) on glucose transport and lipid metabolism of newly-differentiated human fat cells in cell culture. Diabetologia 1995; 38:764-771.

41. Liu LS, Spelleken M, Röhrig K et al. Tumor necrosis factor-α acutely inhibits insulin signaling in human adipocytes. Implication of the p80 tumor necrosis factor receptor. Diabetes 1998; 47:515-522.

42. Orban Z, Remaley AT, Sampson M et al. The differential effect of food intake and β-adrenergic stimulation on adipose-derived hormones and cytokines in man. J Clin Endocrinol Metab 1999; 84:2126-2133.

43. Vgontzas AN, Papanicolaou DA, Bixler EO et al. Elevation of plasma cytokines in disorders of excessive daytime sleepiness: role of sleep disturbance and obesity. J Clin Endocrinol Metab 1997; 82:1313-1316.

44. Fried SK, Bunkin DA, Greenberg AS. Omental and subcutaneous adipose tissues of obese subjects release interleukin-6: Depot difference and regulation by glucocorticoid. J Clin Endocrinol Metab 1998; 83:847-850.

45. Mohamed-Ali V, Goodrick S, Rawesh A et al. Subcutaneous adipose tissue releases interleukin-6, but not tumor necrosis factor-α, in vivo. J Clin Endocrinol Metab 1997; 82:4196-4200.

46. Zhang Y, Proenca R, Maffei M et al. Positional cloning of the mouse obese gene and its human homologue. Nature 1994; 372:425-432.

47. Friedman JM, Halaas JL. Leptin and the regulation of body weight in mammals. Nature 1998; 395:763-770.

48. Considine RV, Sinha MK, Heiman ML et al. Serum immunoreactive-leptin concentrations in normal-weight and obese humans. N Engl J Med 1996; 334:292-295.

49. Heymsfield SB, Greenberg AS, Fujioka K et al. Recombinant leptin for weight loss in obese and lean adults. JAMA 1999; 282:1568-1575.

50. Seufert J, Kieffer TJ, Leech CA et al. Leptin suppression of insulin secretion and gene expression in human pancreatic islets: Implications for the development of adipogenic diabetes mellitus. J Clin Endocrinol Metab 1999; 84:670-676.

51. Girard J. Is leptin the link between obesity and insulin resistance? Diabetes Metab 1997; 23(Suppl 3):16-24.

52. Juhan-Vague I, Alessi MC, Vague P. Increased plasminogen activator inhibitor 1levels. A possible link between in sulin resistance and atherothrombosis. Diabetologia 1991; 34:457-462.

53. Juhan-Vague I, Pyke SDM, Alessi MC et al. Fibrinolytic factors and the risk of myocardial infarction or sudden death in patients with angina pectoris. Circulation 1996; 94:2057-2063.

54. Alessi MC, Peiretti F, Henry M et al. Production of plasminogen activator inhibitor 1 by human adipose tissue: possible link between visceral fat accumulation and vascular disease. Diabetes 1997; 46:860-867.

55. Eriksson P, Reynisdottir S, Lönnqvist S et al. Adipose tissue secretion of plasminogen activator inhibitor-1 in nonobese and obese individuals. Diabetologia 1998;41:65-71.

56. Loskutoff DJ, Samad F. The adipocyte and hemostatic balance in obesity. Studies of PAI-1. Arterioscler Thromb Vasc Biol 1998; 18:1-6.
57. Gottschling-Zeller H, Birgel M, Röhrig K et al. Effect of tumor necrosis factor alpha and transforming growth factor beta 1 on plasminogen activator inhibitor-1 secretion from subcutaneous and omental human fat cells in suspension culture. Metabolism 2000; 49:666-671.
58. Morange P-E, Aubert J, Peiretti F et al. Glucocorticoids and insulin promote plasminogen activator inhibitor 1 production by human adipose tissue. Diabetes 1999; 48:890-895.
59. Karlsson C, Lindell K, Ottosson M et al. Human adipose tissue expresses angiotensinogen and enzymes required for ist conversion to angiotensin II. J Clin Endocrinol Metab 1998; 83:3925-3929.
60. Schling P, Mallow H, Trindl A et al. Evidence for a local renin angiotensin system in primary cultured human preadipocytes. Int J Obesity 1999; 23:336-341.
61. Umemura S, NyuiN, Tamura K et al. Plasma angiotensinogen concentrations in obese patients. Am J Hypertens 1997; 10:629-633.
62. Jones BH, Standridge MK, Taylor JW et al. Angiotensinogen gene expression in adipose tissue: Analysis of obese models and hormonal and nutritional control. Am J Physiol 1997; 273:R236-R242.
63. Auwerx J. PPARγ, the ultimate thrifty gene. Diabetologia 1999; 42:1033-1049.
64. Auboeuf D, Rieusset J, Fajas L et al. Tissue distribution and quantification of the expression of mRNAs of peroxisome proliferator-activated receptors and liver X receptor-α in humans. No alteration in adipose tissue of obese and NIDDM patients. Diabetes 1997; 46:1319-1327.
65. Vidal-Puig AJ, Considine RV, Jimenez-Linan M et al. Peroxisome proliferator-activated receptor gene expression in human tissues. Effects of obesity, weight loss, and regulation by insulin and glucocorticoids. J Clin Invest 1997; 99:2416-2422.
66. Lefebvre AM, Laville M, Vega N. Depot-specific differences in adipose tissue gene expression in lean and obese subjects. Diabetes 1998; 47(1):98-103.
67. Ristow M, Müller-Wieland D, Pfeiffer A et al. Obesity associated with a mutation in a genetic regulator of adipocyte differentiation. N Engl J Med 1998; 339:953-959.
68. Valve R, Sivenius K, Miettinen R et al. Two polymorphisms in the peroxisome proliferator-activated receptor-γ gene are associated with severe overweight among obese women. J Clin Endocrinol Metab 1999; 84:3708-3712.
69. Barroso I, Gurnell M, Crowley VEF et al. Dominant negative mutations in human PPARγ associated with severe insulin resistance, diabetes mellitus and hypertension. Nature 1999; 402:880-883.
70. Caro JF, Dohm LG, Pories WJ et al. Cellular Alterations in liver, skeletal muscle, and adipose tissue responsible for insulin resistance in obesity and type II diabetes. Diabetes/Metab Rev 1989; 5:665-689.
71. Garvey WT, Kolterman OG. Correlation of in vivo and in vitro actions of insulin in obesity and noninsulin-dependent diabetes mellitus: Role of the glucose transport system. Diabetes/Metab Rev 1988; 4:543-569.
72. Friedman JE, Dohm GL, Leggett-Frazier N et al. Restoration of insulin responsiveness in skeletal muscle of morbidly obese patients after weight loss. Effect on muscle glucose transport and glucose transporter GLUT4. J Clin Invest 1992; 89:701-705.
73. Ahmad F, Considine RV, Bauer TL et al. Improved sensitivity to insulin in obese subjects following weight loss is accompanied by reduced protein-tyrosine phosphatases in adipose tissue. Metabolism 1997; 46:1140-1145.
74. Klein S, Luu K, Gasic S et al. Effect of weight loss on whole body and cellular lipid metabolism in severely obese humans. Am J Physiol 1996; 270:E739-E745.
75. Hellstrom L, Reynisdottir S, Langin D et al. Regulation of lipolysis in fat cells of obese women during long-term hypocaloric diet. Int J Obesity 1996; 20:745-752.
76. Geldszus R, Mayr B, Horn R et al. Serum leptin and weight reduction in female obesity. Eur J Endocrinol 1996; 135:659-662.
77. Bastard JP, Hainque B, Dusserre E et al. Peroxisome proliferator activated receptor-gamma, leptin and tumor necrosis factor-alpha mRNA expression during very low calorie diet in subcutaneous adipose tissue in obese women. Diabetes Metab Res Rev 1999; 15:92-98.
78. Bastard JP, Vidal H, Jardel C et al. Subcutaneous adipose tissue expression of plasminogen activator inhibitor-1 gene during very low calorie diet in obese subjects. Int J Obesity 2000; 24:70-74.

79. Frederich RC, Kahn BB, Peach MJ et al. Tissue-specific nutritional regulation of angiotensinogen in adipose tissue. Hypertension 1992; 19:339-344.
80. Millet L, Vidal H, Andreelli F et al. Increased uncloupling protein-2 and -3 mRNA expression during fasting in obese and lean humans. J Clin Invest 1997; 100:2665-2670.
81. Vidal-Puig A, Rosenbaum M, Considine RC et al. Effects of obesity and stable weight reduction on UCP2 and UCP3 gene expression in humans. Obes Res 1999; 7:133-140.

Index